D1271502

FLIGHT IDENTIFICATION OF
RAPTORS

OF EUROPE, NORTH AFRICA AND THE MIDDLE EAST

HELM IDENTIFICATION GUIDES

FLIGHT IDENTIFICATION OF
RAPTORS

OF EUROPE, NORTH AFRICA AND THE MIDDLE EAST

Dick Forsman

CHRISTOPHER HELM
LONDON

All the images taken by the author are of wild, free-living birds of prey.

Christopher Helm
An imprint of Bloomsbury Publishing Plc

50 Bedford Square   1385 Broadway
London   New York
WC1B 3DP   NY 10018
UK   USA

www.bloomsbury.com

BLOOMSBURY, CHRISTOPHER HELM and the Helm logo are trademarks
of Bloomsbury Publishing Plc

First published 2016

A catalogue record for this book is available from the British Library.

Library of Congress Cataloguing-in-Publication data has been applied for.

ISBN (hardback): 978-1-4729-1361-6
ISBN (epdf): 978-1-4729-2554-1
ISBN (epub): 978-1-4729-2555-8

2 4 6 8 10 9 7 5 3

Designed by Julie Dando, Fluke Art
Printed in China by C & C Offset Co. Ltd.

CONTENTS

PREFACE 9

ACKNOWLEDGEMENTS 11

INTRODUCTION 13
    Flight identification of raptors 13
    Plumage variation in raptors 13
    Viewing conditions 13
    Moult 14
    Factors affecting flight 18
    Previous experience 18

TOPOGRAPHY 19

GLOSSARY 22

HAWK-WATCHING IN AND AROUND EUROPE
    by Keith L. Bildstein and Anna Sandor 25

MIGRATION ECOLOGY OF RAPTORS *by Ian Newton* 36

SPECIES ACCOUNTS 49

Osprey *Pandion haliaetus* 49

European Honey-buzzard *Pernis apivorus* 56

Crested Honey-buzzard *Pernis ptilorhyncus* 67

Identification of hybrid honey-buzzards 77

Identification of *Milvus* kites 81

Black Kite *Milvus migrans* 81

Western Black Kite *Milvus migrans migrans* 82

Black-eared Kite *Milvus migrans lineatus* 91

'Eastern Black Kite' – Identifying hybrids and intergrades
    between Western Black Kite and Black-eared Kite 95

Yellow-billed Kite *Milvus aegyptius* 99

Red Kite *Milvus milvus* 107

Identifying hybrids between Black Kite and Red Kite 114

Black-shouldered Kite *Elanus caeruleus* 115

Swallow-tailed Kite *Elanoides forficatus* 119

White-tailed Eagle *Haliaeetus albicilla* 121

Pallas's Fish Eagle *Haliaeetus leucoryphus* 131

Bald Eagle *Haliaeetus leucocephalus* 135

African Fish Eagle *Haliaeetus vocifer* 138

Egyptian Vulture *Neophron percnopterus* 143

Hooded Vulture *Necrosyrtes monachus* 153

Eurasian Griffon Vulture *Gyps fulvus* 157

Rüppell's Griffon Vulture *Gyps rueppellii* 166

African White-backed Vulture *Gyps africanus* 174

Himalayan Griffon Vulture *Gyps himalayensis* 179

Bearded Vulture *Gypaetus barbatus* 184

Black Vulture *Aegypius monachus* 192

Lappet-faced Vulture *Torgos tracheliotos* 199

Short-toed Snake Eagle *Circaetus gallicus* 206

Bateleur *Terathopius ecaudatus* 215

Western Marsh Harrier *Circus aeruginosus aeruginosus* 222

How to separate Black Kite, Marsh Harrier and
  dark morph Booted Eagle 232

Identification of 'ringtail' harriers 233

Hen Harrier *Circus cyaneus* 234

Northern Harrier *Circus hudsonius* 242

Pallid Harrier *Circus macrourus* 246

Montagu's Harrier *Circus pygargus* 254

Identification of hybrid harriers 267

Identification of *Accipiter* hawks 270

Northern Goshawk *Accipiter gentilis* 270

Eurasian Sparrowhawk *Accipiter nisus* 275

Levant Sparrowhawk *Accipiter brevipes* 282

Shikra *Accipiter badius* 287

Dark Chanting Goshawk *Melierax metabates* 293

Gabar Goshawk *Micronisus gabar* 297

Separating Common Buzzard and European Honey-buzzard 302

Identification of *Buteo* buzzards 302

Identification of *Buteo* hybrids 303

Common Buzzard *Buteo buteo buteo* 304

Steppe Buzzard *Buteo buteo vulpinus* 315

Eastern Long-legged Buzzard *Buteo rufinus rufinus* 327

Atlas Long-legged Buzzard *Buteo rufinus cirtensis* 335

Rough-legged Buzzard *Buteo lagopus* 343

Golden Eagle *Aquila chrysaetos* 352

Eastern Imperial Eagle *Aquila heliaca* 359

Spanish Imperial Eagle *Aquila adalberti*   370

Separating Steppe Eagle and Lesser Spotted Eagle   373

Steppe Eagle *Aquila nipalensis*   373

Tawny Eagle *Aquila rapax*   388

Separating Greater Spotted Eagle and Lesser Spotted Eagle   395

Greater Spotted Eagle *Aquila clanga*   395

Lesser Spotted Eagle *Aquila pomarina*   409

Identification of hybrid spotted eagles   418

Verreaux's Eagle *Aquila verreauxii*   421

Bonelli's Eagle *Aquila fasciata*   425

Booted Eagle *Hieraaetus pennatus*   432

Wahlberg's Eagle *Hieraaetus wahlbergi*   442

Differences in flight proportions between juvenile and adult falcons   445

Identification of kestrels   445

Common Kestrel *Falco tinnunculus*   446

Lesser Kestrel *Falco naumanni*   455

American Kestrel *Falco sparverius*   464

Identification of small, hobby-like falcons   467

Red-footed Falcon *Falco vespertinus*   467

Amur Falcon *Falco amurensis*   475

Eurasian Hobby *Falco subbuteo*   481

Eleonora's Falcon *Falco eleonorae*   489

Sooty Falcon *Falco concolor*   498

Merlin *Falco columbarius*   503

Lanner Falcon *Falco biarmicus*   509

Saker Falcon *Falco cherrug*   514

Gyr Falcon *Falco rusticolus*   519

Peregrine Falcon *Falco peregrinus*   526

Barbary Falcon *Falco peregrinus pelegrinoides*   535

REFERENCES   540

INDEX   542

FUNDACIÓN
MIGRES

Migres Foundation is a private non-profit organisation founded in 2003 to promote scientific research on bird migration and to boost sustainable development activities. We are convinced that these policies offer the best tools for biodiversity conservation.

Moreover, Migres Foundation has become a necessary association which reconciles sustainable development and biodiversity conservation, providing solutions for environmental challenges which, if conveniently tackled, become true opportunities of economic and social sustainable growth.

The relationship between Migres Foundation and the Tarifa Wind Power Association has provided the necessary resources for the publication of this guide, which forms part of the suite of environmental measures designed by Migres Foundation to mitigate the impact of wind farms around Tarifa. This book will enhance wildlife conservation through the dissemination of knowledge and research. The Tarifa Wind Power Association was founded in 2000 to coordinate the efforts of their associates in promoting their projects.

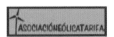

PREFACE

Although focusing on flight identification only, this new book draws strongly on its predecessor *The Raptors of Europe and the Middle East: A Handbook of Field Identification*, which still stands its ground after all these years and contains much useful information not found in this book. In particular, the introductory chapters are well worth reading and are not repeated here. Nevertheless, since its first publication in October 1998, there was always a plan to publish a more user-friendly edition, with a clear focus on flight identification, and this book now fulfils this aim.

With the breakthrough of digital photography in the early 2000s, everything changed. Never before has bird photography been so easy. Suddenly there was a need to capture all the raptors again with the latest digital equipment. Since 2003 my goal has been to find and to document all the raptor species found within the Western Palearctic, capturing their different plumages for this book. This quest has taken me on scores of trips abroad, with more than 60 tours to Spain, Israel and Arabia alone, and additional ventures to other far-flung corners of the Palearctic and African regions.

This book focuses on the identification of flying birds for a reason. Identifying raptors perched and in flight are two completely different games, and the characters and methods used for each are rarely the same. Since raptors are mostly seen in flight, it feels natural that their flight identification is what needs to be addressed.

This book relies strongly on its photographs. They have been chosen to show the different plumages but also to give an idea of known plumage variation. Most identification challenges could probably be dealt with by just consulting the images and their captions, the main text being there as a back-up reference for more difficult situations.

Raptors will still be tough to identify even after the publication of this book. Their geographic forms, complex taxonomy and bewildering plumage variation will see to that. With this in mind, it is important to understand that no book can ever replace time spent in the field. There is no substitute for personal field experience!

Juvenile Steppe Eagle. Oman, 6.11.2004 (DF)

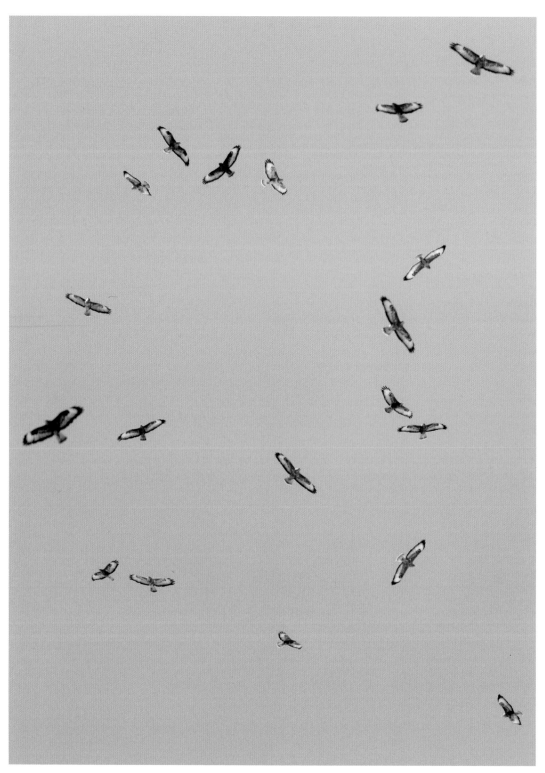

A flock of soaring Steppe Buzzards on spring migration over Eilat, Israel. 28.3.2008 (DF)

ACKNOWLEDGEMENTS

A book like this would not be possible to produce without help from others. During my travels I have met a lot of people with whom to discuss the various points of raptor identification. Some of the acquaintances were brief, while others led to a long and lasting friendship. Nevertheless, they all contributed to this book in one way or another. I extend a collective thank you to everybody for crossing paths with me.

Some people have played a major role and must be specifically acknowledged. Mark Constantine and The Sound Approach must be thanked for constant support and encouragement during the entire project. I am most grateful to Fundación Migres, in particular Miguel Ferrer, Antonio-Roman Muñoz, Eva Casado and Luis Barrios, for their cooperation and for granting financial support for my work.

I am also sincerely grateful to two of the biggest names in raptor research, Ian Newton and Keith Bildstein, for participating in this publication with their instructive and authoritative articles on raptor migration.

The following persons have helped me in various ways; some by joining me in the field, others by providing hints regarding literature, contacts and accommodation, or by providing help in some other way. I am greatly indebted to all of you; it has been a great pleasure knowing you: Dan Alon, Janos Bagyura, Arnoud van den Berg, Patrick Bergier, Stavros Christodoulides, José Luis Copete, Paul Dernjatin, Gerald Driessens, Marc Duquet, Javier Elorriaga, Hanne and Jens Eriksen, David Erterius, Wouter Faveyts, Waheed al Faziri, Juan Antonio "Kiko" Gil, Sundev Gombobaatar, Miguel Gonzales, Barak Granit, Tom Gullick, José "Pepe" Guzman, Thomas Hadjikyriakou, Teppo Helo, Vesa Hyyryläinen, Pentti Kallio, Igor Karyakin, Aleksei Koshkin, Aleksi Lehikoinen, Petteri Lehikoinen, James Lidster, Jan Lontkowski, Dani Lopez Velasco, Patricia Maldonado, Björn Malmhagen, Jonathan Meyrav, Jozef Mihók, Istvan Moldovan, John Muddeman, Killian Mullarney, Antonio-Roman Muñoz, Alejandro Onrubia, Gerald Oreel, Carlos Pacheco, Vanesa Palacios, Clairie Papazoglou, Vittorio Pedrocchi, Jari Peltomäki, Yoav Perlman, Kostas Pistolas, René Pop, Pekka Pouttu, Mátyás Prommer, Colin Richardson, Magnus Robb, Josele Saíz, Yeray Seminario, Itai Shanni, Kari Soilevaara, Matti Suopajärvi, Magnus Ullman, Ülo Väli, Machiel Valkenburg, Aarne Vattulainen and Pim Wolf. Please forgive me if I have inadvertently omitted any names from this list.

I am grateful to Dale Forbes of Swarovski Optik and to Hans Dahlgren of Swarovski Optik Nordic for their support in obtaining appropriate field optics. A sincere thank you must also go to all my clients who have travelled with me over the years – thanks for your patience when I left you standing to chase after yet another raptor. A special thank you must also go to all the photographers who participated and kindly let me use their material for the book: Ralph Buij, Richard Crossley, Peter Csonka, Javier Elorriaga, Jaakko Esama, Stewart Finlayson, Annika Forsten, Hannu Huhtinen, Johann Knobel, James Lidster, Jerry Liguori, Tom Lindroos, Bruce Mactavish, Rishad Naoroji, Gabor Papp, René Pop, Janne Riihimäki, Markku Saarinen, Yeray Seminario, Martti Siponen, Ülo Väli and Brian K Wheeler.

I would like to thank Nigel Redman at Bloomsbury Publishing for commissioning this book and for seeing it through to publication, and Julie Dando at Fluke Art for turning the text and photos into a beautiful work.

Finally, I must thank all those hundreds of people who over the years have been in touch with me, either in the field or through letters and emails, usually asking for help with the identification of particular birds. I have learnt a lot from those contacts, but I hope the gain has been mutual. Thank you.

A cocktail of eagles: juvenile Greater Spotted top two, subadult Steppe Eagle in the middle and adult Eastern Imperial Eagle bottom. Oman, 12.12.2012 (DF)

INTRODUCTION

FLIGHT IDENTIFICATION OF RAPTORS

Identifying raptors in flight can feel like a daunting task for the beginner; the species are many, the variation is often overwhelming, depending on age, sex or just individual variation, and the viewing conditions may vary from one extreme to the other, depending on distance and light. To identify a close-flying raptor one needs to deploy a different set of characters compared to identifying the same bird from a distance, when only major feather tracts rather than plumage details are visible. The approach of this book is to show the birds at close range, showing the necessary details needed for a positive identification. Once you are familiar with the details they can always be converted to be used on more distant birds.

PLUMAGE VARIATION IN RAPTORS

Raptors as a group are notoriously variable. Just think of the European Honey-buzzard as a species, with endless individual variation in every age-class and on top of this the difference between the sexes in the adults. The variation may appear bewildering, but focusing on the right feather tracts can help a lot. In the case of the honey-buzzard one can ignore the variable body plumage completely and focus on the less variable pattern of the flight feathers, and suddenly it all becomes much easier: juveniles are easily separated from the adults, and adult males and females are equally easy to tell apart. The same applies to a lot of raptor species: if the markings of the underwing flight feathers can be made out reasonably well, the species can usually be identified with certainty by this single character alone. Also, the geographical variation can be considerable, not only between subspecies of different geographical origin, but also within one subspecies. For instance, the nominate subspecies of the Common Buzzard has several subpopulations in Europe which differ in size and average coloration, yet they are all considered to belong to the same subspecies.

VIEWING CONDITIONS

The viewing conditions have a great effect on how we perceive a bird, ambient light being the single most important factor of all. The impression of a bird will change dramatically depending on the quality and direction of the light. Even reflected light, or the lack of it, can add a dimension, which sometimes can feel puzzling. To take an example, a Common Buzzard against a blue sky will show some plumage details, but as it drifts against a light cloud it suddenly changes to a featureless black silhouette. If it is circling above a landscape covered in snow, or even light-coloured sand, its colours appear almost surreal, because of the intense light bouncing off the ground. Conversely, if the bird flies over a green forest or a green field it will look dark, as very little light is reflected from these surfaces. It is also important to realise that the quality of light changes during the course of the day. Around noon, when the sun is at its highest in the sky, more light is reflected onto the underparts of a bird from the ground, while the same bird will look dark underneath in the morning and afternoon because of the low angle of the sun, when light is no longer reflected upwards.

The images for this book have been chosen to show the various identification features as clearly as possible. Many of the selected images have been taken in special light conditions, with snow or sand reflecting light onto the underparts. Conditions may not always be as favourable as shown, but the intention has been to show as much detail of the bird as possible, knowing that the light in most field situations is less favourable. To get a more life-like feeling of the images one can hold out the book at an arm's length or look at the images while squinting your eyes; both methods help to give a better idea of the bird in the field.

Figure 1. Adult female Eurasian Hobby photographed one hour after sunrise. Although the head and body catch the direct morning sun the underwings and tail remain dark due to lack of reflected light. In similar lighting conditions Eurasian Hobbies could easily be mistaken for an Eleonora's Falcon, and adults of Steppe or Lesser Spotted Eagles may look practically identical to an adult Greater Spotted Eagle, and so on. Finland, 5.8.2015 (DF)

MOULT

Moult is the single phenomenon responsible for the greatest changes in a bird's appearance. As a nestling every raptor grows its first full set of feathers, the juvenile plumage, which, as a rule, is carried for approximately the first year of life. Since all feathers have been grown at the same time this plumage appears very uniform, in terms of its condition. Soon after the breeding season it is fresh, often with clearly visible lighter tips and margins to each feather. During the first year it gradually wears and fades, but is still rather uniform in appearance. There are, however, exceptions to this general rule. Some small species, mostly long-distance migrant species of falcons and harriers, have a partial body moult in their first winter, resulting in a transitional plumage comprising both juvenile and new, moulted feathers. The main rule is that if a raptor is moulting its flight feathers in late summer/autumn, it cannot be a juvenile, but is at least one year old.

After about one year from fledging the first complete moult starts and the entire plumage is moulted, but both the extent and the timing of moult vary from species to species, sometimes even between populations of the same species. Most small and medium-sized resident or short-distance migrant species

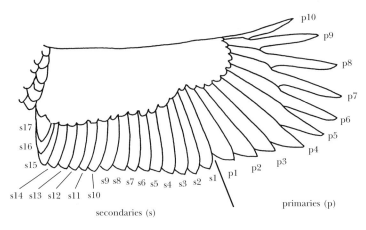

Figure 2. Wing of juvenile White-tailed Eagle showing the numbering of the remiges.

covered by this book moult during the breeding season, replacing the entire plumage in one moult cycle. The long-distance migrants, wintering in sub-Saharan Africa, may or may not commence their moult while still on the breeding grounds, but most of the moulting takes place on the wintering grounds in Africa.

The large eagles and vultures have a different moulting strategy. Because of their long flight feathers and the rather slow feather growth they are not capable of replacing the entire set of flight feathers in one season. In order to retain their flying ability they moult just a part of their plumage in a season, and the process of replacing the entire juvenile plumage takes several years. Knowing the moulting strategy, and the timing and extent of each moult, immatures of these large species can be aged by the number of retained juvenile feathers in the wing, for some species up to the age of 3–4 years. As these large raptors get older they gradually develop a stepwise moult, in which neighbouring sections of flight fathers are moulted simultaneously, thus reducing the time needed to complete a moult cycle. The stepwise moult also explains why adult eagles and vultures appear to have a comparatively fresh-looking plumage, rarely showing the very faded and worn flight feathers that are so commonly seen in immatures of the same species.

Moult can also have implications on species identification. Several closely related species can be safely identified by the wing-formula (the relation between the tips of the longest primaries forming the wingtip), which is species-specific, often used for separating species pairs or groups like Common and Lesser Kestrels, Shikra and Levant and Eurasian Sparrowhawks, Hen and Pallid/Montagu's Harriers, and so on. However, as the primary moult reaches the long primaries, missing and growing feathers will make this character useless. A moulting Hen Harrier in Aug–Sep may show a wingtip reminiscent of a Pallid/Montagu's and a Eurasian Sparrowhawk will at the same time have a wingtip which may recall a Shikra or a Levant Sparrowhawk, and so on.

Since the onset of moult in most raptor species is so strictly linked to the breeding season, different populations can sometimes be told apart just by checking the status of moult. In Ospreys, birds breeding in Arabia and the Red Sea start to breed around New Year, while Scandinavian migrants begin their breeding season some 4–5 months later. The different breeding season is then reflected in the start of the wing-moult in juveniles, with southern birds dropping their innermost primary some four months prior to their migratory relatives from the north. A similar situation is also found in Peregrines, where

Figure 3. Male Montagu's Harrier in a late 2nd cy summer transitional plumage, illustrating the progress of the flight feather moult in an *Accipitridae* hawk. In the primaries p1–p6 have been replaced by adult-type feathers and p7 is more than half-grown, while p8–p10 are retained juvenile feathers. In the secondaries only s1, s5 (with a black band) and the innermost s11 have been moulted, all others being retained juvenile feathers. Although p7 is growing the bird has suspended its moult and is not actively moulting; the remaining juvenile feathers will be replaced on the wintering grounds. Spain, 3.9.2013 (DF)

resident birds from the Mediterranean complete their complete flight feather moult by late October, while their migratory relatives from the far north are not even halfway through with theirs.

Moult in hawks and eagles *Accipitridae*

The primary moult in hawks and eagles follows a simple basic pattern, where the primaries are shed in a sequence starting with the innermost p1 and finishing with the outermost p10. In most species with a normal complete annual moult the replacement of primaries starts at the onset of breeding, in females usually at the time of laying, in males up to several weeks later. The entire moulting process of the flight feathers starts with the dropping of p1 and only after this does the moult of the secondaries and tail commence.

The secondary moult is simultaneously active at several different points, but usually starts with the dropping of either s1 or s5, quickly followed by the other, and from these foci the moult progresses inwards, towards the body. Soon after the activation of the foci at s1 and s5, a third focus is also activated at the innermost secondary, from where the moult progresses outwards, away from the body. Towards the end of the moult cycle the waves that started at s5 and s11, moving in opposite directions, will meet halfway along, usually around s8–s9, which is the place to look for the last remaining old feathers. Sometimes these secondaries are left unmoulted and may offer important clues for ageing, particularly in the case of retained juvenile feathers.

The basic pattern explained above applies to the smaller species, such as accipiters and harriers, while larger species have developed a more complex strategy. *Buteo* buzzards, for instance, regularly do not complete their first moult, but retain some outer juvenile primaries and keep them until their second moult in their 3rd cy summer. When the second moult commences these outer primaries are shed simultaneously with the inner primaries of a new moult cycle, starting from the innermost primary p1. This creates a seemingly irregular pattern, with two simultaneous moult fronts chasing each other

Figure 4. Older immature Eastern Imperial Eagle clearly showing the stepwise moult typical of older immatures and adults of larger eagles and vultures. Note how waves of freshly moulted (dark) feathers are chasing each other, separated by more worn (lighter) feathers in between. In this particular case the difference between the new and old feathers is emphasised by the fact that the older feathers are of the lighter immature type, while the more recently replaced feathers are of the darker adult type. Since new moult waves start from the active points (foci) before previous waves have been completed, several parallel moult fronts are active at the same time, enabling the bird to moult a greater number of feathers per moult cycle compared to a simpler moult strategy with just one front active at a time. Oman, 7.12.2010 (DF)

in the primaries. These follow the bird through life and older *Buteo* buzzards regularly show two or three simultaneous moult fronts in the primaries. This kind of moult is known as stepwise moult, and is typical of all larger raptors.

With age all large eagles and vultures develop a stepwise moult, but the details vary depending on the size of the bird, but also depending on the species' ecology and migration habits. Most of these large species only replace a few inner primaries in their first moult, while the rest of the plumage remains juvenile. In the next moult the following year the primary moult is resumed where it stopped the year before, but again only a few flight feathers are replaced and many of the outer primaries are still juvenile feathers. During the second moult many species start to moult their inner primaries again, for a second time, thus showing two simultaneous moult fronts in the primaries. As the first and second waves gradually progress towards the outer primaries additional new waves are activated at p1, eventually resulting in several simultaneously active moult fronts in the primaries. Birds following the above strategy include all large species of eagle and vulture, while the smaller spotted eagles and Egyptian Vulture are capable of replacing more feathers per moult cycle, partly because they also moult extensively in the winter quarters and are thus more similar in this respect to the *Buteo* buzzards.

Moult in falcons *Falconidae*

The flight feather moult in falcons differs greatly from the moult described above for hawks and eagles. The primary moult starts by dropping one of the median primaries, usually p4 (sometimes p5), after which the moult progresses both ways, one wave proceeding towards the wingtip and the other towards the innerwing. Halfway through the moult the primaries show a block of fresh median feathers flanked by worn inner and outer primaries. The last primaries to be shed are p1 and p10, and usually the latter terminates the wing moult, being the last feather to reach its full length.

Figure 5. This female Lesser Kestrel, captured at an early stage of its primary moult, has dropped pp3–6 (pp4–6 on the left wing) in rapid succession, and also s5. The dark tip of a new p5 can already be seen in both wings, while p4 and p6 are still in pin. Spain, 30.6.2015 (DF)

The secondary moult starts with s5 (sometimes s4) from where the moult proceeds both ways, outwards and inwards. After a while another moult wave starts from s11 and proceeds outwards, eventually meeting the wave from s5 somewhere in the middle. The innermost two secondaries, or the tertials, are moulted separately during the course of the secondary moult. The last secondary to be replaced is usually the outermost, s1. Occasionally the odd median secondary may be retained after moult is completed, but this happens more rarely in falcons compared to hawks of a similar size.

Resident and short-distance migrant species, like Gyr Falcon, Common Kestrel and Merlin, and also resident populations of Peregrine, complete their flight feather moult in one go, except perhaps for a short temporary break during breeding, while long-distance migrants such as Eurasian Hobby, Sooty and Eleonora's Falcons, and Red-footed and Amur Falcons only moult a few or no remiges at all while breeding, as most of the moult is undertaken on the wintering grounds.

FACTORS AFFECTING FLIGHT

Although many species can be identified, with experience, by the way they fly or move through the air there are several factors affecting the flight, which need to be taken into account.

Moult is one factor, which may considerably change the way a bird flies. Missing feathers have a direct impact on the bird's wing-loading (the relation between the bird's weight and its wing area), by hampering the carrying capacity of the wings. Moult does not only affect the bird's flight, it also alters the silhouette and may in some cases change the look of a bird to remind one of another species. Wind conditions are also of importance to how we perceive the flight of a bird. Strong or light winds, and head- or tail-winds all have a bearing on the flight. Raptors can sometimes be seen carrying prey, which also affects the wing-loading and hence the normal flight. Even a full crop may change the flight of a bird, as the point of gravity is shifted forward and to counterbalance itself the bird has to push its wings further forward. This is particularly obvious in large vultures, in which a full crop also extends their necks, changing the entire flight silhouette. Also, the purpose of the flight makes a huge difference to the wing-action and the behaviour of the bird. Raptors can be soaring, gliding, flapping, stooping, hovering, and so on, not to mention various sorts of flight displays and mutual interactions. With some experience it is easy to see different modes, even of the normal powered flapping flight depending on the situation. The normal powered flight on migration is more relaxed than the very determined and purposeful wing-action of a bird taking up a chase after prey or fending off an intruder from its territory.

PREVIOUS EXPERIENCE

When it comes to identifying raptors, the simple truth is that there is no substitute for experience. The more you see and the more often you see the same birds over and over again, the more confident you will get. Books are good for checking plumage details, but other important clues, like learning the different flight modes, the ever varying silhouettes, and tricks of light and wind, can only be learnt through repeated and careful observation. Start with whichever species is the commonest around you and never miss an opportunity to look and learn, over and over again. Once you know the common birds any vagrant will stand out and set the alarm bells ringing.

TOPOGRAPHY

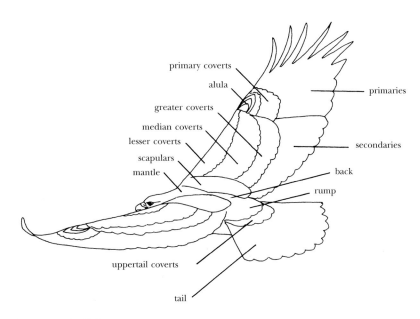

primary coverts
alula
greater coverts
median coverts
lesser coverts
scapulars
mantle
primaries
secondaries
back
rump
uppertail coverts
tail

Feather tracts of upperparts in flight

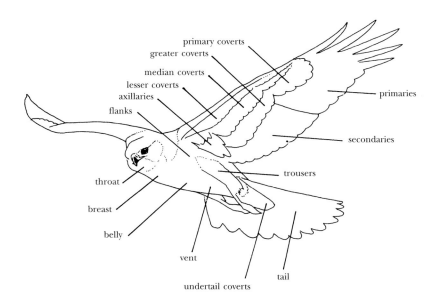

primary coverts
greater coverts
median coverts
lesser coverts
axillaries
flanks
primaries
secondaries
trousers
throat
breast
belly
vent
undertail coverts
tail

Feather tracts of underparts in flight

19

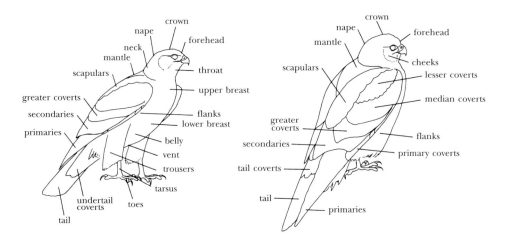

Feather tracts of a perched bird

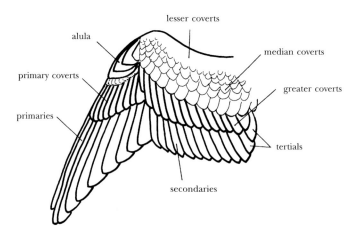

Upperwing

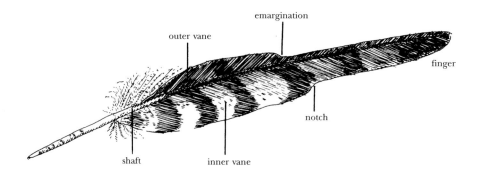

A fingered primary

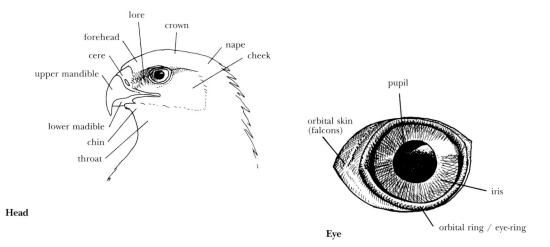

Head

pupil

orbital skin
(falcons)

iris

orbital ring / eye-ring

Eye

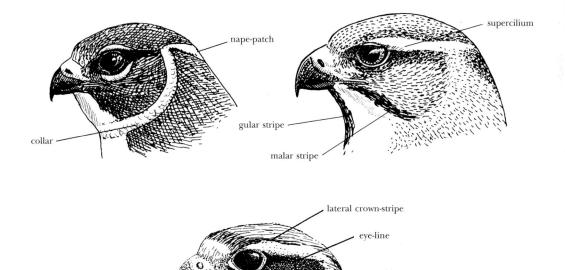

nape-patch

supercilium

collar

gular stripe

malar stripe

lateral crown-stripe

eye-line

auricular patch

moustache

Head markings

GLOSSARY

Adult (ad; plumage) The final plumage of a bird.

Anterior Towards the front or the head of the bird.

Ascendant moult (moulting sequence) When the moult wave in the wing progresses towards the body (away from the wingtip).

Auricular (spot or patch) Plumage mark on the ear coverts, sometimes also including part of the cheeks.

Axillaries The elongated feathers of the 'armpit'.

Bare parts The parts of a bird not covered with feathers (usually cere, bill, eyes, eye-rings and feet).

Body plumage All contour feathers of a bird, excluding the remiges and the rectrices.

Carpal (carpal-patch, -area, -comma, etc.**)** The underwing primary coverts or the area close to it.

Centre (moult centre) The position from where the moult of the flight feathers starts (cf. Focus).

Cere The unfeathered, waxy skin covering the base of the upper mandible in birds of prey.

Cheek The side of the head below the eye.

Collar Specially shaped feathers surrounding the facial disk of harriers *Circus*.

Contour feather All (visible) feathers covering the bird, as opposed to down.

cy (calendar year) A bird is in its 1st cy from its birth until 31st December of that same year, when overnight it becomes 2nd cy, and so on. The 'birthday' is thus on 1st January every year.

Descendant moult (moulting sequence) When a moult wave progresses away from the body (towards the wingtip).

Definitive adult The final adult plumage of a bird which will not change further with age. The term 'full adult' is often used for this final plumage stage.

Dihedral Wing position of a bird in flight, with wings held above the horizontal plane in a shallow V.

Dimorphic A bird showing two distinct (colour) morphs.

Distal Towards the periphery; away from the bird's body; opposite of Proximal.

Divergent moult Wing-moult starting from a focus but progressing simultaneously inwards and outwards; typical of remex-moult in falcons.

Dorsal The upper surface (upperparts) of a bird; opposite of Ventral.

Eye-line Usually a dark line in front of and/or behind the eye.

Eye-ring The bare skin around the eye; rather prominent in falcons *Falco*, for example.

Feather tract A set of feathers that grow and wear together, and are moulted at roughly the same time.

Fingers The emarginated tips of the longest primaries.

First-adult plumage In many medium-sized raptors a first-adult plumage can be separated from the definitive adult plumage at close range.

First-summer plumage The plumage worn during the 2nd cy summer, at an age of *c.* one year; this is usually a mix of juvenile and adult feathers.

Flight feathers The remiges and retrices.

Focus (plural foci) The point where a moult wave starts, indicating the first feather to be shed in a given sequence.

Gular stripe A longitudinal streak on the throat.

Immature (imm; plumage) In this book given the meaning 'not mature' or 'non-adult' and includes all the plumages except for the adult plumage (also the juvenile if not specified more exactly).

Juvenile A young bird in its first year of life, still carrying its juvenile remiges.

Juvenile plumage (juv; plumage) The first complete plumage acquired in the nest and during the fledging stage. Most raptor species have this plumage until their first complete moult in the 2nd cy spring, but some (smaller) species undergo a partial moult of the body feathers during the first winter.

Leading edge of wing The anterior edge of the wing.

Lores The area between the eyes and the cere. Usually covered in bristles, but in honey-buzzards covered in scale-like feathers.

Malar stripe Usually a dark stripe running from the gape along the lower edge of the cheeks.

Mantle The feather tract between and in front of the scapulars, sometimes also given the meaning of the upperparts more generally, including the scapulars.

Monotypic Species with only one (the nominate) subspecies.

Morph A distinct (plumage) variant of a species.

Moustache A prominent dark mark below the eye, especially in falcons.

Nape The upper hindneck, to the rear of the crown.

p = primary (pp = primaries) Used for numbering the primary flight feathers: p1, p2 etc; numbered descendantly from the innermost to the outermost.

Patagium The fold of skin between the wrist and the body (which equates to the forearm).

Phase An age-related colour form of a species, often incorrectly applied to Morph.

Plumage All the feathers a bird has at any one time.

Plumage-type Indicates a separable, age-linked plumage, although exact ageing may not be possible.

Polymorphic A species with two or more distinct morphs, usually differing in colour.

Polytypic A species with two or more subspecies (cf. Monotypic).

Posterior Towards the rear end of the bird.

Primaries The outer flight feathers of the wing attached to the metacarpus and digits, forming the 'hand'.

Primary projection The tips of the primaries that project beyond the tertials on a closed wing.

Proximal Towards the base of a structure (tail, feather, etc.); opposite to Distal.

Rectrix (plural Rectrices) Tail feather(s).

Remex (plural Remiges) Wing feather(s), used for the primaries and secondaries together.

Ruff The elongated neck feathers in some vultures.

s = secondary (ss = secondaries) Used for numbering secondary flight feathers: s1, s2 etc; numbered ascendantly, from the outermost to the innermost.

Scapulars Dorsal feathers on each side of the mantle, covering most of the upperparts on a perched bird.

Secondaries The inner flight feathers of the wing, attached to the ulna (and humerus), forming the 'arm'.

Serially descendant moult When a second moult starts before the first has finished, resulting in two (or three) active moult centres in a wing.

Sp. Species (plural **spp.**).

Ssp. Subspecies.

Subadult (subad; plumage) The plumage(s) preceding the adult plumage.

Subterminal band (on feather) A dark band just inside the tip of a feather.

Supercilium A contrasting (usually pale) line running above the eye.

Suspended moult When the moult is interrupted e.g. for migration or during breeding, but resumed shortly after.

Terminal band (on feather) A band at the tip of a feather.

Tertials The innermost secondaries, varying in number between two and three in raptors.

Trailing edge The posterior edge of the wing, formed by the tips of the secondaries and inner primaries.

Trousers The elongated feathers of the thighs.

Underbody/underparts The feathering of the underside of the body and the underwing-coverts.

Underwing The lower surface of the wing.

Upperparts The feathering of the entire upperside including body, wings and tail.

Upperwing The upper surface of the wing.

Ventral The underparts or lower surface of the bird or a structure; opposite to Dorsal.

Wing-formula The relative distances between the primary tips forming the wingtip; mostly compared with the tip of the longest primary.

Wingspan The measurement between the extended wingtips.

Wrist The joint between the hand and the arm in the wing. Also called the carpal joint or the bend of the wing.

Adult Eurasian Griffon Vulture. Spain, 23.11.2010 (DF)

HAWK-WATCHING
IN AND AROUND EUROPE

Keith L. Bildstein and Anna Sandor

Field guides are the backbones of successful raptor-watching. World-class guides like this one allow the uninitiated to quickly acquaint themselves with a region's birds of prey. Raptors, in particular, can be notoriously difficult to distinguish in the field and the present offering will be of great value in helping raptor-watchers, and raptor biologists for that matter, to correctly identify the birds that they see. Furthermore, seeing lots of raptors with this field guide in-hand will speed the rate at which raptor-watchers build their identification skills.

Unfortunately, because they are secretive and wide-ranging for most of the year, raptors tend to be difficult to see, let alone to identify. This general rule changes for many species during migration, however, offering numerous opportunities for seeing large numbers of birds of prey at close range. Although raptors often migrate across broad fronts – particularly at the beginning and end of their migratory journeys – many predictably aggregate in large numbers along well-established migration corridors, sometimes congregating by the tens of thousands at specific geographic features that include mountain ridges and passes, narrow coastal plains, isthmuses and peninsulas. For raptors, traditional migration corridors usually occur along what are known as 'leading lines', geographic or topographic features such as mountain ranges that are orientated along or near the preferred direction of travel and that provide updrafts for low-cost soaring flight, as well as along 'diversion lines', including land–water interfaces and lowlands next to high mountain ranges, where migrants concentrate not because they are attracted to them but because they are trying to avoid what lies beyond (Bildstein 2006).

Because many raptors hesitate or refuse to cross bodies of water that are wider than 25 km, raptor migration 'bottlenecks' occur in areas that allow migrants to forego or reduce the lengths of extensive flights over water. Water avoidance in the Europe–Africa migration system is best seen at places such as the Strait of Gibraltar at the western end of the Mediterranean, the Bosphorus at the north-eastern corner, and the Strait of Messina between Sicily and south-westernmost peninsular Italy in the central Mediterranean (Bildstein 2006).

The daily passage of migrating raptors by the hundreds and even thousands of individuals along leading lines and diversion lines, and at migration bottlenecks, has intrigued scientists for hundreds of years. French zoologist Pierre Belon described spectacular movements of Black Kites over the Bosphorus thus: "If they had continued for a fortnight in the same strength as on that day we could have surely said that they were in greater number than all men living on the earth..." (Nisbet & Smout 1957). Historically, such concentrations increased the vulnerability of many populations of raptors to human threats including shooting and trapping (Zalles & Bildstein 2000, Bildstein 2006). More recently, well-timed visits to established watch-sites along traditional migration corridors have provided raptor enthusiasts and hawk-watchers with excellent opportunities for seeing large numbers of migrating raptors at close range, creating some of the best situations for using this field guide to hone one's identification skills.

The hobby of raptor-watching has some of its deepest roots in and around Europe, and many of the world's most famous raptor migration watch-sites are found there. With this new field guide in hand, watching birds of prey at such sites offers one of the best ways to develop an appreciation for the many rules and complexities of both raptor identification and raptor migration. Well-timed visits to such sites provide opportunities for seeing large numbers of birds of prey and in the company of other enthusiasts who can help you identify them. Below we describe several of these sites and the raptors that one is likely to see at them. The real challenge in putting together these descriptions has been to reduce to eight the number of 'representative' watch-sites described. Several of the sites were chosen largely for their history, others for their diversity of species or the incredible magnitude of their migrations, and others still for their geographic settings and landscapes. Dozens of additional great European migration watch-sites are described in Zalles & Bildstein (2000).

FALSTERBO BIRD OBSERVATORY, SWEDEN

Located between 55º and 56º N, the Falsterbo Bird Observatory in south-western Sweden ranks as one of the northernmost raptor migration watch-sites in the world. It is also one of the world's oldest continually active sites. The hammer-shaped, 7-km long, flat and sandy peninsula that juts into the Baltic Sea, 24 km across the Öresund Strait from western Denmark, is a natural collecting point and major autumn migration bottleneck for many of western Scandinavia's land-restricted migrants (Zalles & Bildstein 2000, Karlsson 2004).

The site, which was first described as migratory bottleneck in the early 1800s by zoologist Sven Nilsson, gained fame as a raptor migration hotspot on the heels of Gustav Rudebeck's painstaking investigations of migration geography in the region in the 1930s and 1940s. Rudebeck backtracked, on a bicycle, the sources of individual streams of migrants and mapped their movements towards the site through south-western Sweden. His empirical approach allowed him to develop a model for the concentrated nature and location of the migration that used a combination of the site's coastal diversion lines, predominant winds, and a series of habitat-associated leading lines inland to explain each day's migration (Rudebeck 1950). His accurate observations also laid the foundation for understanding the seasonal progression of the migration, starting with the acute passage of most of the site's European Honey-buzzards over the course of several days in late August–early September, and concluding with the decidedly more protracted passage of Northern Goshawks in October and early November.

Falsterbo has attracted the attention of numerous writers and researchers over the years. Anyone planning a visit to the watch-site is recommended to read the colour-illustrated and informative account *Wings over Falsterbo* (Karlsson 2004). Ulfstrand *et al.* (1974) provide an additional useful overview of the site and the visible migration to be seen there. Animal ecologist Nils Kjellén has focused his attention on age and sex differences in the migration at Falsterbo, and has linked both annual fluctuations in age and sex ratios, and age and sex differences in the timing of the migration, to critical aspects of the population biology and ecology of the species involved (Kjellén 1992, 1994, 1998). Kjellén (1997) has also assessed the degree to which Falsterbo serves as a regional bottleneck by comparing the magnitude of migration there with published estimates of Swedish populations of migratory raptors. His analysis suggests that, on average, the site records the movements of 38% of Sweden's Red Kites and 33% of its European Honey-buzzards, as well as 10–20% of its Marsh Harriers, Eurasian Sparrowhawks, Common Buzzards and Peregrine Falcons.

Thirty-one species of raptors are recorded as migrants at Falsterbo; 15 species are regular migrants. Approximately 40,000 raptors are seen in autumn. On average, more than 50% of the migration is made up by Eurasian Sparrowhawks (>17,000; 45,000 in 2012). The second most numerous raptor is the nominate subspecies of the Common Buzzard (30% of the migration; >10,000). Other significant migrants include European Honey-buzzard (>4000), Red Kite (>800), Rough-legged Buzzard (>800), Common Kestrel (>500), Osprey (>200), and Merlin (>200). Other birds seen in high numbers include (2012 total count/average): Barnacle Goose (>180,000), Common Eider (>90,00), Wood Pigeon (490,000), and Chaffinch/Brambling (>2,400,000).

Together with several sites in neighbouring Denmark, including Stevns Klint, directly across the Baltic Sea from Falsterbo, and Stigsnæs on the south-west coast of Sjaelland (Zalles & Bildstein 2000), Falsterbo offers the best opportunity for observing concentrated raptor movements in all of Scandinavia, especially during the peak passage around the first and second weeks of October.

Full-season counts have been conducted at Falsterbo by the Swedish Environmental Protection Agency since 1973. From 2001, counts have been made at Nabben at the southern end of the peninsula daily from 1 August to 20 November (Kjellén 2001). The Agency also operates the Falsterbo Bird Observatory in the town of Falsterbo on the peninsula, and they maintain an active website (www.falsterbofagelstation. se) that provides useful information about raptor migration and other ornithological events in the area.

COL D'ORGANBIDEXKA, FRANCE

Since 1980, Organbidexka Pass, which is believed to be the most important raptor migration bottleneck in the western Pyrenees, has been a *col libre*, or 'shooting-free zone'. In the late 1990s, there were more than 6,000 shooting hides along 200 km of ridges of the Atlantic Pyrenees. Although most hunters target passerines and pigeons, raptors are often also killed (Devisse 2000). The French League for the Protection of Birds (LPO Aquitaine) monitors the movements of the raptors at the site (Urcun, pers. comm.).

The watch-site, situated in the northern Basque country of southern France, 1,283 m above sea level, in the heart of the Irati Europe's most extensive beech forest, offers one of the best opportunities in Europe for observing large numbers of both Red and Black Kites.

Although avian biologists have long speculated on the extent of visible migration in the Pyrenees, little had been published on the movements of birds through this sparsely populated part of Europe until the early 1950s when David and Elizabeth Lack reported the results of three autumns of migration 'reconnaissance' tours (Lack & Lack 1952). Anticipating a concentrated coastal migration, the two biologists instead discovered extensive large-scale inland movements among the north–south 'cols' or mountain passes in the western Pyrenees that funnelled raptors and other diurnal migratory species into concentrated streams of migrants.

The Lacks also uncovered the ancient Basque pastime of *La chasse de la palombe* or Wood Pigeon hunting, in which teams of hunters hurling a series of 'zimbelas', or small white wooden discs, below flocks of low-flying Wood Pigeons and Stock Doves, elicited an alighting response in the birds that caused them to fly lower still, either into a series of nets, or onto the ground, at which time a horn was sounded and the hunters 'let fly in every direction'. The so-called sport remains popular today, and estimates of the annual kill range into the tens of thousands of pigeons and doves. The hunters are not particularly specific in their targeting, and many raptors are also slaughtered in autumn.

This is the backdrop for Organbidexka Col Libre (OCL), which has been monitoring the movements of and protecting trans-Pyrenean migrants since it was founded in 1979. In addition to Le Col d'Organbidexka, which is located on the northern slopes of Pic d'Orhy, halfway between Saint-Jean-Pied-de-Port and Tardets, OCL conducts counts and protects raptors at two other important watch-sites in the western Pyrenees, Le Col de Lizarrieta, on the border with Spain approximately 20 km south of Ascain, and Lindux, a Napoleonic fortification, 30 km south of Saint-Jean-Pied-de Port and several km north-west of Col de Roncevaux, where Roland was slain in AD778 while conducting a rear-guard action for Charlemagne.

The daily count is conducted annually from 15 July until 15 November, from sunrise until sunset.

Fourteen species of raptors are recorded as regular migrants at Organbidexka, and another seven are irregular migrants, although there are days when 20 species can be seen. Approximately 50,000 migrants are seen in autumn. Three species, European Honey-buzzard (23% of the overall flight; >12,000), Black Kite (67%; >35,000), and Red Kite (8%; >4,500), together make up 95% of all raptors seen. Most migrants at the site fly relatively low while crossing the pass, making them easy to see and identify. Opportunities for seeing large numbers of migrants occur mid- to late August during the peak movements of Black Kites and European Honey-buzzards. October is the best time to see Red Kites, which are rare migrants elsewhere in Europe, and Booted Eagles, which rarely occur this far north.

Common Cranes (>20,000) also migrate at the site. One additional notable aspect of the site is the late October migration of well over 100,000 Wood Pigeons and Stock Doves, both of which are legal game on migration. The autumn passage of these two species attracts thousands of hunters, many of which operate near the watch-site. Col d'Organbidexka is one of only a few raptor migration sites that records the numbers of gunshots heard as well as the numbers of migrants seen, and the current single-day record of 25,360 shots heard on 28 October 1982 highlights the degree to which shooting, much of it indiscriminate, persists in the region.

Although all of OCL's watch-sites are easy to find, the best place to begin a visit is at OCL headquarters at 11 Rue Bourgneuf, F-64100 Bayonne, France. One can find useful information on the www.migraction. net website as well.

STRAIT OF GIBRALTAR, SPAIN AND GIBRALTAR

The narrow body of water that connects the western end of the Mediterranean Sea to the Atlantic Ocean is also one of the most identifiable landscapes on Earth, and one of its most prominent migration bottlenecks, a 14 km-wide sea passageway across which hundreds of thousands of central and western European raptors travel en route to wintering areas in West Africa.

The fact that large numbers of raptors concentrated in southern Iberia before streaming into West Africa has been known to science since the late 1700s (Nisbet *et al.* 1961). British officer and ornithologist H. L. Irby was the first to detail the timing of the migration, as well as the role that cross winds played in determining the likelihood and location of the passage (Irby 1875). However, it was not until the 1960s that the magnitude of the passage became known (Evans & Lathbury 1973, Bernis 1980). Since then a number of researchers, most notably Gibraltar native Clive Finlayson (1992), have studied and reported on the migration there.

The narrow channel that separates Europe from Africa provides ample evidence for the extent to which raptors differ in their willingness to fly over water. Indeed, although the Strait appears to be a minor hindrance for some, it presents a formidable barrier for others. Ospreys, European Honey-buzzards, harriers, Peregrine Falcons, and other migrants with powerful flapping flight regularly cross the Strait almost immediately upon reaching land's end, whereas Eurasian Griffons, Black Kites, Booted Eagles and other obligate soaring migrants spend up to a week or more milling about in southernmost Iberia waiting for the right weather. Although such species' hydrophobia plays havoc with counting at the site, it also creates first-rate opportunities for birding there.

In general, migrants approach southernmost Spain and Gibraltar across a something of a broad front. With few exceptions, however, most cross the Strait in *streams* in a series of narrow-frontal movements, with the exact point of departure along the length of the passageway depending upon predominant winds. Movements tend to be greater overall and the migration heaviest at the eastern end of the Strait when 'Poniente' or westerly cross winds prevail, and lesser overall and heaviest towards the middle of the Strait when the far more dangerous 'Levante' or easterly cross winds prevail, the latter substantially increasing the likelihood of migrants being blown out into the Atlantic.

Thirty-four species are recorded as regular migrants at the Strait of Gibraltar, 25 of which are regular migrants. Approximately 230,000 and >260,000 migrants are seen in spring and autumn, respectively. Black Kites (50% of the overall total; >130,000) and European Honey-buzzards (30%; >85,000), make up more than two-thirds of the overall migration. Four other species, Short-toed Snake Eagle (7%), Booted Eagle (6%), Eurasian Griffon (3%), Egyptian Vulture (<1%) and Eurasian Sparrowhawk (<1%), migrate by the thousands to tens of thousands. Aside from being a great place to see migrating Old World vultures, the Strait of Gibraltar is also a great place to see three species of European harriers (Marsh, Hen and Montagu's), as well as the occasional Lesser Kestrel, Common Kestrel, Eurasian Hobby and Peregrine Falcon. Rarities, including Rüppell's Griffon Vulture and Eleonora's Falcon, can been seen among the huge flocks of more common species (Harris 2013).

The late-summer and early-autumn movements of Black Kites and European Honey-buzzards are truly spectacular. Most kites pass in July and August, whereas most honey-buzzards do so during a brief period at the end of August and early September. Kites tend to cross along the central and western parts of the Strait, whereas honey-buzzards, which typically approach along coastal Mediterranean flight lines, are far more likely to depart from the Rock of Gibraltar at the eastern end of the Strait.

In addition to soaring raptors, tens of thousands of White Storks and thousands of Black Storks can be seen migrating at the site in July and August and in September and October, respectively.

Equally systematic, albeit proportionately smaller, counts of spring migrants occur along the coast of southernmost Spain between the towns of Tarifa and Algeciras, at a time when birds crossing from Morocco on their return migration often pass at close range while flying close to the surface of the water.

The principal stream of migration shifts along the more than 20 km length of the Strait depending upon the wind, and migration at the site is all-but-impossible to track from a single lookout. Hawk-watching is best from a series of watch-sites, including the Rock of Gibraltar, as well as from a number of points in southern Spain between Punta del Carnero, south of Algeciras, west to Punta Marroquí, near Tarifa. The watch-site on the Rock of Gibraltar is operated by the Gibraltar Ornithological and Natural History Society; those in Spain are coordinated by Programa Migres, in cooperation with the Government of Andalucía. Both organisations maintain active websites (www.gonhs.org and www.fundacionmigres.org). Programa Migres has produced a useful *Guide to the Common Birds of the Strait of Gibraltar*. The guide and maps of the watch-sites are available at the Huerta Grande Visitor Centre for Los Alcornocales National Park, on the main coastal highway several miles west of Algeciras, which is the best place to begin a visit to the site. Clive Finlayson's *Birds of the Strait of Gibraltar* (1992) provides an excellent history of the site and well as useful details regarding the timing, ecology and geography of each species' migration.

THE STRAIT OF MESSINA, SICILY–CALABRIA

This spring coastal mountain watch-site, which is one of the most important bottlenecks for migrants using the Central Mediterranean Corridor, overlooks the 3 km-wide Strait of Messina, the body of water that separates the 'toe' of the Italian peninsula (Calabria) from north-eastern Sicily. Counts are made from both sides of the Strait along the slopes of Monti Peloritani in Sicily, as well as along the coast of Calabria. Raptors cross from Sicily to Calabria along a relatively broad front with the greatest concentrations occurring along 20 km at the northern part of the Strait, with the exact crossing depending on wind and general weather conditions. The presence of clouds and fog is strictly connected with the time of the passage. In early morning and late afternoon there are fewer thermals and birds pass over in flapping flight. In autumn, most of the birds of prey are concentrated in the mountains of Aspromonte National Park in Calabria.

Sadly, this site is well known among bird protection associations for the raptor shooting that has been happening there for decades. In the late 1970s, raptors and storks were being shot from dozens of elaborate, multi-person concrete bunkers in the Peloritani Mountains in spite of regional and national prohibitions. Raptor conservation protest camps were initiated in the region in 1981, and counts of both migrants and numbers of shots heard started in 1984. By the late 1990s, many shooting bunkers had been abandoned (Giordano *et al.* 1998). Shooting continued regularly on the Calabrian side of the Strait as recently as 2002. Despite the activities of conservationists in Calabria shooting currently remains an episodic problem there, although this activity is now rare in Sicily.

In 1984 Sicilian environmental associations established a study and surveillance camp to help protect migratory birds, still held by the Mediterranean Association for Nature in collaboration with WWF and NABU (Germany). Since then 37 species of raptors have been recorded on migration – no single site in Europe has recorded such a large variety of raptors. Two distinctive subspecies, rare elsewhere in Central–Western Europe, are regularly observed at the Strait as well: the *calidus* race of Peregrine Falcon (Corso 2001) and Steppe Buzzard, the *vulpinus* race of the Common Buzzard. The first documented observation for Italy of a Crested Honey-buzzard (*Pernis ptilorhyncus*) was made in May 2011 (A. Scuderi in Janni & Fracasso 2013). Vagrants include Black-shouldered Kite, Black Vulture, Levant Sparrowhawk, Steppe Eagle, Eastern Imperial Eagle, Greater and Lesser Spotted Eagles, and Amur Falcon. Yet another highlight of the Strait of Messina is the possibility of seeing the rare and local *feldeggii* race of the Lanner Falcon (Corso 2000).

The Strait of Messina is now recognised as one of the best places in the Western Palearctic to see the spring migration of the globally endangered Pallid Harrier (Corso 2004), together with the endangered Lesser Kestrel. Occasional irruptions of Red-footed Falcons also occur there (A. Giordano pers. comm.).

Annual spring raptor counts increased from 3,198 in 1984 to more than 40,000 in recent years, including 35,000 European Honey-buzzards. Other common species include Marsh Harrier (>3,000) followed by Black Kite (*c*.1,000), Common Kestrel (>900), and Montagu's Harrier (>800). Red-footed Falcon (a record number of >7,000 in 1992) is also common, along with Eurasian Hobby (>200) and Lesser Kestrel (>200). Egyptian Vulture, Lesser Spotted Eagle, Booted Eagle, Saker Falcon and Peregrine Falcon are less common regular migrants.

April typically has high species diversity, including some rarities, but low numbers (one day in mid-April more than 100 raptors were seen, but of 17 different species). Harriers are the earliest migrants, peaking as early as the first half of that month. The first two weeks in May have high overall numbers (up to 3,000/5,000 a day) but fewer species, with European Honey-buzzards making up the overwhelming majority of the passage.

The first autumn count occurred in the National Parks of Aspromonte in 2000 (Guglielmi *et al.* 2003), and 20,000 raptors were counted there between 11 August and 10 October 2012 (Grasso *et al.* 2012). More information can be found on the website of the Mediterranean Association for Nature, which is involved in studying and protecting migratory birds over the Strait of Messina (www.migrazione.it).

BURGAS, BULGARIA

Reports on the migration of soaring raptors using the Via Pontica (west coast of the Black Sea) migration corridor at the Bosphorus were published in the late nineteenth century (Alleon & Vian 1869, 1870). In Bulgaria, the Black Sea coast and the nearby hills play an important role in the orientation of these migrant birds; and the nature of the coastline near the town of Burgas creates a significant bottleneck, which is the best-known watch-site on the Via Pontica (Zalles & Bildstein 2000). From Cape Emine, the coast turns sharply west for 15 km, along the northern edge of Burgas Bay, and then continues south and west in a stepped fashion for a further 20 km. The Black Sea reaches its westernmost extremity near the town of Burgas, which is an ideal place for observing autumn migration. Beyond Burgas the route diverges: some birds follow the coast, while others continue inland (Michev *et al.* 2011).

Counts at Burgas Bay started in 1978 (Michev & Simeonov 1981) from a dyke on the NE shore of Lake Atanasovsko. Research at the site is part of International Counts of Soaring Birds in the Western Palearctic. Although the site is decidedly 'low key', it remains an important concentration point along this significant migration corridor in eastern Europe.

Around the city of Burgas, several lakes and wetlands host migrant waterfowl, pelicans, waders, gulls and terns in spring and autumn as well. The shallow, salty Lake Atanasovsko, just north of the city, records more than 300 bird species. One of the world's rarest birds, the Slender-billed Curlew, was last reliably seen here in 1993 (Harris 2013).

The Kableshkovo Hills, located a few miles north of Burgas, seem to benefit from the most concentrated passage of soaring birds: the average count (2004–2011) is 150,000 individuals in spring and 270,000 in autumn, made up mainly of White Pelicans, Lesser Spotted Eagles, Common Buzzards and White Storks (Harris 2013).

Thirty-three species of raptors were recorded at Burgas between 1979 and 2003. The mean number of migrating raptors observed in autumn is close to 40,000, with a clear trend towards increasing numbers in recent years (Michev *et al.* 2011). Count data suggest that Burgas is one of the most important sites for monitoring the migration of soaring birds in Europe, and the most important for the autumn migration of several species, including White and Dalmatian Pelicans, White Storks, harriers, Levant Sparrowhawks, Lesser Spotted Eagles and Red-footed Falcons. One of the most numerous migrants is the European Honey-buzzard with the main period of migration occurring between late August and early September (8,779 individuals on 4 September 1996), the same time at which numbers of Black Kites and White Storks peak (Michev *et al.* 2011). Two subspecies of Common Buzzards migrate at the site, the nominate race and the Steppe Buzzard, with the latter being far more common. Lesser Spotted Eagle migration peaks between late August and early October.

More information can be found on the website of the Central Laboratory of General Ecology, Institute of Biodiversity and Ecosystem Research, Bulgarian Academy of Sciences (www.ecolab.bas.bg).

BOSPHORUS, TURKEY

The Via Pontica migration corridor along the western coastline of the Black Sea is a major route for raptors in the region (Bijlsma 1987). In both spring and autumn, raptors migrating along the corridor are funnelled by the narrow bridge that links Europe and Asia at the Bosphorus (Cramps & Simmons 1980), where more than 167,000 migrants were observed from 22 September to 13 October 2008, (Milvus Group, unpublished.). Most of the world population of Levant Sparrowhawks and Lesser Spotted Eagles may migrate at the site (Bijlsma 1987).

The Bosphorus is the 30 km-long, 1.5 km-wide, N–S strait between the Sea of Marmara and the Black Sea, separating European from Asian Turkey. Migrants can be seen throughout the strait, which includes Istanbul and its suburbs. Locations of the flight-lines change according to wind direction and time of day. Autumn watch-sites include two hilltops, Büyük Çamlica and Kuçuk Çamlica, on the Asian side of the strait. A hilltop near Sariyer, 15 km north of Istanbul on the European side, is good in both spring and autumn (Zalles & Bildstein 2000). The site is not regularly monitored.

Twenty-nine species appear to be regular migrants in spring; 31 or 32 are regular migrants in autumn (Zalles & Bildstein 2000). The largest movements of birds of prey occur on days of light northeasterly winds. The main species found migrating were (with total numbers recorded in brackets) White Stork (207,145), Black Stork (6,194), European Honey-buzzard (8,997), Common/Steppe Buzzard (12,949), Greater/Lesser Spotted Eagle (4,309) and Eurasian/Levant Sparrowhawk (5,224) (Porter & Willis 1968).

According to the Milvus Group's count data, >149,000 raptors passed through here during the 19 days of counting. Most of the birds were Common Buzzards (>79,000, with a peak of >8900 individuals on the 29 September), and Lesser Spotted Eagles (>60,000, with a peak of >23,000 birds on the 30 September). Other species observed in high numbers were Short-toed Snake Eagle (>4800), Levant Sparrowhawk (>3000), Eurasian Sparrowhawk (>1700), Booted Eagle (194) and European Honey-buzzard (178). The third most numerous migrant during the same period was Black Stork (>15,000).

The following website offers additional details: http://milvus.ro/raptor-migration-watch-site-at-bosphorus/307.

BATUMI, GEORGIA

This major bottleneck along the southeastern coast of the Black Sea has only recently been discovered as the most important autumn flyway in Eurasia. The watch-site provides an opportunity to see large numbers of migratory raptors during their outbound and return migrations. What makes Batumi special is the close juxtaposition of the Black Sea coast and the formidable Greater Caucasus mountain range several km to the east, which creates a narrow passageway through which hundreds of thousands of southbound migrants funnel each autumn. Depending upon the weather the passage can pass just a few metres overhead, making the movement spectacular and the birds easy to identify. Unfortunately this also makes the birds vulnerable to shooters and trappers. Eurasian Sparrowhawks have been trapped for falconry for a long time in Georgia, particularly along the western Black Sea coast (van Maanen *et al.* 2001). By the late 1990s Georgian law 'protected' migrants from such actions, but enforcement has remained insufficient in many areas, up to the present day. The Georgian Center for the Conservation of Wildlife began monitoring the movements of raptors and conducting education programmes near Batumi in autumn 2000 (van Maanen *et al.* 2001). Presently Batumi Raptor Count (BRC), a young and dynamic international team, conducts annual autumn counts, works to promote tourism and raises environmental awareness in the surrounding villages and schools, as well as working to put an end to illegal shooting at the site. In addition to counting the numbers of migrants, counters also record the numbers of gunshots heard.

The watch-sites are not easy to find, so the best way to prepare for a visit is via the BRC's website (www.batumiraptorcount.org), where visitors can find all the necessary information regarding travel, accommodation and guided tours.

Every autumn, from 15 August to 15 October, an international team of volunteer counters records a visible passage of more than 1,000,000 birds of prey, together with hundreds of European Bee-eaters, European Rollers and Black Storks. The counts are conducted in two small villages, Sakhalvasho and Shuamta, just north of Batumi, near the southeastern coastline of the Black Sea.

Thirty-five species of raptors have been recorded by the Batumi Raptor Count. Two species make up the majority of the passage: European Honey-buzzards (about 400,000 each autumn, or approximately 50% of the overall passage) and Steppe Buzzards (>250,000, or about 30% of the overall passage). Daily peak counts for both species reach 60,000 to 80,000 individuals. Other numerous species include Black Kite (>50,000), Marsh Harrier (*c.* 4,000), Montagu's Harrier (*c.* 3,000), Pallid Harrier (*c.* 1,000), Eurasian Sparrowhawk (>4,000), Levant Sparrowhawk (*c.* 4,000), Booted Eagle (*c.* 4,000), and Lesser Spotted Eagle (*c.* 4,000). Osprey (>50), Short-toed Snake Eagle (>600), Long-legged Buzzard (*c.* 100), Greater Spotted Eagle (>100), Steppe Eagle (>300), Red-footed Falcon (*c.* 300), Lesser Kestrel (>200), Common Kestrel (>300), and Eurasian Hobby (<500) also are seen in large numbers. Red Kite, Crested Honey-buzzard, White-tailed Eagle, Egyptian Vulture, Eurasian Griffon Vulture, Black Vulture, Hen Harrier, Rough-legged Buzzard, Eastern Imperial Eagle, Golden Eagle, Merlin, Lanner Falcon, Saker Falcon and Peregrine Falcon are observed in lower numbers.

The peak period of passage of European Honey-buzzards is at the end of August–beginning of September (with 179,000 individuals on 3 September 2012). At this time of the season large flocks, often of several thousand honey-buzzards mixed with Black Kites, can be seen daily. Late September into October is the best time to see large flocks of Steppe Buzzards, and high numbers of Lesser and Greater Spotted Eagles. Booted Eagles, Short-toed Snake Eagles and juvenile Steppe Eagles regularly join the spectacular passage.

EILAT, ISRAEL

Each spring, millions of birds of prey that have over-wintered in Africa set out from that continent for breeding areas in Europe and Asia. Some leave Africa via the Strait of Gibraltar at the western end of the Mediterranean, others via the Sicilian Channel in the central Mediterranean. Most, however, do so via the Middle East, where, in most years, the elliptical geography of the Eurasian–East African Flyway positions the bulk of the spring passage over the outskirts of the coastal resort city of Eilat, in southernmost Israel, at the head of the Gulf of Aqaba.

One's first impression of the rugged mountains west of Eilat is that this is not a place to watch migrating raptors. Aside from a blazing blue sky above, the Negev Desert landscape looks more lunar than earthly, and, indeed, Eilat may be the only migration watch-site on Earth where, while scanning for raptors, one can see more countries – four, namely Israel, Jordan, Saudi Arabia and Egypt – than trees. What Eilat lacks in plant life, however, it more than makes up for in migrating raptors. The site is one of only three raptor watch-sites in the world that has counted more than a million raptors in a single season, and it has accomplished this in spring, during the inherently 'thinner' return migration. With reports of 38 species recorded there, Eilat is also one of the world's most biologically diverse raptor watch-sites.

Published accounts of raptor migration in the region date from Tristram's mid-19th century reports of large-scale movements in Palestine and Sinai (Tristram 1865–1868). Sporadic counts in and around Eilat date from the late 1950s (Safriel 1968). Systematic, full-season counts date from 1977 when Christensen *et al.* (1981) reported more than 750,000 migrants at the site between February and May.

The most thorough description of migration at the site is Hadoram Shirihai and David Christie's 1992 paper in *British Birds*. *Raptor migration in Israel and the Middle East* provides a brilliant overview of 30 years of fieldwork in the region (Shirihai *et al.* 2000). The International Birding and Research Center (IBRC) maintains an active ringing station for raptors and other birds at and around its field station and headquarters north of Eilat. IBRC organises the counts, which, in most years, are made by one or more groups of volunteers both from within and outside of Israel. Those interested in visiting the site should contact the International Birding and Research Center at Eilat (www.arava.org/birds-eilat/).

Twenty-two species are recorded as regular migrants at Eilat; another 16 species are irregular or rare migrants. Approximately 830,000 and 20,000 migrants are seen in spring and autumn, respectively. Two species, the European Honey-buzzard and the Steppe Buzzard, together make up about 85% of the passage. Other numerically significant migrants include Black Kite (3% of the passage), Steppe Eagle (3%) and Levant Sparrowhawk (2%). Eilat is clearly one of the world's best sites for viewing the latter, a small *Accipiter* whose migratory tactics, including obligate flocking and soaring migration, as well as nocturnal movements, belie its taxonomic affinities. Eilat is also an excellent spot for viewing significant movements of the one of the world's largest long-distance raptor migrants, the Steppe Eagle, as well as a great place to see migrating Egyptian Vultures and Eurasian Griffons. Notable residents at the site include Verreaux's Eagle and the 'Barbary Falcon' race of the Peregrine. The return passage begins with Steppe Eagles in late February–early March, and continues with Black Kites and Steppe Buzzards in late March and early April; it concludes with the decidedly more acute flights of Levant Sparrowhawks in late April–early May, and European Honey-buzzards in early May. Although the passage can be high, on most days good numbers of the birds can be seen at close range, and almost always in good light. Counts are conducted from a large gravel car park in the shadow of Mount Yoash, several kilometres west of Eilat on the road to Egypt. A substantial, although far less studied, autumn migration also occurs at the site (Shirihai & Christie 1992).

NORTHERN VALLEYS, ISRAEL

The Northern Valleys is a lowland watch-site in the agricultural valleys north of the Samarian Hills and south of the Galilee Hills, along a 75 km E–W transect between the Mediterranean Sea and the Jordan River (Zalles & Bildstein 2000).

The migration count that occurs between early August and mid-October at the site has been underway since 1988. Before that date, between 1982 and 1987, it was conducted from the Kfar Kasem migration corridor. The count is the joint project of the Israeli Ornithological Center and the Israeli Air Force (D. Alon, pers. comm.).

The watch-site records 27 raptor species as regular migrants, with an average of 470,000 individual birds seen annually. The most numerous species is the European Honey-buzzard (>300,000, with a peak of 544,215 in 1997) (Alon *et al.* 2004) that often appears in endless streams on a very broad front, covering large portions of the sky around the third week of August and peaks towards the first days of September. On hot days they can migrate at very high altitudes. Levant Sparrowhawks (>40,000, with a peak of 70,311 in 2008) (D. Alon pers. comm.) arrive in huge flocks around 10 September, reaching their peak at the end of the month. Towards the third week of September Lesser Spotted Eagles (>90,000) appear gradually with their peak of 48,000 birds around 29 September–5 October.

In addition to raptors, White Storks (>230,000, with a record of 540,000 in 1997, peaking at the end of August) and White Pelicans (*c.* 40,000, peaking in mid-September) are seen migrating at the site (D. Alon pers. com.).

This chapter provides but a brief introduction to a handful of the better-known raptor migration watch-sites in and around Europe. Although most raptor watchers spend most of their time at a single, favourite site, those that make it a point to visit other watch-sites regularly quickly realise that much is to be learned in doing so. In addition to consulting the general references listed below, the Hawk Migration Association of North America's website at www.hmana.org provides additional information on these and other watch-sites in the region.

References

ALLEON, A. & VIAN, J. 1869. Les migrations des oiseaux de proie sur le Bosphore de Constantinopole. *Rev. et Mag. Zool.* 21(2): 258–273, 305–315, 342–348, 369–374, 401–409.

ALLEON, A. & VIAN, J. 1870. Les migrations des oiseaux de proie sur le Bosphore de Constantinopole. *Rev. et Mag. Zool.* 22(2): 81–86, 129–138, 161–165.

ALON, D., GRANIT, B., SHAMOUN-BARANES, J., LESHEM, Y., KIRWAN, G. M. & SHIRIHAI, H. 2004. Soaring bird migration over northern Israel in autumn. *Brit. Birds* 97: 160–182.

BERNIS, F. 1980. *La Migración de las Aves en el Estrecho de Gibraltar.* Vol. 1: Aves Planeadors. Univ. Computense de Madrid, Madrid.

BIJLSMA, R. G. 1987. *Bottleneck areas for migratory birds in the Mediterranean region: an assessment of the problems and recommendations for action.* Internatl. Coun. Bird Preserv., Cambridge, UK.

BILDSTEIN, K. L. 2006. *Migrating raptors of the World: their ecology and conservation.* Cornell University Press, Ithaca & London.

CHRISTENSEN, S., LOU, O., MÜLLER, M. & WOHLMUTH, H. 1981. The spring migration of raptors in southern Israel and Sinai. *Sandgrouse* 3: 1–42.

CORSO, A. 2000. Identification of European Lanner. *Birding World* 13: 200–213.

CRAMP, S. & SIMMONS, K. E. L. 1980. *Handbook of the Birds of Europe, the Middle East and North Africa: the Birds of the Western Palearctic.* Vol. 2. Oxford Univ. Press, Oxford.

DEVISSE, J.-S. 2000. *Organbidexka Col Libre et al chasse aux migrateurs au Pays Basque en guise de conclusión. Oiseaux migrateurs chassés en mauvais état de conservation et "points chauds" européens.* Organbidexka Col Libre, Bayonne, France.

EVANS, P. R. and LATHBURY, G. W. 1973. Raptor migration across the Straits of Gibraltar. *Ibis* 115: 572–585.

FINLAYSON, C. 1992. *Birds of the Strait of Gibraltar.* T. & A. D. Poyser, London.

GIORDANO, A., RICCIARDI, D., CELESTE, S., CANDIANO, G. & IRRERA, A. 1998. Anti-poaching on the Straits of Messina: results after 15 years of activities. In *Holarctic Birds of Prey* (R. D. Chancellor, B.-U. Meyburg, & J. J. Ferrero, eds.). World Working Group on Birds of Prey and Owls, Berlin.

GRASSO, E., SICLARI, A., MUSCIANESE, E., PANUCCIO, M., POLICASTRESE, M., AGOSTINI, N., SCUDERI, A., DUCHI, A.,

MARTINO, G., CAMELLITI, G., CIULLA, A., POLIMENI, F., SIGNORINO, G., CALABRO, M. & CUMBO, G. 2012. La migrazione post riproduttiva nel Parco Nazionale d'Aspromonte (RC) settembre 2012. *Infomigrans* 30.

GUGLIELMI, R., REPACI, E., MORABITO, N. 2003. La migrazione post-nuzuale di Accipitriformi e Falconiformi in Aspromonte. *Avocetta* 27: 69.

HARRIS, T. 2013. *Migration Hotspots.* Bloomsbury Publishing, London.

IRBY, L. H. 1875. *The Ornithology of the Straits of Gibraltar.* R. H. Porter, London.

KARLSSON, L. 2004. *Wings over Falsterbo.* Falsterbo Bird Observatory, Falsterbo.

KJELLÉN, N. 1992. Differential timing of autumn migration between sex and age groups in raptors at Falsterbo, Sweden. *Ornis Scand.* 23: 420–434.

KJELLÉN, N. 1994. Differences in age and sex ratio among migrating and wintering raptors in southern Sweden. *Auk* 111: 274–284.

KJELLÉN, N. 1997. Importance of a bird migration hot spot: proportion of the Swedish population of various raptors observed on autumn migration at Falsterbo 1986–1995 and population changes reflected by the migration figures. *Ornis Svecica* 7: 21–34.

KJELLÉN, N. 1998. Annual variation in numbers, age and sex ratios among migrating raptors at Falsterbo, Sweden from 1986–1995. *J. für Ornithol.* 139: 157–171.

KJELLÉN, N. 2012. Migration counts at Falsterbo in the autumn of 2011. *Fåglar i Skåne* 2011: 5–44.

LACK, D. & LACK, E. 1953. Visible migration through the Pyrenees: an autumn reconnaissance. *Ibis* 95: 271–309.

MICHEV, T. & SIMEONOV, P. 1981. [Studies on the autumn migration of some waterfowl and birds of prey near Bourgas.] *Ecology, Sofia* 8: 43–48. (In Bulgarian, with English summary)

MICHEV, T., PROFIROV, L., NYAGOLOV K. & DIMITROV, M. 2011. The autumn migration of soaring birds at Burgas Bay, Bulgaria. *Brit. Birds* 104: 16–37.

NISBET, I. C. T. and T. C. SMOUT. 1957. Autumn observations on the Bosphorus and Dardanelles. *Ibis* 99: 483–499.

NISBET, I. C. T., EVANS, P. R. & FEENY, P. P. 1961. Migration from Morocco into southwest Spain in relation to weather. *Ibis* 103: 349–372.

PORTER, R. and WILLIS, I. 1968. The autumn migration of soaring birds at the Bosphorus. *Ibis* 110: 520–536.

RUDEBECK, G. 1950. Studies on bird migration based on field studies in southern Sweden. *Vår Fågelvårld,* Suppl. 1: 5–49, 74–85, 147–148.

SAFRIEL, U. 1968. Bird migration at Eilat Israel. *Ibis* 110: 283–320.

SHIRIHAI, H. & CHRISTIE, D. A. 1992. Raptor migration at Eilat. *Brit. Birds* 85: 141–186.

SHIRIHAI, H., YOSEF, R., ALON, D., KIRWAN, G. M. & SPAAR, R. 2000. *Raptor migration in Israel and the Middle East.* Internatl. Bird. Res. Center, Eilat; Soc. Protect. Nat. Israel, Tel Aviv.

TRISTRAM, H. B. 1865–1868. Notes on the ornithology of Palestine. *Ibis* 1: 67–83, 241–263; 2: 59–88, 280–292; 3: 73–97, 360–371; 4: 204–215, 321–335.

ULFSTRAND, S., ROOS, G., ALERSTAM, T. & ÖSTERDAHL, L. 1974. Visible bird migration at Falsterbo, Sweden. *Vår Fågelvårld,* Suppl. 8.

VAN MAANEN, E., GORADZE, I., GAVASHELISHVILI, A. & GORADZE, R. 2001. Trapping and hunting of migratory raptors in western Georgia. *Bird Conserv. Internatl.* 11: 77–92.

ZALLES, J. I. & BILDSTEIN, K. L. 2000. *Raptor Watch: A global directory of raptor migration sites.* BirdLife International, Cambridge, UK; Hawk Mountain Sanctuary, Kempton, PA.

MIGRATION ECOLOGY OF RAPTORS

Ian Newton

Migration in birds is often defined as a regular return movement between breeding and wintering areas. It occurs not just among birds that breed at high latitudes, with warm summers and cold winters, but also among species that breed in the tropics, with regular wet and dry seasons. In fact, migration occurs in any seasonal environment in which bird food supplies change predictably from abundant to scarce during the course of each year (Newton 2008, 2010).

Migration is assumed to have evolved wherever birds benefit from leaving their breeding areas for a period than by staying there year-round (Lack 1954). The usual reason that breeding areas become unsuitable during part of the year is lack of food, as plant growth stops for part of the year, and many prey-animals die off, hibernate or become inaccessible under snow and ice. In the case of the fish-eating Osprey, for example, lakes in the breeding areas freeze over in winter and fish retreat to deeper water, so become inaccessible; and in the case of the European Honey-buzzard, the bee and wasp grubs which form the diet are dormant or dead in winter. At high latitudes, days also shorten to such an extent in winter that many diurnal birds would have insufficient time to get enough food, even if food were available. So the reason why many birds move to lower latitudes in autumn is fairly obvious.

A more difficult question is why they bother to return in spring, when their wintering areas – many in Africa – seem perfectly capable of supporting them year-round. But by returning north in spring, migrants can exploit the seasonal abundance of food at high latitudes, as well as the longer summer days, and probably raise more young than if they stayed in their low latitude wintering areas, and competed with the birds resident there.

So whereas the advantage of autumn migration to lower latitudes can be seen as improved winter survival, the advantage of spring migration to higher latitudes can be seen as improved breeding success. Migration is assumed to occur when the net benefits of moving both ways outweigh the costs of staying in the breeding areas year-round. But these are just plausible ideas, practically impossible to test experimentally, but for which there is considerable circumstantial evidence.

GEOGRAPHICAL TRENDS

Among birds as a whole, the proportions of migrants among local breeding bird species increase with increasing latitude, as winters become more severe (Newton & Dale 1996, Newton 2008). Even in the mild climes of Morocco, 30% of all breeding bird species are completely migratory, wintering further south. This proportion increases with latitude, reaching more than 90% in the far north, in places such as Svalbard. The same trend holds among diurnal raptors (Figure 1), and the only species which winters in the far north, with at most only a few hours of twilight in mid-winter, is the Gyr Falcon, dependent mainly on Ptarmigan as prey. However, conditions are such that even the majority of Gyr Falcons move out of the high arctic for the winter. The important point is that a geographical trend in migratory behaviour occurs among birds (including raptors), which corresponds to the severity of winter.

This geographical trend is evident not just from comparing different species, but also within species. Many widespread raptor species are wholly resident in the southern parts of their breeding range and wholly migratory in the north, while in between they are partial migrants, with some individuals staying in their breeding areas year-round while others move out. Examples include the Common Kestrel and Common Buzzard. In general, the proportion of migratory individuals within a species increases with distance northwards, roughly corresponding to the level of reduction of food supply in winter. Among such partial migrants, migration is often more frequent in one sex than in the other, and among juveniles than adults (Newton 2008). Among short-distance migrants that winter within Europe, juveniles tend to leave before adults, and migrate further, as in the Common Kestrel, Common Buzzard, Red Kite, Eurasian Sparrowhawk and Northern Goshawk; but in long-distance migrants that winter in Africa south of the Sahara, adults tend to leave before juveniles, as in the Osprey, Eurasian Hobby and European Honey-buzzard (Kjellén 1992, 1994a, Newton 2008). Among Marsh Harriers tracked on migration from Sweden, in which juveniles migrated more slowly and shorter distances than adults, most wintered in West Africa (Strandberg *et al.* 2008).

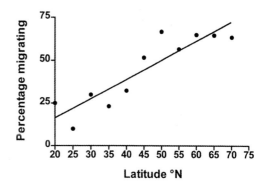

Figure 1. Proportion of raptor species breeding at different 5° latitudes in western Europe which migrate south for the winter. For this analysis, only species which move out completely from any given latitude were classed as migratory there. From Newton 1998.

RELATIONSHIP BETWEEN MIGRATION AND DIET

Whether particular species leave particular latitudes relates to their diets, and whether the foods they eat remain available at that latitude in winter. If one divides European raptors according to whether they feed primarily on warm-blooded prey (birds and mammals, which remain active and available in winter at high latitudes) and cold-blooded prey (reptiles, amphibians and insects, which become inactive and unavailable in winter), differences in migration are apparent (Figure 2). Within each group, the proportion of migrant species increases with latitude, following the general trend among birds as a whole (Newton & Dale 1996). At any one latitude, however, a larger proportion of species that eat cold-blooded than warm-blooded prey leave for the winter, while species with mixed diets are intermediate (Figure 2A). Furthermore, the cold-blooded feeders generally move longer distances than the warm-blooded feeders (Figure 2B). The reasons for this difference are fairly obvious, in that species that feed on cold-blooded prey and breed at high latitudes must winter in the tropics or the southern hemisphere if they are to have access to the same types of prey year-round.

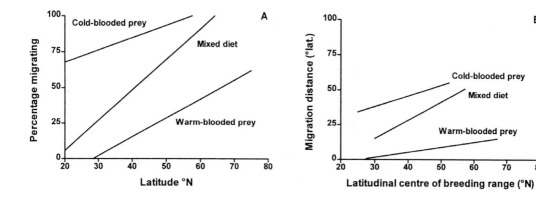

Figure 2. Migration in relation to diet in European raptors. A: Proportion of breeding species which migrate from different 5° latitudes. B: Distances moved, as measured by the latitudinal shift between the latitudinal centres of breeding and wintering ranges. Lines calculated by regression analyses. From Newton 1998.

Table 1. Wintering areas of Western Palaearctic raptors in relation to diet.

Wintering area	Main prey types		
	Warm-blooded	**Mixed**	**Cold-blooded**
North of Sahara	16	0	0
North and south of Sahara	5	8	2
South of Sahara	10	4	

Significance of variation between categories (examined by Monte Carlo randomisation test): $x^2_4 = 35.9$, P<0.001.

Of the 22 species of Western Palaearctic raptors that eat mainly warm-blooded prey, 16 spend the northern winter entirely within Eurasia, five partly in Eurasia and partly in Africa, and only one entirely in Africa (Table 1). Of the nine species that eat mainly cold-blooded prey, none winter entirely in Eurasia, two partly in Eurasia and partly in Africa, and seven entirely in Africa. Moreover, of the six insectivorous species, two winter partly and four entirely south of the equator, where the seasons are reversed. So, through migration, these latter species live in almost perpetual summer, thereby gaining access to plentiful insect food year-round. The 12 species with mixed diets show intermediate patterns. Such patterns again underline the link between migration and the seasonal changes in specific food sources (Newton 1979).

Food also seems to influence the timing of raptor migration. At the end of summer the first animals to disappear with the onset of cold weather at high latitudes are large insects, followed by reptiles and amphibians, while fish retreat to deeper water. Somewhat later, many small birds begin to migrate, and mammals begin to disappear, some hibernating and others spending increasing periods in sheltered sites where they are unavailable to raptors. Eventually the landscape is snow-covered, rendering small mammals even less accessible. Given this pattern, it is not surprising that insect-eaters, such as the European Honey-buzzard, are first to leave their breeding areas, followed by Ospreys, and later by bird-eating and mammal-eating species. During spring, the situation is reversed, with mammals appearing first in the environment and large insects last. The Steppe Eagle, which eats chiefly mammals, is the first to migrate north, passing through Israel mainly in early March, while the honey-buzzard, which eats insects, migrates last, mainly in May. The Steppe Eagle spends six months in its breeding areas, passing south through Israel mainly in mid October to mid November, while the honey-buzzard spends only three months in its breeding areas, passing through Israel mainly in early September. The eagle also has a longer breeding cycle than the honey-buzzard, with longer incubation, nestling and post-fledging periods (Leshem & Yom-Tov 1996a). But the general pattern is that raptor species which are the first to leave their breeding areas in late summer are also among the last to return there next spring, while species which are among the last to leave their breeding areas in autumn are among the first to return there next spring.

MODE OF MIGRATION

While most bird species travel on migration by using energy-demanding flapping flight, many raptors are able to travel by soaring and gliding, which is much less energy demanding. It is in the larger species that this ability is best developed, as their spread wings span a large area relative to body weight, and can thereby provide good lift in rising air currents. The larger raptors share this ability with some other large birds, such as storks and pelicans.

Raptor species that use passive soaring–gliding flight are not sharply separated from those that use powered flapping flight. Different species form a continuum between the two extremes, depending on their body size and wing shape, and in all soaring species the ratio of flapping to gliding can vary with air conditions at the time. Vultures and eagles are most dependent on soaring–gliding, followed in descending order by *Buteo* buzzards, *Milvus* kites, *Accipiter* hawks and *Pernis* honey-buzzards, and then by *Circus* harriers and *Pandion* Ospreys (Kerlinger 1989, Newton 2008). Falcons are more active fliers, less dependent on updrafts, but make use of them when available. This order of listing broadly follows the sequence of wing loading, from lowest to highest, and the variation in wing shape from long and broad, with slotted primary feathers, to narrow and pointed, with little or no slotting. It also reflects the dependence of the various species on updrafts, and hence the extents to which they take advantage of land-based topographic features and avoid long sea crossings. It is chiefly the falcons that regularly make long (<100 km) flights over water, and occasionally migrate at night, but other species (*Pandion*, *Pernis, Butastur* and others) do so in some parts of the world. The most extreme is the Amur Falcon, which each autumn crosses the Indian Ocean between India and East Africa on a journey sometimes exceeding 4,000 km. This over-water flight is assisted by prevailing winds. On the return spring migration, winds over the sea are less favourable, and migration is largely overland, up the east side of Africa and eastward across Asia.

Soaring species migrating overland gain lift from rising air, whether updrafts or thermals. Updrafts are formed when the wind, striking a slope or cliff, is deflected upwards, enabling 'slope soaring'. Long mountain ridges thus provide excellent flyways for soaring migrants in those places where the ridge lies roughly north-south, in a direction appropriate for migration. In such places, birds can glide for tens of kilometres without flapping (Kerlinger 1989). The Rift Valley, extending from the Middle East into Africa, provides such conditions over parts of its length.

Thermals are localised columns of rising air created through the uneven heating of the ground by the sun. These columns rise to high elevation, until they have cooled to the temperature of the surrounding air, where they often produce a cumulus cloud, marking their position. They begin each morning once the ground has heated sufficiently, but gather strength during the day. They climb gradually faster and higher, often reaching more than 1,000 m at noon, and then wane in the evening as the ground cools. Birds progress on migration by circling in one thermal to gain height, and then gliding with loss of height to the next thermal where they rise again, repeating the process along the route (Figure 3). This enables birds to travel across country at around 30–50 km per hour, depending on the rate and extent of rise within thermals and the distance covered between thermals (in turn dependent on the 'glide coefficient', which is the ratio between the horizontal distance covered by the bird and its altitude loss over that distance). Species with light wing-loadings ascend more rapidly so spend less time in each thermal; they also lose less height per unit distance in a glide, so they can travel further before having to climb again. They can also get underway earlier in the morning and continue later into the evening than species with heavier wing-loadings that are restricted to a shorter period each day when the thermals are strongest. The flight times of the smaller species migrating through Israel on thermals

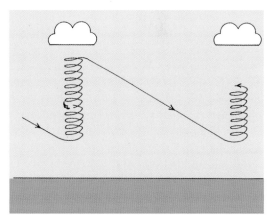

Figure 3. Soaring–gliding bird migration, indicating soaring within thermals, gaining height, and gliding between thermals, losing height. Thermals are often topped with cumulus clouds.

typically extend over 8–10 hours each day (beginning around 09.00) and the larger ones over 6–7 hours (beginning around 10.00). All species tend to make more rapid progress in the middle part of the day, when climbs are fastest and highest, and glides are longest. Particular species can travel across country twice as fast around noon than in the morning or evening (Spaar & Bruderer 1996).

In Steppe Eagles studied in Israel, cross-country speed was related to the climb rate in thermals which averaged 1.9 m s^{-1} over the whole daily migration period, but reached up to 5.0 m s^{-1} around noon. Mean gliding airspeed between thermals was 56 km per hour which, allowing for climb times, gave a mean cross-country speed of 45 km per hour. The upper limit of migration was about 1,600 m above ground, but was mostly below 1,000 m (Spaar & Bruderer 1996). Smaller European Honey-buzzards and Steppe Buzzards achieved lower cross-country speeds, at 37 and 35 km per hour. However, in a 12-hour day, Steppe Eagles soared for only 6 hours and covered 270 km, whereas the two smaller species migrated for 10 hours and covered 360 km. Hence, although the largest species travelled faster between thermals, it did not cover more kilometres per day (Spaar & Bruderer 1996). In cooler climates, where thermals occur during a smaller part of each day, migration times are shorter, and larger raptors, such as Golden Eagles, travel for at most a few hours per day when using thermals, mostly between 12.00 and 14.00 hours. But when using mountain updrafts, which continue through the day, large raptors can migrate over longer periods. As expected, weather conditions also influence progress, as they affect the formation of updrafts and thermals, while rain can stop migration altogether.

Partly because of their dependence on thermals, all soaring landbird species migrate primarily by day. Taking advantage of topography, they often form into concentrated streams, using the same routes year after year, and crossing seas at the narrowest points. Crossings at the Bosphorus or Gibraltar are best accomplished in one long descending glide, in order to avoid laborious flapping flight. Early in the day, when conditions are not ideal, birds sometimes start a crossing, and having reached a few kilometres from shore, turn back to try again later. On their migrations, soaring raptors usually travel low enough to be seen with the naked eye and, with large numbers passing predictable places year after year, they can be counted by ground-based observers in ways that other, higher-flying or night-flying birds cannot. The seasonal timing of their migrations can thereby be assessed accurately and day-to-day passage can be related to weather and other conditions. Moreover, in some species, the different sex and age groups can be distinguished, enabling the movements of these different groups to be examined separately.

Well-known observation points for soaring migrants in western Eurasia include Falsterbo in Sweden, Gibraltar and the Bosphorus at either end of the Mediterranean Sea, the Black Sea coast in northeast Turkey, various localities in the Rift Valley in Israel and Suez in Egypt (see Bildstein & Sandor in this book). At these points, large numbers of soaring migrants pass in spring or autumn, with total numbers varying between tens of thousands and hundreds of thousands, depending on the site. At certain sites in the New World, much larger numbers can be seen, and several millions of soaring raptors and others cross between North and South America at the narrow land bridge at Panama (Bildstein 2006). In contrast to the soaring species, raptors and other birds that migrate primarily by flapping flight tend to migrate more on a broad front, as they are less dependent on topography, and do not concentrate to the same extent at short sea crossings or narrow land bridges. They can more readily cross larger stretches of water.

Many birds, whether soaring or flapping species, take somewhat different routes in spring and autumn, depending on wind and other conditions, as mentioned above for the Amur Falcon (Newton 2008, 2010). At least two major loops have been described for migration between Europe and Africa. In one, the southward movement occurs through Gibraltar and the northward one through Sicily–Italy (an anti-clockwise loop). In the other, a southward route occurs across Arabia, down the east side of the Red Sea and crossing to Africa at Bab el Mandeb, and the northward route occurs up through the eastern Sahara and up the west side of the Red Sea to cross from Africa at the Gulf of Suez (a clockwise loop). Both loops are used by many raptors, as well as by passerines and other kinds of birds. They have been demonstrated by counts on the different flyways, by ringing recoveries and by the satellite-based tracking of radio-marked birds (see Meyburg *et al.* 2003 for Steppe Eagle). A less marked clockwise loop was noted in individual Marsh Harriers which, in crossing the western Sahara, travelled northward in spring to the west of their southward route in autumn, associated with different wind conditions (Klaassen *et al.* 2010).

WEATHER EFFECTS

Birds in general prefer to migrate under clear skies with following winds. Clear skies facilitate navigation because the sun or stars – which are the main compass cues – are clearly visible; and following winds hasten the birds on their journeys, reducing their energy costs and also the risks of being blown off-course. In contrast, rain, mist and heavily overcast conditions, or strongly opposing winds, normally deter birds from setting off on migration, or cause birds already on the wing to settle if they are overland. These generalisations also apply to raptors, but soaring species operate under even greater constraints. To create an updraft from a ridge or mountainside, the wind has not only to be sufficiently strong, but must also strike the ridge at an angle, for if it flows in the same direction as the ridge, no significant updraft is created. In these conditions, a following wind may be of little or no help. The development of thermals is heat-dependent, so such up-currents are likely to operate for longer each day, and achieve greater heights, in warmer regions than in cold ones. Even in warm regions, however, strong winds from any direction can prevent thermals from developing, a situation which, in the absence of other updrafts, suppresses raptor migration. In the northern latitudes of Europe, migration of any kind is often stopped for up to several days at a time by rain or other adverse weather, but from southern Europe southward to the tropics, thermals can usually develop for part of every day in the migration seasons.

This does not necessarily mean that migration will be visible from the same locations every day, for as in other birds, migration streams are often shifted laterally by cross-winds. Thus, migrating raptors may pass localities up to several tens of kilometers apart on different days, depending on the strength and direction of the wind. To a large extent, the likelihood of strong migration on particular days during migration seasons can be predicted from examination of synoptic weather maps. But because birds differ in body size, flight mode and other aspects, they are affected by adverse weather to different extents, some species being able to fly in conditions that would ground others.

POSSIBLE SOCIAL INFLUENCES

Soaring species migrating along favourable routes often form long drawn-out flocks. One presumed advantage of migrating in this way is that it makes finding thermals easier, thus conserving energy. By watching the birds ahead that are already circling upward, a bird can head for them without wasting time and energy in thermal location. Use of radar in Israel revealed that, on peak migration days, the lines formed by flocks extended up to 200 kilometres or more, so that most individuals had before them a continuous route marked out by their predecessors (Leshem & Bahat 1999). However, it is difficult with raptors to tell whether the birds migrate in flocks simply because they share the same narrow migration route, and the same thermals and updrafts within it, or whether they are attracted to one another for other reasons. When leaving the top of a thermal, birds seem to depart individually. Not surprisingly, the biggest flocks are seen in relatively numerous species which migrate within a short time period, such as the European Honey-buzzard and Levant Sparrowhawk; but some insectivorous raptors, such as Red-footed and Amur Falcons, seem always to associate in flocks, whether on migration or not.

ENERGY NEEDS AND FEEDING

While birds that travel by flapping flight depend on internal energy for migration, those that use soaring–gliding flight depend largely on external energy, derived from the atmosphere. Our knowledge of the extent to which raptors feed on migration, and how much body fat they accumulate for each stage of the journey, is somewhat limited. In theory, migrant raptors could forage in the early mornings when conditions are unsuitable for soaring, and suffer no reduction in overall migration speed; but how much they do so is unknown. Those that winter in Africa migrate partly through desert or other terrain offering little food, and often birds travel in such large numbers over such narrow routes that most would have little chance of picking up a meal.

Nevertheless, many raptor species can be seen to feed on migration, or to fly with a full crop, especially those which hunt or locate their prey in flight. Bird-eating falcons and accipiters migrate at the same time as their prey, and could in effect survive by eating their fellow travellers. Ospreys have also been observed to divert from their paths to visit wetlands and search for and catch fish before moving on. Various harriers have been seen to migrate and hunt at the same time, and aerial insectivorous species have been seen to pause and forage if they encounter concentrations of suitable insects en route. The term 'fly-and-forage' has been used to describe this 'eat-on-the-road' strategy (Strandberg & Alerstam 2007). Eagles and buzzards normally hunt by a sit-and-watch strategy; and those that migrate long distances feed more episodically, on definite stopovers, and probably make large parts of their journeys without eating. This is also true of honey-buzzards, which seem to make little attempt to feed over most of their route, although they frequently descend to drink. These groups are among the raptors known to accumulate migratory fat (up to 30% of body weight in Steppe Buzzards, Gorney & Yom-Tov 1994). In North America, museum labels on skins of Swainson's Hawks *Buteo swainsoni* and Mississippi Kites *Ictinia mississippiensis* collected at migration times often carried the note 'very fat' (W. S. Clark). Bald Eagles *Haliaeetus leucocephalus* have been tracked on migration for up to 12 days without being seen to feed (Harmata 2002), and long-distance fasting has also been proposed for various other soaring raptors, which show long uninterrupted travel steps, implying the use of stored reserves (Meyburg *et al.* 1995, Kjellén *et al.* 1997, Håke *et al.* 2003). Even though they may make much of their journey on energy-saving soaring–gliding, therefore, it is clear that some raptors accumulate and use migratory fat like other birds. Whether there is a clear division in migratory fat levels between species that hunt on the wing and those that do not must await further research.

SATELLITE TRACKING

Recent satellite-based radio-tracking of individual raptors has provided much new information on the speed and duration of migration, on the routes and stopover sites used and on their behaviour en route. The use of Google Earth images has enabled exploration of the habitats used on migration. Most of these tracked birds made long journeys, some from one continent to another, and their need for thermals also led some to take indirect routes to avoid long sea-crossings. Their journeys were thereby lengthened by up to 50% over the shortest (great circle) routes.

Among birds as a whole, the flight speeds of different species need not correlate well with their migration speeds, because a large part of most migratory journeys is spent resting or foraging. In general, larger bird species fly faster than small ones, but partly because of their slower food-processing rates, large species take longer to accumulate the body reserves necessary to fuel the journey (Newton 2008). This size-related difference may be less marked among soaring raptors than most other birds because their cheaper flight mode reduces the time they must spend feeding to accumulate fuel. In addition, some raptor species can pick up food as they travel, as mentioned above.

One of the earliest satellite-based studies involved a Short-toed Snake Eagle, which was monitored every night on its autumn journey between France and Niger in West Africa (Meyburg *et al.* 1998). Every roost location was thus recorded, as were the distances travelled each day throughout the journey. These daily distances varied from 17 to 467 km (mean 234 km), and the whole migration of 4,685 km was accomplished in 20 days (Figure 4). Other raptors for which sufficient data were obtained have occasionally moved more than 500 km per day, with 746 km recorded from an Osprey through Europe (Kjellén *et al.* 2001), 663 km for a Black Kite through Europe (Meyburg & Meyburg 2009b), and 537 km for a Lesser Spotted Eagle through Africa (Meyburg *et al.* 2001). But these are extreme values, and daily distances of 150–250 km per day would be more typical. Inevitably, some non-stop sea crossings

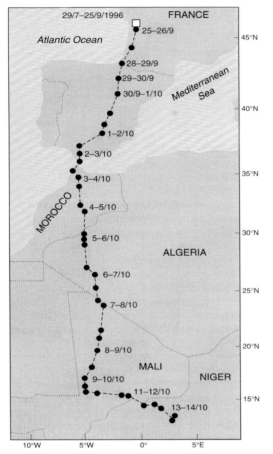

Common Buzzards involved spent 4–44 travel days on their autumn journeys and 6–10 travel days over their spring journeys over distances of 452–1,449 km. Similarly, among 24 juvenile Steller's Sea Eagles *Haliaeetus pelagicus* tagged at nests in various parts of the breeding range, mean speed per day increased with length of journey (as measured by degrees of latitude travelled) (McGrady *et al.* 2003). Again, however, most of these movements could be considered short distance, at less than 2,000 km.

Not surprisingly, those species tracked between Europe and Africa travelled more rapidly over the Sahara (which offered excellent soaring conditions but no food) than over other parts of the journey where feeding stops were possible (for Short-toed Snake Eagle see Meyburg *et al.* 1998; for Osprey see

made by Amur and other falcons must be longer.

Comparing records from different individuals, the time spent on migration increased with the length of the journey, a trend apparent both between and within species, but with considerable individual variation (Newton 2008). Within species, the trend was most apparent among Peregrines and Ospreys, in which large numbers of birds were tracked from different parts of the breeding range (Newton 2008, Burnham *et al.* 2012). Overall, among long-distance migrant raptors journeys of around 2,000 km took, on average, around 20 days, and every additional 2,000 km added another ten days (Figure 5). It seemed that, on journeys exceeding 2,000 km, migration speed showed no tendency to increase with length of journey. This was presumably because soaring species required much less food to fuel the flights, and did not need to stop for long feeding periods during the journey (although some individuals clearly did). However, examination of short-distance Common Buzzards, migrating within Europe, showed that they generally moved more slowly, with some steps of their journey triggered by adverse changes in weather and food supplies. Adding these data to Figure 5 altered the relationship between duration and distance, with duration increasing approximately in proportion to the square root of distance (Strandberg *et al.* 2009b). The 12

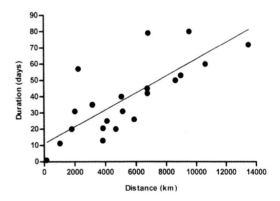

Kjellén *et al.* 2001; for European Honey-buzzard see Håke *et al.* 2003; for Egyptian Vulture see Meyburg *et al.* 2004; for Lesser Spotted Eagle see Meyburg & Meyburg 2009; for Eurasian Hobby see Strandberg *et al.* 2009c). Elsewhere some species broke their autumn journeys for up to several days at a time, apparently when they encountered good feeding areas. Once in the winter quarters, individuals of some species remained in one locality throughout their stay, and those individuals studied in successive years returned to the same breeding and wintering localities each year (for Osprey see Alerstam *et al.* 2006, for European Honey-buzzard see Håke *et al.* 2003, for Marsh Harrier see Strandberg *et al.* 2008). Like some other birds, some tracked raptors used different areas in Africa at different times in the non-breeding period, moving successively further south during their stay, apparently in association with southward shifts in food availability, related to previous rainfall. This behaviour was recorded in 30 Montagu's Harriers from Europe which used four different African areas in succession, each separated by more than 200 km, to which they also showed high fidelity in successive years (Trieweiler *et al.* 2013). Similarly, an adult Black Kite tracked from Germany to West Africa used three main areas in succession during the course of a single winter (Meyburg & Meyburg 2009b). Some Lesser Spotted Eagles wandered over many thousands of square kilometres in southern Africa, frequently making long moves during the course of their stay (Meyburg *et al.* 1995, 2004). This species was known from previous observations to move around according to rainfall, concentrating temporarily in areas of abundant food, such as recently emerged termites (Newton 1979). However, not all tracked eagles of this species moved over such large areas (Meyburg & Meyburg 2009a).

Among various long-distance raptors, no consistent difference was apparent between the duration of autumn and spring journeys, either from the mean values calculated for different populations or using the values for those individuals of each species tracked on both outward and return journeys (Figure 6). In some populations (or individuals) the autumn journey took longer, in others the spring journey. The biggest difference was recorded in a White-tailed Eagle which took markedly different routes at the two seasons, so that the autumn journey lasted 57 days (39 km per day) and the spring journey 67 days (81 km per day) (Ueta *et al.* 1998). However, even including this bird, the regression line relating the duration of autumn and spring journeys in different species was not significantly different from 0.5, which implies equality in the mean duration of both journeys. The longest-distance travellers between northern Europe and southern Africa, or between northern North America and southern South America, spent up to 152 days per year on migratory journeys (42% of each year).

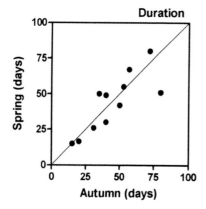
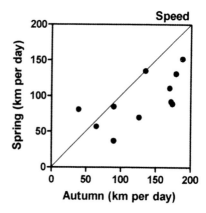

Figure 6. Relationship between the durations and speeds of the autumn and spring migrations of different raptor populations, based on the mean values from different studies listed in Newton 2008: Table 8.3 (but excluding a White-tailed Eagle which took markedly different routes in autumn and spring). The line on each graph shows the relationship expected if the two migrations were of equal duration and speed. Overall, no significant difference was apparent in the duration or speed between seasons (on a paired t-test for duration, t = 0.08, P = 0.94; for speed, t = -0.93, P = 0.37).

The large individual variation in the duration of migration within populations was mentioned above. Among Ospreys that bred in Sweden and wintered in West Africa, the total length of journey varied between 5,813 and 7,268 km (longest journey 1.25 times longer than the shortest), but this journey took between 14 and 55 days to complete (greatest value 3.9 times longer than shortest) in a sample of 13 individuals (Kjellén et al. 2001). Mean speeds varied between 108 and 431 km per day. Among Honey-buzzards over the same route, the journey varied between 6,299 and 7,091 km (longest journey 1.13 times longer than the shortest), and took between 34 and 70 days to complete (greatest value 2.06 times longer than shortest) in a sample of nine individuals (Håke et al. 2003). Mean speeds varied between 93 and 209 km per day. In other species, equivalent journeys took up to three times longer in some individuals than others (Newton 2008: Table 8.3).

Over the same journey, no difference was apparent in the mean autumn migration speeds of juvenile and adult Ospreys. In European Honey-buzzards the average speeds of adults and juveniles on travelling days were similar, at about 170 km per day in Europe, 270 km per day across the Sahara and 125 km per day in Africa south of the Sahara. However, as the adults had fewer stopover days en route, they maintained higher overall speeds and completed their migration in a shorter time (42 days) than juveniles (64 days) (Håke et al. 2003). The spring journeys of several eagle species were slower in juveniles than in adults, mainly for the same reason (Newton 2008: Table 8.3). Although Peregrines and Ospreys often progress by flapping flight when soaring is not possible, their mean migration speeds were within the range of values recorded for other soaring species (Newton 2008: Table 8.3). In general, within regional populations, those adults that spent the longest periods on migration spent the shortest periods in their wintering areas, while no effect was apparent on the amount of time spent in the breeding areas.

Satellite tracking has also shown that individuals of some species, such as Osprey and European Honey-buzzard, start on migration as soon as they finish parental care or, in the case of juveniles, as soon as they become independent. But in other species some individuals (adults or young) move to a pre-migratory area where they spend some weeks before departing on their migration. This last behaviour has been recorded among Marsh Harriers in Sweden, Montagu's Harriers in Spain and Peregrines in Greenland (Strandberg et al. 2008, Limiñana et al. 2008, Burnham et al. 2012).

Tracking studies have also taught us more about the hazards facing migrating birds. One study involved more than 90 trans-Saharan journeys by four species of raptors (Osprey, European Honey-buzzard, Marsh Harrier and Eurasian Hobby) from Sweden (Strandberg et al. 2009a). Aberrant behaviour occurred on 40% of these crossings, indicating difficulties or hazards for the migrants. These events included changing course abruptly, slow travel speeds, interruptions, aborted crossings followed by retreat migration out of the desert, and failed crossings due to death. The mortality rate associated with Saharan crossings was significantly higher among juveniles (31% of crossing attempts, first autumn migration) than among adults (2% per crossings attempts, autumn and spring combined). Mortality associated with desert crossings made up about half of the total annual mortality recorded for juveniles. Among the survivors, aberrant behaviours resulted in late arrival in the breeding areas and increased the probability of breeding failure.

In some species, ringing and satellite-based radio-tracking have shown that different birds from the same breeding locality can migrate to widely-separated wintering places, and conversely that birds from a single wintering locality can migrate to widely separated breeding places. Individuals may return year after year to their own breeding and wintering sites, but have different sets of neighbours at the two seasons. For example, Peregrine Falcons caught wintering on a 50 km stretch of coast in eastern Mexico were tracked to breeding areas that lay across much of North America and western Greenland, with a west–east spread of more than 5,000 km (McGrady et al. 2002). Similarly, Peregrines from any one breeding area wintered at sites scattered over a wide range of latitudes from southern North America to northern South America, mixing with Peregrines from other breeding areas (Fuller et al. 1998, Burnham et al. 2012). Again, individuals tracked in more than one year showed fidelity to their own breeding and wintering sites, confirming trapping results from ringed birds (Newton 2008, Burnham *et al.* 2012). In some species, even breeding partners tracked by satellite wintered in areas separated by more than 1,000 km, as shown in Ospreys, Greater Spotted Eagles and others (Kjellén *et al.* 1997, Meyburg & Meyburg 1998).

Finally, satellite tracking has for several species confirmed the implications from ring recoveries that, in raptor species in which individuals migrate between Europe and Africa and do not breed until they are two or more years old, at least most young birds stay in Africa for the whole of their first year or more, and do not return to their breeding areas until they are approaching breeding age (for Osprey see Dennis 2008; for European Honey-buzzard see Hake *et al.* 2003; for Egyptian Vulture see Meyburg *et al.* 2004). This means that individuals in full juvenile plumage are seen in Europe chiefly during the few weeks between leaving the nest and leaving the continent on their first southward migration. However, among a small number of tagged Short-toed Snake Eagles, some individuals remained in Africa in their second summer while others migrated north to the breeding range, arriving later and departing earlier than breeding adults (Yanez *et al.* 2014). Based on observations of migrants in southern Italy, the same holds for Honey-buzzards (Panuccio & Agostini 2006).

Summary

Migration is assumed to occur in species in which individuals survive better by wintering at lower latitudes than they breed, and breed at higher latitudes than they winter: when the benefits of the two-way journey outweigh the costs of staying in the same place year-round.

The proportion of breeding raptor species that are migratory increases from south to north within Europe. Also, many widespread species are totally resident in the southern parts of their breeding range and totally migratory in the north, and partial migrants in between.

As well as latitude, diet influences migratory habits. In general, species that eat cold-blooded insects, amphibians and reptiles are more migratory than species that eat warm-blooded birds and mammals. The former also in general migrate longer distances, arriving in their breeding areas later in spring and departing earlier in autumn.

Raptors form a continuum from species that migrate almost entirely by soaring–gliding flight to those that migrate almost entirely by powered flapping flight. The former is less energy-demanding, but largely restricts the birds to travelling by day, at relatively low altitude and overland, avoiding all but relatively short water crossings. Soaring species often follow the same routes from year to year, concentrating to pass through specific bottlenecks where they can be readily counted.

The tracking of radio-tagged individuals has provided much new information on the speed and duration of migratory journeys, on the routes taken, and on the behaviour and mortality occurring en route. Raptors typically travel overland at speeds of 150–250 km per day. In species that migrate from Europe to Africa, most mortality occurs in the Sahara Desert, especially among juveniles on their first journey. In some species, individuals return not only to the same breeding areas in successive years, but also to the same non-breeding areas. Individuals of some species which migrate to Africa use a succession of widely separated areas during the course of a single non-breeding season to which they may also return in subsequent years.

In raptor species in which individuals migrate between Europe and Africa and do not breed until they are two or more years old, at least most young birds stay in Africa for the whole of their first year or more, and do not return to their breeding areas until they are approaching breeding age.

References

Alerstam, T., Håke, M. & Kjellén, N. 2006. Temopral and spatial patterns of repeated migratory journeys by Ospreys. *Anim. Behav.* 71: 555-66.

Bildstein, K. L. 2006. *Migrating birds of prey of the world; their ecology and conservation.* Cornell University Press, Ithaca, New York.

Burnham, K. K., Burnham, W. A., Newton, I., Johnson, J. A. & Gosler, A. G. 2012. The history and range expansion of Peregrine Falcons in the Thule area, northwest Greenland. Museum Tusculanum Press, Copenhagen.

Dennis, R. 2008. *A Life of Ospreys.* Whittles, Dunbeath.

Gorney, E. & Yom-Tov, Y. 1994. Fat, hydration condition, and moult of Steppe Buzzards *Buteo buteo vulpinus* on spring migration. *Ibis* 136: 185–192.

Håke, M., Kjellén, N. & Alerstam, T. 2003. Age-dependent migration strategy in Honey Buzzards *Pernis apivorus* tracked by satellite. *Oikos* 103: 385–396.

Harmata, A. R. 2002. Vernal migration of Bald Eagles from a southern Colorado wintering area. *J. Raptor Res.* 36: 256–264.

Kerlinger, P. 1989. *Flight Strategies of Migrating Hawks.* Chicago University Press, Chicago.

Kjellén, N. 1992. Differential timing of autumn migration between sex and age groups in raptors at Falsterbo, Sweden. *Ornis Scand.* 23: 420–434.

Kjellén, N. 1994. Differences in age and sex-ratio among migrating and wintering raptors in southern Sweden. *Auk* 111: 274–284.

Kjellén, N., Hake, M. & Alerstam, T. 1997. Strategies of two Ospreys *Pandion haliaetus* migrating between Sweden and tropical Africa as revealed by satellite tracking. *J. Avian Biol.* 28: 15–23.

Kjellén, N., Håke, M. & Alerstam, T. 2001. Timing and speed of migration in male, female and juvenile Ospreys *Pandion haliaetus* between Sweden and Africa as revealed by field observations, radar and satellite tracking. *J Avian Biol.* 32: 57–67.

Klaasson, R. H. G., Strandberg, R., Hake, M., Olofsson, P., Tottrup, A. P. & Alerstam, T. 2010. Loop migration in adult Marsh Harriers *Circus aeruginosus* is explained by wind rather than by habitat availability. *J. Avian Biol.* 41: 200–207.

Lack, D. 1954. The natural regulation of animal numbers. Clarenden Press, Oxford.

Leshem, Y. & Bahat, O. 1999. *Flying with the Birds.* Chemed Books, Israel.

Leshem, Y. & Yom Tov, Y. 1996a. The magnitude and timing of migration by soaring raptors, pelicans and storks over Israel. *Ibis* 138: 188–203.

Leshem, Y. & Yom Tov, Y. 1996b. The use of thermals by soaring migrants in Israel. *Ibis* 138: 667–674.

Limiñana, R., Soutullo, A., López-López, P. & Urios, V. 2008. Pre-migratory movements of breeding Montagu's Harriers *Circus pygargus. Ardea* 96: 81–90.

McGrady, M. J., Maechtle, T. L., Vargas, J. J., Seegar, W. S. & Pena, M. C. P. 2002. Migration and ranging of Peregrine Falcons wintering on the Gulf of Mexico coast, Tamaulipas, Mexico. *Condor* 104: 39–48.

McGrady, M. J., Ueta, M., Potapov, E. R., Utekhina, I., Masterov, V., Ladyguine, A., Zykov, V., Cibor, J., Fuller, M. & Seegar, W. S. 2003. Movements by juvenile and immature Steller's Sea Eagles *Haliaeetus pelagicus* tracked by satellite. *Ibis* 145: 318–328.

Meyburg, B.-U., Ellis, D. H., Meyburg, C., Mendelsohn, J. M. & Scheller, W. 2001. Satellite tracking of two Lesser Spotted Eagles, *Aquila pomarina*, migrating from Namibia. *Ostrich* 72: 35–40.

Meyburg, B.-U., Gallardo, M., Meyburg, C. & Dimitrova, E. 2004. Migrations and sojourn in Africa of Egyptian Vultures (*Neophron percnopterus*). *J. Orn.* 145: 273–280.

Meyburg, B.-U. & Meyburg, C. 1998. Satellite tracking of Eurasian raptors. *Torgos* 28: 33–48.

Meyburg, B.-U. & Meyburg, C. 2009a. Annual cycle, timing and speed of migration of a pair of Lesser Spotted Eagles (*Aquila pomarina*) – a study by means of satellite telemetry. *Populationsökologie Greifvogel- und Eulenarten* 6: 63–85.

Meyburg, B.-U. & Meyburg, C. 2009b. GPS-Satelliten-Telemetrie bei einem adulten Schwarzmilan (*Milvus migrans*): Aufenthaltsraum während der Brutzeit, Zug und Überwinterung. *Populationsökologie Greifvogel- und Eulenarten* 6: 243–284.

Meyburg, B.-U., Meyburg, C. & Barbraud, J.-C. 1998. Migration strategies of an adult Short-toed Eagle *Circaetus gallicus* tracked by satellite. *Alauda* 66: 39–48.

Meyburg, B.-U., Meyburg, C., Belka, T., Šreibr, O. & Vrana, J. 2004. Migration, wintering and breeding of a Lesser Spotted Eagle (*Aquila pomarina*) from Slovakia tracked by satellite. *J. Orn.* 145: 1–7.

Meyburg, B.-U., Paillat, P. & Meyburg, C. 2003. Migration routes of Steppe Eagles between Asia and Africa: a study by means of satellite telemetry. *Condor* 105: 219–227.

Meyburg, B.-U., Scheller, W. & Meyburg, C. 1995c. Migration and wintering of the Lesser Spotted Eagle (*Aquila pomarina*) - a study by means of satellite telemetry. *J. Orn.* 136: 401–422.

Newton, I. 1979. *Population ecology of raptors*. T. & A. D. Poyser, Berkhamsted.

Newton, I. 1998. Migration Patterns in West Palaearctic Raptors. Pp. 603-612 in *Holarctic Birds of Prey*. eds. R. D. Chancellor, B.-U. Meyburg & J. J. Ferrero. ADENEX-WWGBP, Badajoz, Spain.

Newton, I. 2008. *The Migration Ecology of Birds*. Academic Press, London.

Newton. I. 2010. *Bird Migration*. HarperCollins, London.

Newton, I. & Dale, L. 1996. Relationship between migration and latitude among west European birds. *J. Anim. Ecol.* 65: 137–146.

Panuccio M. & Agostini N. 2006. Spring passage of second-calendar-year Honey-buzzards at the Strait of Messina. *Brit. Birds* 99: 95–96.

Spaar, R. & Bruderer, B. 1996. Soaring migration of Steppe Eagles *Aquila nipalensis* in southern Israel: flight behaviour under various wind and thermal conditions. *J. Avian. Biol.* 27: 289–301.

Strandberg, R. & Alerstam, T. 2007. The strategy of fly-and-forage migration, illustrated for the Osprey (*Pandion haliaetus*). *Behav. Ecol. Sociobiol.* 61: 1865–1875.

Strandberg, R., Klaasson, R. H. G. & Alerstam, T. 2009a. How hazardous is the crossing of the Sahara desert for migratory birds? Indications from satellite tracking of raptors. *Biology Letters* 6: 2297–2300.

Strandberg, R., Alerstam, T., Hake, M. & Kjellen, N. 2009b. Short distance migration of the Common Buzzard *Buteo buteo* recorded by satellite tracking. *Ibis* 151: 200–206

Strandberg, R., Klaasson, R. G. H., Hake, M., Olofssonj, P. Thorup, K. & Alerstam, T. 2008. Complex timing of Marsh Harrier *Circus aeruginosus* migration due to pre- and post-migratory movements. *Ardea* 96: 159–171.

Strandberg, R., Klaasson, R. H. G., Olofsson, P. & Alerstam, T. 2009c. Daily travel schedules of adult Eurasian Hobbies *Falco subbuteo* – variability in flight hours and migration speed. *Ardea* 97: 287–295.

Trierweiler, C., Mullie, W. C., Drent, R. H., Exo, K.-M., Komdeur, J., Bairlein, F., Harouna, A., de Bakker, M. & Koks, B. J. 2013. A Palaearctic migratory raptor species tracks shifting prey availability within its wintering range in the Sahel. *J. Anim. Ecol.* 82: 107–120.

Ueta, M., Sato, F., Lobkov, E. G., Naga Lisa, M. 1998. Migration route of White-tailed Sea Eagles *Haliaetus albicilla* in northeastern Asia. *Ibis* 140: 684–696.

Yáñez, B., Muñoz, A.-R., Bildstein, K. L., Newton, I., Toxopeus, A. G., & Ferrer, M. 2014. Individual variation in the over-summering areas of immature Short-toed Snake Eagles *Circaetus gallicus*. *Acta Ornithol.* 49: 137–141.

OSPREY
Pandion haliaetus

VARIATION Occurs as nominate *haliaetus* in the region, but resident birds from the Red Sea and Arabia may warrant taxonomic recognition (as 'Arabian Osprey'). They are whiter and less marked below and the upperparts are often heavily worn and bleached as a result of exposure to intense sunlight and flying sand. There are two records of the North American ssp. *carolinensis* from the region (Iceland and the Azores; Strandberg 2013).

DISTRIBUTION Cosmopolitan, with a nearly worldwide distribution. In our region confined to temperate parts of W, C and N Europe, with additional populations in Arabia, the Red Sea, W Mediterranean, Portugal and on some Atlantic islands. Reintroduced and now breeding again in S Spain.

BEHAVIOUR Often seen hovering over shallow waters, both salty and fresh, looking for fish. Perches prominently on posts and tops of dead trees; often also on rocks by the water.

MOULT The adults moult in the breeding area as well as on the wintering grounds, with moult suspended for the migration. Adults can thus be growing flight feathers during any month of the year.

The Osprey is the only species in the region, except for Black-shouldered Kite, to start its complete wing-moult in its 1st cy. By late Nov–Dec migrant juveniles start their wing-moult by dropping their innermost primary, after which the moult is a more or less continuous process for the rest of the bird's life, save for temporary stops during migration and part of breeding season. Resident populations of the Red Sea and Arabia moult c. 4–6 months earlier in the season compared to migratory populations, mirroring their earlier breeding season. Juveniles of the Arabian population moulting for the first time can thus be told from migrant juveniles wintering in the same area by their more advanced moult at any given date.

SPECIES IDENTIFICATION A rather straight-forward species to identify, thanks to its diagnostic plumage and shape, and limited plumage variation. Even from a distance, or in poor light, when the underwings may look just dark, the white body and head in combination with the diagnostic, long and narrow, clearly angled wings and rather short and narrow tail are enough to identify an Osprey. Powered flight with wings pushed forward and angled at the carpal, with most of wing-action typically occurring below horizontal level. Glides with wings smoothly arched, reminiscent of large gull.

PLUMAGES Normally only two plumages can be separated in the field, juveniles and adults, although, given good views, a transitional plumage can also be recognised by its retained juvenile secondaries. Owing to the strong wear and bleaching of the upperside, juveniles already lose their distinctive pale feather tips during their first winter, becoming much more difficult to separate from older birds by spring.

Juvenile

Fresh autumn juveniles are slimmer in outline than adults, with narrower wings in particular. Seen against a dark backdrop the white trailing edge to the wings and the white tail-tip are diagnostic as is the distinct pale scaling of the dark upperparts, although this is often difficult to discern from any distance, except for a wider whitish elbow-patch, which is visible from afar on the inner upperwing.

Seen from below, the barring of tail and remiges differ from that of the adults by being more distinct and even, and also by the lack of a broad subterminal band. The greater underwing-coverts are clearly barred, compared with the dark-looking greater coverts of adult birds, which combines to give the whole underwing a lighter and more distinctly and loosely patterned appearance. The white forearm is stained yellowish-buff, a further difference from the adults. The crown is dark-streaked in juveniles and the dark breast-band, when present, appears scaly, although this is discernible only at close range. Iris is orange, not light yellow as in adult.

Transitional plumage

The pale tips and margins of the upperparts soon wear off, and may be lost already by Nov in the 1st cy. At the same time the moult of the primaries also starts. First-winter birds would thus be moulting their inner primaries, while the rest of the plumage remains juvenile. Immatures can be aged well into their first moult by their remaining juvenile secondaries and greater underwing-coverts, showing the juvenile-type barring (see under Juvenile above). Also the pale tips of the upper primary coverts are often retained until later, giving a clue about the bird's age. As long as juvenile secondaries are retained, the immatures can be aged by the diagnostic barring of the juvenile feathers, but by late 2nd cy autumn at least some birds would probably have replaced all their juvenile remiges, making exact ageing impossible.

Adult

Adults are uniformly brown above, save for irregularities owing to variously worn feathers, the crown is shining white and the tail, when fanned, is darker in the middle while the paler sides show rather distinct sparse barring and a broad diffuse darker subterminal band ('wheatear-pattern'). Seen from below the body is gleaming white, with a variably clear, darker breast-band (see under Sexing). The underwing has a diagnostic and constant pattern, with white forearm, a darker line of greater coverts and a rectangular dark carpal patch. The flight feathers are brownish-grey with rather faint barring, the tips of the long primaries being darkest. The darker tips of the secondaries create a broad but faint dark trailing edge to the arm and the greater underwing-coverts appear all-dark, both features different from the juvenile plumage.

SEXING Often possible but since many of the characters are overlapping, multiple features need to be seen. The white forewing looks clean in males, with at the most just one row of small dark spots, while females often show several lines of bigger spots; the underwing thus looks more 'messy' compared to males. In males, the dark carpal patch is restricted mostly to the forewing, while in females it also stretches to the greater coverts. As a rule, males have a paler and less conspicuous breast-band than females, but difficult intermediates occur. Males are also slimmer in silhouette than females, with narrower wings and a slimmer body, and the head is rounder and bill smaller, but appreciating these differences requires former experience or a chance to compare the two side-by-side. Juveniles can be sexed following the principles given above, but the differences may be less pronounced.

CONFUSION RISKS Not possible to mistake for any other raptor when seen well, thanks to diagnostic shape, with long narrow and angled wings and distinct plumage. The Osprey is the only raptor which, from a distance, is vaguely reminiscent of a large gull.

NOTE The population of Arabia and the Red Sea breeds up to six months earlier in the season compared to the migratory birds from further north, hence their moult also starts earlier in the year. Apart from being overall whiter and less marked below, they are also more heavily worn and more bleached above compared to migrants from the north, which may occur in the same area on migration and in winter. This difference in plumage condition is particularly obvious in young birds during autumn and first winter, when migrants appear in a fairly good plumage while the local birds are extremely faded and worn. By the time the migrant juveniles commence their first moult the local birds' moult is already well underway, showing up to 4–5 fresh inner primaries.

REFERENCES

Strandberg, R. 2013. Ageing, sexing and subspecific identification of Osprey, and two Western Palearctic records of American Osprey. *Dutch Birding* 35: 69–87.

1. Osprey *Pandion haliaetus*, adult female on migration, showing typical silhouette of the species, with long, narrow, distinctly bent wings and rather small tail. White body and wing linings, together with dark carpals, greater coverts and outer hand, make the species rather unmistakable. Israel, 25.3.2013 (DF)

2. Osprey *Pandion haliaetus*, juvenile male, wintering. Aged as a juvenile by yellowish wash to underwing-coverts, distinct and uniform barring of secondaries, and streaked crown; identified as a male by slim build, faint breast-band and clean underwing-coverts. Oman, 7.11.2013 (DF)

3. Osprey *Pandion haliaetus*, juvenile (male), wintering. Note more distinctly barred secondaries and more loosely patterned greater coverts compared to adult. Weak breast-band, clean wing-coverts and small, broken carpal patch indicate a male. Oman, 7.11.2013 (DF)

4. Osprey *Pandion haliaetus*, juvenile female, wintering. Females are broader-winged, heavier-bodied and stronger-billed than males, even as juveniles. Note also more distinct breast-band and spotted wing-linings of female, and compare with 2. Oman, 12.11.2013 (DF)

5. Osprey *Pandion haliaetus*, juvenile female, wintering (same as 4). Fresh juveniles feature pale feather-tips on entire upperparts. Oman, 12.11.2013 (DF)

6. Osprey *Pandion haliaetus,* juvenile of the resident Arabian population. Note whiter crown and poorly defined dark eye-mask but, above all, the heavily worn plumage; compare with the wintering migrant in **5**, photographed only three days later. Inner primaries and central tail are moulting. Oman, 9.11.2013 (DF)

7. Osprey *Pandion haliaetus,* juvenile of the resident Arabian population. Note pale overall impression with poorly defined dark carpal patch. Oman, 21.11.2014 (DF)

8. Osprey *Pandion haliaetus,* juvenile of the resident Arabian population (same as **7**). Extremely worn plumage is typical of autumn juveniles of this population; note advanced moult compared to migrant juveniles. Oman, 21.11.2014 (DF)

9. Osprey *Pandion haliaetus,* juvenile female, wintering. Moult of inner primaries and central tail has commenced, but timing is about three months later compared to locally bred juveniles (cf. **6**). Note distinct breast-band, spotted underwing-coverts and solid dark carpal area, indicative of a female. Oman, 15.2.2013 (DF)

10. Osprey *Pandion haliaetus*, juvenile male, wintering. Primary moult has only just commenced; sexed by clean underwing-coverts. Oman, 10.2.2014 (DF))

11. Osprey *Pandion haliaetus*, first-winter juvenile (migrant?). Compare pattern of retained juvenile and moulted adult-type secondaries, and note scaly appearance of breast-band, typical of a juvenile. Israel, 28.3.2012 (DF)

12. Osprey *Pandion haliaetus*, adult male, breeding. Diffuse breast-band and white underwing-coverts suggest male, as does the shape of the carpal patch, being darkest along the leading edge of the wing. Finland, 20.4.2013 (DF)

13. Osprey *Pandion haliaetus,* adult male, breeding (same as **12**). Note almost uniformly darkish secondaries and greater coverts, and compare with juvenile. Finland, 20.4.2013 (DF)

14. Osprey *Pandion haliaetus,* adult female, breeding. Heavy build, broad wings and spotting to white underwing-coverts identify this bird as a female. Finland, 23.4.2014 (DF)

15. Osprey *Pandion haliaetus*, adult female, migrant. Obvious breast-band, heavily spotted underwing-coverts and a rather solid dark carpal patch are all indications of a female. Israel, 19.3.2007 (DF)

16. Osprey *Pandion haliaetus*, adult male, resident. Adults are brown above, often clearly variegated as here, save for white crown and pale sides to tail. Cape Verde, 2.2.2006 (DF)

17. Osprey *Pandion haliaetus*, adult male, resident. Note overall whiteness and poorly defined markings of this local Red Sea form, except for darker wingtips. Egypt, 28.10.2008 (DF)

EUROPEAN HONEY-BUZZARD
Pernis apivorus

Other names: Western or Eurasian Honey Buzzard/
Honey-buzzard

VARIATION Monotypic.

DISTRIBUTION Breeds over much of Europe,
reaching the Arctic Circle in the north, and W Siberia
in the east. A long-distance migrant wintering in sub-
Saharan Africa. Autumn passage from mid-Aug to
mid-Oct, with peak of adults in late Aug/early Sep,
juveniles 1–2 weeks later. Spring migration mostly
from late Apr to late May.

BEHAVIOUR A skulking forest-dweller. During
breeding season mostly seen in flight only, either
when moving between nest-site and hunting grounds,
or in diagnostic aerial display flight, which can be
seen throughout summer. On migration follows
major flyways around the Mediterranean to and from
wintering grounds in sub-Saharan Africa.

SPECIES IDENTIFICATION Adults can be identified
by diagnostic wing- and tail-barring and rather
distinctive silhouette, while juveniles are more similar
to Common Buzzard in proportions, but plumage
details are diagnostic. Soars on flat wings and glides
on gently arched, yet flattish-looking wings, giving a
diagnostic flat head-on profile, with body appearing
small in relation to wingspan compared to Common
Buzzard. When soaring the bulging arm is typically
broader than the hand, wings are clearly pinched-in at
the base and the whole wing is pushed more forward
compared to Common Buzzard. Flapping flight more
relaxed than Common Buzzard's, with deeper and
slower wing-beats, small head often lifted slightly to
create cuckoo-like impression, as if the bird is looking
back over its shoulder. Juveniles have dark eyes and
an *extensively yellow gape, cere and base of bill, with only
tip of bill narrowly black*. Adults have a yellow iris and a
greyish cere and bill.
   Difficult to tell from Crested Honey-buzzard (which
see), particularly in juvenile plumage.

MOULT in adults starts in the breeding area, but is
suspended for the autumn migration. By then usually
up to four inner primaries have been replaced, 0–3
in males and 2–4 in females. Moult is resumed in the

winter quarters and finished by Feb–Mar, well in time
before the spring migration.
   Juveniles stay in Africa for almost their first two
years of life and knowledge about their moult while
in Africa is scant. The body plumage is replaced from
the first winter, and the moult of the flight feathers
starts during the 2nd cy summer. 2nd cy birds seem to
complete their first moult sooner than adults, possibly
already by their 2nd cy autumn, 2–3 months ahead of
the adults.
   On autumn passage adults show fewer moulted
primaries compared to Crested Honey-buzzards.

PLUMAGES In Europe juveniles and adults occur
together only in autumn, and since young birds remain
in Africa for the next two years all spring migrants
should be adults. Only exceptionally do young birds
return to Europe before their second spring (in 3rd cy).
   The European Honey-buzzard is one of the most
variable of all raptor species. However, the wing- and
tail-barring is less prone to variation, being diagnostic
in all plumages, while body plumage of underparts
varies from white, through streaked or barred to all-
dark. Juveniles and adults differ in plumage, even adult
males and females differ on average but overlapping
types occur. Juveniles and adults also show highly
different flight silhouette, with juveniles being slimmer-
winged and shorter-tailed than adults. This explains
why juveniles are so often misidentified, even by people
familiar with adult European Honey-buzzards.

Juvenile

Regardless of colour of body plumage the pattern of
the flight feathers remains distinctive, and is alone
sufficient to distinguish juveniles from any confusion
risks: *both primaries and secondaries show 3–4 evenly
spaced broad bands, the secondaries always appear dusky,*
contrasting against the paler bases of the inner
primaries, the fingers are extensively dark, making the
entire wingtip of juvenile European Honey-buzzards
more extensively dark compared to any *Buteo* buzzard.
The tail-banding is variable, often just sparsely but
irregularly barred brown from above, while from below
it looks pale, sometimes with two or three well-spaced
'honey-buzzard bands' visible. Although juveniles
come in many different plumage-types, uniformly dark

brown birds predominate and account for about 90% of all juveniles. The remaining birds are lighter below, from sandy buff to pure white, variably streaked or mottled, often with a pale or whitish head, with showy dark 'sunglasses' in many. The underwing-coverts vary in accordance with the general plumage colour and pattern, but in darker forms *the greater underwing-coverts usually stand out as a paler band*, opposite to *Buteo* buzzards, where they often are clearly patterned and darker than the adjacent median coverts.

The upperparts also vary according to the general coloration of the plumage, but unlike in *Buteo* buzzards the *uppertail-coverts stand out as a pale U on most*. Lighter forms often show distinct pale tips/margins to their upperwing-coverts, sometimes forming a pale wing-patch across the median coverts, recalling Black Kite or even Booted Eagle.

First-summer

Although the majority of young birds stay in Africa for their first summer, a few individuals have been reliably recorded in Europe (pers. obs.). Prior to the onset of the wing-moult first-summers can be identified by their juvenile flight and tail feathers, although their body plumage may already be moulting. Birds in wing-moult are quite conspicuous, as the new tail feathers and primaries are much longer than the juvenile feathers and protrude clearly beyond the old and extremely worn feathers. The new inner primaries show the adult pattern with a broad black subterminal band and the new central tail feathers are equally strongly marked, and reach well beyond the tip of the old tail. By autumn the inner 4–5 primaries have been replaced giving the wing a very strange shape, being widest across the carpal area. The silhouette is like no other raptor, with a longish tail, narrow wing-base, very broad inner hand and a narrower, rounded tip.

Adult male

Often easily identified from below by its 'clean-looking' flight feathers, with a broad black trailing edge to the wing and with 'finger-tips dipped in ink', while the rest of the barring is confined to the basal portions of the remiges, just outside the coverts. Typically, there is a long jump from the dark trailing edge to the next wing-band, and the same pattern is repeated in the tail, where the median tail-band is barely visible behind the tips of the undertail-coverts. The band across the underwing secondaries, if at all present, disappears behind the coverts well before reaching the body.

The upperparts are normally clearly grey, with black bands visible along the trailing edge of the wing and tail, and additional dark bands on the greater coverts and inner primaries. The underbody and underwing-coverts vary from nearly pure white to all-dark, with a barred plumage-type being the most common. The change from one morph to another is gradual with all possible intergrades occurring, making some individuals difficult to assign to a certain plumage type. *The head looks uniformly grey* and pigeon-like, and the iris is yellow. Adult males have slightly narrower wings and a comparatively longer tail compared to adult females, giving them a rather diagnostic shape.

Adult female

Females have more patterned flight feathers than males, and the bands are finer, less distinct and run further out on the feathers. The dark trailing edge is narrower and less conspicuous compared to adult males, and the fingered primaries darken gradually towards the tips. The wing-bands of the secondaries run further out on the feathers than in males, with less of a gap between the trailing edge and the barring. The upperparts are browner and more variegated compared to the males, often with a pale area on the inner hand, and the uppertail-coverts are pale in many. In females the darker bands of the upperwing are browner and less distinct compared to the distinct black bands of adult males. The head is often pale, or of same colour as the rest of the bird, streaked or mottled, and the grey colour, if present, is restricted to the sides of the head or just to the area around the eye. However, some (old?) females can show a grey head and male-like underwing banding and may thus be wrongly sexed if no other features are consulted. The structural differences between the sexes can be used in situations like this, but this requires extensive previous experience of the species: the females are broader-winged (in particular having a broader hand) and shorter-tailed, thus are more compact and heavier-looking, compared to adult males which appear slimmer, with a narrower hand and a longer tail.

CONFUSION RISKS *Juvenile European Honey-buzzards are probably the most often misidentified birds of prey in Europe.* This is partly explained by their extreme plumage variation, while their proportions are rather similar to Common Buzzard, and they are surprisingly different from the rather conspicuously shaped adults. Very similar to Crested Honey-buzzard (which see).

NOTE Some adults can be difficult to sex as they seem to show partly male and partly female characters. These are probably old females, judging by their female-like proportions, which have acquired some male-type characters. The head is largely grey and they show rather male-like underwing- and tail-barring, with the dark wing-bands running further in on the feathers compared to the average adult female. However, they still retain brownish upperparts lacking the contrasting upperwing markings seen in adult males and the secondaries are more strongly marbled between the main wing-bands compared to the cleaner look of adult males; most importantly, they show female proportions with broad wings and a shorter tail compared to adult males.

See also Crested Honey-buzzard.

18. European Honey-buzzard *Pernis apivorus*, juvenile, migrant. The uniformly dark brown plumage is by far the most common form in juveniles. Note the diagnostic flight-feather pattern, with extensively dark wingtips, dusky secondaries and a few heavy and well-spaced wing-bars. Finland, 14.9.2013 (DF)

19. European Honey-buzzard *Pernis apivorus*, juvenile, migrant. A slightly more mottled dark juvenile. Finland, 25.9.2010 (DF)

20. European Honey-buzzard *Pernis apivorus*, juvenile, migrant. Some dark juveniles may appear almost black. Sweden, 6.9.2014 (DF)

21. European Honey-buzzard *Pernis apivorus*, juvenile, migrant. Note the diagnostic sparse wing-barring and compare with Common Buzzard. Finland, 22.9.2006 (DF)

22. European Honey-buzzard *Pernis apivorus*, juvenile, migrant. A lighter brown juvenile. Note the extensively yellow bill-base and dark eye of all juveniles. Finland, 5.9.2011 (DF)

23. European Honey-buzzard *Pernis apivorus*, juvenile, migrant. A mottled lighter brown juvenile. Finland, 15.9.2012 (DF)

24. European Honey-buzzard *Pernis apivorus*, juvenile, migrant. Fewer than 10% of juveniles belong to the largely white form. They often have a dark mask around the eye. Finland, 27.9.2005 (DF)

25. European Honey-buzzard *Pernis apivorus*, juvenile, migrant. Most of the white birds are variably streaked below, but birds with all-white body plumage do occur. Finland, 13.9.2013 (DF)

26. European Honey-buzzard *Pernis apivorus*, juvenile, migrant. Most dark juveniles are fairly uniform above, apart from pale uppertail-coverts, which is a diagnostic character compared to Common Buzzard. Note also the fine bill, which is black only at the very tip, and the dark eye; compare with juvenile *Buteo* buzzards. Finland, 9.9.2009 (DF)

27. European Honey-buzzard *Pernis apivorus*, juvenile, migrant. The pale uppertail-coverts are visible from afar, as is the extensively yellow bill. Finland, 22.9.2006 (DF)

28. European Honey-buzzard *Pernis apivorus*, juvenile, migrant (same as **22**). Lighter birds often show a pale area on the upperwing-coverts, resembling Black Kite or even Booted Eagle. Finland, 5.9.2011 (DF)

29. European Honey-buzzard *Pernis apivorus*, adult male, spring migrant. Unlike juveniles, adults have a yellow iris and a greyish bill and cere, but only adult males are grey-headed. Spain, 1.5.2011 (DF)

30. European Honey-buzzard *Pernis apivorus* adult male, spring migrant. A typical adult male, showing clean-looking flight feathers, a broad black trailing edge to the wing and a grey head. Spain, 5.5.2011 (DF)

31. European Honey-buzzard *Pernis apivorus*, adult male, autumn migrant. Note that the dark carpal patches, which are obvious in all but dark morph birds, are oval in shape in honey-buzzards, but round in *Buteo* buzzards. Spain, 4.9.2013 (DF)

32. European Honey-buzzard *Pernis apivorus*, adult male, spring migrant. Spain, 1.5.2011 (DF)

33. European Honey-buzzard *Pernis apivorus*, adult male, autumn migrant. A more heavily barred male. Spain, 5.9.2013 (DF)

34. European Honey-buzzard *Pernis apivorus* adult male, spring migrant. An almost white male with reduced dark markings on the body plumage. The growing primary is not regular moult, but is replacing an accidentally lost feather. Spain, 5.5.2011 (DF)

35. European Honey-buzzard *Pernis apivorus* adult male, spring migrant. A very dark bird, classified as belonging to the dark morph. Despite the great variations in body plumage between individuals, the remiges retain their diagnostic pattern. Spain, 1.5.2011 (DF)

36. European Honey-buzzard *Pernis apivorus*, adult male, autumn migrant. Adult males typically appear rather grey above with distinct dark wing-bands. Spain, 4.9.2013 (DF)

37. European Honey-buzzard *Pernis apivorus* adult female, spring migrant (with adult male above). Adult females are broader-winged and comparatively shorter-tailed than males; their secondaries often appear duskier and the wing- and tail-barring runs further out on the feathers compared to adult males. Spain, 5.5.2011 (DF)

38. European Honey-buzzard *Pernis apivorus* adult female, spring migrant. A typical adult female with regard to wing- and tail-barring. Note also that females tend to have duskier fingers compared to the black-tipped primaries of adult males. Spain, 6.5.2011 (DF)

39. European Honey-buzzard *Pernis apivorus*, adult female, autumn migrant. Note the brownish head and the dusky fingers, both typical female traits. Finland, 25.8.2005 (DF)

40. European Honey-buzzard *Pernis apivorus* adult female, spring migrant. Some (old?) females may attain a male-like plumage and are best sexed by their different proportions. This female has retained the typical female-type barring on its underwings and tail. Spain, 28.4.2011 (DF)

41. European Honey-buzzard *Pernis apivorus* adult female, spring migrant. This female shows broad dark fingers and dusky secondaries. It is birds like this that account for many of the claims of 2nd cy immatures from Europe. Spain, 6.5.2011 (DF)

42. European Honey-buzzard *Pernis apivorus*, adult female, autumn migrant. This bird shows the broad-winged female proportions and also typical underwing-barring. Spain, 10.9.2012 (DF)

43. European Honey-buzzard *Pernis apivorus* adult female, spring migrant. A dark bird with typical adult female-type barring on the underwing remiges. Spain, 5.5.2011 (DF)

44. European Honey-buzzard *Pernis apivorus* adult female, spring migrant. Another dark female, which could easily be mistaken for an immature by the unaware. Spain, 5.5.2011 (DF)

45. European Honey-buzzard *Pernis apivorus*, adult female, autumn migrant. Adult females appear rather uniformly brown above, lacking the greyness and the black bands typical of adult males. Instead they show darker secondaries contrasting clearly with the pale inner hand. Spain, 10.9.2012 (DF)

CRESTED HONEY-BUZZARD
Pernis ptilorhyncus

Other names: Oriental or Eastern Honey Buzzard/Honey-buzzard

VARIATION Occurs in the region as migratory ssp. *orientalis*, with another five races (some possibly independent species) in SE Asia. In the light of recent observations apparently hybridises extensively with European Honey-buzzard in western part of range (pers. obs.).

Ssp. *ruficollis* from India may occur as a vagrant to the Arabian Peninsula, based on sightings from Oman, where non-moulting adults have been observed in winter, at a time when adult *orientalis* are all actively moulting (pers. obs.), but further studies are required.

DISTRIBUTION A rare but regular passage migrant and winter visitor to the Middle East. The migrant ssp. *orientalis* breeds from W Siberia eastwards to NE China, Japan and Korea, with the majority wintering in SE Asia and the Indonesian islands. Breeding range probably reaches further west than previously thought. In the west mixes with European Honey-buzzards on migration, and some, both pure *orientalis* but also hybrids and intergrades, migrate through the Middle East each autumn and spring.

BEHAVIOUR A skulking forest-dweller spending most of the time hidden in the canopy. Birds wintering in the Middle East prefer palm plantations and similar evergreen forest.

SPECIES IDENTIFICATION Extremely variable and superficially similar to European Honey-buzzard, the two species being particularly difficult to tell apart in certain juvenile plumages. To people with experience of European Honey-buzzard, Crested stands out as *broader-winged and shorter-tailed, more eagle-like in proportions* and the wingtip is broader and blunter, with six clearly fingered primaries. As a rule, *in Crested the tail is shorter than the width of the wing-base, while in European the tail is longer than the width of the wing-base.* Hybrids are intermediate in this respect, with wing-width equalling tail length. Many Cresteds (but not all) feature a dark gorget around the pale throat and they *always lack the dark carpal patch* of European Honey-buzzard, but neither of these diagnostic characters are apparent in dark morph birds. Pale cinnamon-coloured birds are rather common in both adult and juvenile Crested, while this colour form is lacking in European. *The feet of Crested are clearly bigger than in European and the claws are strikingly long and straight,* often clearly visible in flight, unlike in European, where they are usually hidden in the undertail-coverts.

MOULT Annual moult complete, but odd remiges, particularly median secondaries, may be left un-moulted. Adults start moulting during breeding in mid-summer, but the moult is suspended for the autumn migration. During autumn migration females show 4–5 moulted inner primaries, compared to 2–3 in males, while both are ahead of the respective sex of European Honey-buzzard at the same time. Moulting resumes in the winter quarters and finishes on average earlier than in European, by Dec–Feb.

Non-breeding 2nd cy birds probably start moulting earlier in the season than adults, showing a more advanced moult at any given time compared to adults, and they are possibly able to complete their moult by December.

PLUMAGES Highly variable, with body plumage ranging from light buff or cinnamon to all-dark, but the strongly barred plumages, common in European, appear to be lacking. Normally, adult male, adult female and first-years can be separated, and also second-autumn birds until moult is completed. Among winter visitors to western part of range (Middle East) dark birds appear to dominate.

Juvenile

Similar to juvenile European Honey-buzzard, but wings broader and wingtip blunter, with six clearly fingered primaries. *Most birds show more bars and denser barring to secondaries (4–6 bands), compared to 3 in most juvenile European.* Tail either evenly but coarsely barred, or showing three rather prominent dark, sparsely spaced bars, much as in European. Fingers appear all dark, but close views may reveal sparse barring in some. Body plumage of dark birds is mottled or uniformly dark brown to almost blackish, with paler throat. Paler birds variably streaked or mottled below, often with rather finely barred flanks and contrasting dark gorget around pale throat (never in juvenile European), and *always lacking distinct dark carpal patch of pale juvenile European.* Upperparts much like in juvenile European, dark brown, often with prominent pale feather edges and pale tips to upperwing-coverts, and a striking pale U across the uppertail-coverts. Cere and base of bill are extensively yellow, only tip of bill is dark. Iris is dark brown to begin with, turning greyish, then yellowish during first autumn and winter.

Transitional plumage

From juvenile to adult can be separated for as long as juvenile flight feathers are retained, normally until at least Nov–Dec in 2nd cy, sometimes until Feb in 3rd cy. Retained juvenile secondaries are narrower and somewhat shorter (less difference compared to European), very worn and *lack the broad black subterminal band of the newly moulted feathers of adult type.* The same difference in pattern is also seen in the tail. Iris as in adults, yellow or dark, depending on sex, but cere is still yellowish, as in juveniles.

Adult male

This is the most distinctive plumage, with greyish head and deep red (dark) eyes, and diagnostic, sparse underwing barring and a broad black trailing edge to the wings. Tail pattern is equally diagnostic, with a *broad pale central band sandwiched between broad dark subterminal and inner tail-bands.* Body plumage of underparts varies from pale sandy-ochre (most common) to all dark, with pale forms lacking the carpal patch of European. Upperparts grey compared to adult female, with black tips of greater coverts and remiges forming distinct wing-bands, much as in adult male European. Underwings diagnostic with outer secondary band reaching body clear of underwing-coverts, and with primary barring running further out than in adult male European, with outermost band spanning across base of fingers. *The deep red iris is diagnostic* (appears dark in the field); bill and cere are dark greyish, bill-tip dark.

Adult female

Has less grey to head and a yellow or orange-yellow iris, while *underwing shows evenly but sparsely banded secondaries and primaries, lacking the gap in the barring typical of European.* Tail-banding often similar to adult male European Honey-buzzard. Body plumage on underparts variable, similar to adult male, while upperparts are browner and more mottled compared to males, often with a paler U across the uppertail-coverts and a pale window to the upper primaries; sexual dimorphism of upperparts thus similar to European. Bill and cere grey, tip of bill dark.

SEXING Adults can be sexed by their different tail and underwing banding and different iris colour (see above). Sexes are not known to differ in juveniles.

CONFUSION RISKS Very similar to European Honey-buzzard, especially in juvenile plumage, and great care is needed for specific identification. Apart from differences in size and structure, the barring of the flight feathers is the most important feature to note. The two species are known to hybridise frequently (see below).

NOTE Hybrids between European and Crested Honey-buzzards appear to be much more common than previously thought. It is important to keep this in mind when attempting to identify a presumed Crested. This is particularly relevant when identifying birds well west of the species' normal range, where the chances of hybrids are higher. Photographs of birds from Israel and Arabia, claimed as Crested, show that many are in fact hybrids and that the number of published records of Crested in the region may be too optimistic – in fact, hybrids could be more common than the real thing (pers. obs.). It is worth noting, that some of the birds showing intermediate proportions may in fact belong to the Indian subspecies *ruficollis*, which is somewhat smaller with comparatively shorter wings and longer tail compared to *orientalis*. Birds like this are regularly seen in Arabia and since they are not moulting at the same time as *apivorus* or *orientalis*, they are likely to represent a third population with a different breeding and moulting season. Further studies are required.

REFERENCES

Faveyts, W., Valkenburg, M. & Granit, B. 2011. Crested Honey Buzzard: identification, western occurrence and hybridisation with European Honey Buzzard. *Dutch Birding* 33: 149–162.

Forsman, D. 1994: Field identification of Crested Honey Buzzard. *Birding World* 7: 396–403.

46. Crested Honey-buzzard *Pernis ptilorhyncus*, light juvenile, migrant. Very similar to European Honey-buzzard, but note lack of dark carpal patch and broad wingtip with six fingers. Thailand, 17.10.2010 (Martti Siponen)

47. Crested Honey-buzzard *Pernis ptilorhyncus*, light juvenile, wintering migrant. Note distinct black malar mark and broad hand; compare with European Honey-buzzard. Oman, 9.12.2012 (DF)

48. Crested Honey-buzzard *Pernis ptilorhyncus*, cinnamon-buff juvenile, wintering migrant. Broad wings and lack of dark carpal are diagnostic. Juvenile Crested often (but not always) shows four bands on the secondaries, compared to three in European Honey-buzzard. Oman, 7.12.2013 (DF)

49. Crested Honey-buzzard *Pernis ptilorhyncus*, juvenile, wintering migrant. Distinct black gorget and lack of dark carpal indicates Crested. Oman, 12.12.2010 (DF)

50. Crested Honey-buzzard *Pernis ptilorhyncus*, dark juvenile, wintering migrant. Typical black gorget, broad hand and densely banded secondaries are all Crested characters. Oman, 12.12.2010 (DF)

51. Crested Honey-buzzard *Pernis ptilorhyncus*, dark juvenile, migrant. Similar to dark juvenile European Honey-buzzard, but note six fingers and more densely barred secondaries. Thailand, 6.10.2012 (Martti Siponen)

52. Crested Honey-buzzard *Pernis ptilorhyncus*, dark juvenile, migrant. Very much like a dark juvenile European Honey-buzzard, except for broader wings, especially the hand. Thailand, 22.10.2010 (Martti Siponen)

53. Crested Honey-buzzard *Pernis ptilorhyncus*, light juvenile, migrant. Light juveniles have pale uppertail-coverts and pale fringes to the upperwing-coverts, just like European Honey-buzzard, but note broad hand. Thailand, 11.10.2010 (Martti Siponen)

54. Crested Honey-buzzard *Pernis ptilorhyncus*, light immature male moulting from juvenile to second plumage, 2nd cy autumn, wintering migrant. Note different length, colour and condition of retained juvenile feathers compared to moulted ones. Oman, 6.12.2012 (DF)

55. Crested Honey-buzzard *Pernis ptilorhyncus*, cinnamon immature male moulting from juvenile to second plumage, 2nd cy autumn, wintering migrant. This more advanced male has nearly completed its moult, with only one juvenile secondary left (s4). Oman, 7.12.2012 (DF)

56. Crested Honey-buzzard *Pernis ptilorhyncus*, adult male, migrant. Note grey head, dark eye and broad black tail-bands. Compare underwing barring with adult female, and with both sexes of European Honey-buzzard. Thailand, 3.10.2011 (Martti Siponen)

57. Crested Honey-buzzard *Pernis ptilorhyncus*, cinnamon adult male, migrant. This particular plumage colour does not occur in adult European Honey-buzzard. Thailand, 7.10.2012 (Martti Siponen)

58. Crested Honey-buzzard *Pernis ptilorhyncus*, dark adult male, migrant. Note typical tail and underwing pattern. Thailand, 3.11.2011 (Martti Siponen)

59. Crested Honey-buzzard *Pernis ptilorhyncus*, finely barred adult male, wintering migrant. This male shows intermediate characters between European and Crested Honey-buzzards, with a red-orange iris, a narrower rather rounded hand and clean-looking underwings; possibly a hybrid/intergrade. Oman, 11.2.2014 (DF)

60. Crested Honey-buzzard *Pernis ptilorhyncus*, dark adult male, migrant. Greyish face with dark eye, broad tail-bands and grey colour above are all typical features of adult male. Also note status of suspended wing-moult. Thailand, 11.10.2010 (Martti Siponen)

61. Crested Honey-buzzard *Pernis ptilorhyncus*, adult male, migrant. Thailand, 4.10.2012 (Martti Siponen)

62. Crested Honey-buzzard *Pernis ptilorhyncus*, dark adult male, wintering migrant. Typical grey upperparts of a male with bold black bands. Oman, 15.2.2013 (DF)

63. Crested Honey-buzzard *Pernis ptilorhyncus*, finely barred adult male (same as **59**). This male shows rather European-type barring above, although much bolder and more like a Crested; possibly an intergrade European x Crested. Oman, 11.2.2014 (DF)

64. Crested Honey-buzzard *Pernis ptilorhyncus*, adult female, wintering migrant. Note very broad, almost eagle-like wings with underwing barring covering the remiges entirely. Oman, 12.12.2010 (DF)

65. Crested Honey-buzzard *Pernis ptilorhyncus*, adult female, migrant. Very similar to female European Honey-buzzard, except for larger size and broader wings. Thailand, 4.10.2012 (Martti Siponen)

66. Crested Honey-buzzard *Pernis ptilorhyncus*, adult female, migrant. Note the longer and deeper bill of ssp. *orientalis*, compared to European Honey-buzzard. Thailand, 4.10.2012 (Martti Siponen)

67. Crested Honey-buzzard *Pernis ptilorhyncus*, cinnamon-type adult female, migrant. Wings of *orientalis* Crested Honey-buzzards are evenly broad without the clearly narrower hand and bulging arm of European Honey-buzzard. Note also distinct gorget and lack of dark carpal patch. Thailand, 5.10.2012 (Martti Siponen)

68. Crested Honey-buzzard *Pernis ptilorhyncus*, light adult female, migrant. Females are typically brown, not greyish above, the iris is yellow and the head is not grey. Thailand, 6.10.2012 (Martti Siponen)

69. Crested Honey-buzzard *Pernis ptilorhyncus*, adult female, migrant. Thailand, 10.10.2010 (Martti Siponen)

70. Crested Honey-buzzard *Pernis ptilorhyncus*, adult female, migrant. Thailand, 2.11.2011 (Martti Siponen)

71. Crested Honey-buzzard *Pernis ptilorhyncus*, dark adult female, migrant. Some females show grey around the eyes, but much less than males. Thailand, 6.10.2012 (Martti Siponen)

Identification of hybrid honey-buzzards

The Crested Honey-buzzard is thought to be a rare but regular passage migrant and winter visitor to the Middle East, joining migrating flocks of European Honey-buzzards on their way to and from Africa. However, photographs of alleged Cresteds have revealed that a considerable number of the claimed birds actually show a combination of features from both species, including some intermediate characters. Birds featuring mixed characters are likely to be of hybrid origin, originating from a probably wide overlapping zone in W and C Siberia. These presumed intergrades are intermediate in silhouette and wing-formula between the bulkier Crested and the slimmer and smaller European Honey-buzzard; the wing-barring is equally intermediate between the two species, and they may or may not show a darker gorget and/or a hint of a dark carpal patch. The more boldly marked adult male intergrades can easily be mistaken for pure Cresteds, as the tail-bands are often broader compared to European Honey-buzzard, approaching the pattern of adult male Crested Honey-buzzard.

The intergrades are highly variable, which is normal in situations with geographically extensive hybridisation zones (McCarthy 2006). Some birds look like Crested Honey-buzzards, but may show a carpal patch, an intermediate underwing barring and a lighter build. Others are more similar to European Honey-buzzard, but may show a tail-pattern more similar to male Crested, intermediate wing-banding, a longer sixth finger or a hint of a dark gorget. From a limited sample it appears that flight proportions and wing-formulae are diagnostic for Crested, European and hybrids (pers. obs.). There is no reason to believe that honey-buzzards would be any different from hybridising spotted eagles, where the hybrids themselves are fertile, and can reproduce with either of the parental species or with another hybrid. This multitude of possible combinations would also explain the great variability in the phenotype of intergrade honey-buzzards. Juveniles of the two species are very similar anyway, except for different proportions, and great care should be taken when identifying a juvenile Crested Honey-buzzard in the Western Palearctic.

In order to be able to separate between hybrids/ intergrades on the one hand and Crested or European Honey-buzzards on the other, good quality photographs are by far the best way. Important details to focus on are: wing shape and wing-formula (the internal relationship between the fingered primaries of the wingtip), type of barring of remiges and colour and pattern of underwing-coverts (carpal area) and body, colour of iris, size and structure of feet and bill, and status of wing moult.

The possibility of a hybrid/intergrade should always be kept in mind when confronted with a suspected Crested Honey-buzzard in the Western Palearctic. Any honey-buzzard in an odd place or at an odd date should be scrutinised carefully and, if possible, photographed.

References

Faveyts, W., Valkenburg, M. & Granit, B. 2011. Crested Honey Buzzard: identification, western occurrence and hybridisation with European Honey Buzzard. *Dutch Birding* 33: 149–162.

Forsman, D. 1994: Field identification of Crested Honey Buzzard. *Birding World* 7: 396–403.

McCarthy, E. M. 2006. *Handbook of Avian Hybrids of the World.* Oxford University Press.

Vaurie, C. 1965. *The Birds of the Palearctic Fauna: Non-passeriformes.* Witherby, London.

72. Possible hybrid/intergrade European x Crested Honey-buzzard *Pernis apivorus x ptilorhyncus*, juvenile, wintering. Note intermediate structure with bulging secondaries but fingered p5, hint of gorget but also hint of darker carpal area. In this juvenile iris has already turned yellow. Oman, 14.2.2013 (DF)

73. Possible hybrid/intergrade European x Crested Honey-buzzard *Pernis apivorus x ptilorhyncus*, juvenile, wintering (same as **72**). Very similar to European Honey-buzzard and not possible to identify by plumage. Oman, 12.2.2013 (DF)

74. Possible hybrid/intergrade European x Crested Honey-buzzard *Pernis apivorus x ptilorhyncus*, juvenile, wintering. Structurally rather similar to European Honey-buzzard, small-winged with broader arm and narrower hand, but note six fingered primaries, lack of dark carpal and hint of dark gorget. Oman, 7.2.2013 (DF)

75. Presumed hybrid/intergrade European x Crested Honey-buzzard *Pernis apivorus x ptilorhyncus*, adult male, wintering. A striking bird with European-like barring on wings and tail, but lacks dark carpal and shows strong dark gorget like Crested, while iris is deep orange-yellow, not matching either parental species. Oman, 12.2.2013 (DF)

76. Possible hybrid/intergrade European x Crested Honey-buzzard *Pernis apivorus x ptilorhyncus* (or ssp. *ruficollis*), wintering adult (sex uncertain). A rather European-like bird in structure, but hand too broad, while wingtip is too rounded and tail too long for Crested; underwing barring fits neither species. This bird has an orange iris. Oman, 12.12.2010 (DF)

77. Possible hybrid/intergrade European x Crested Honey-buzzard *Pernis apivorus x ptilorhyncus* (or ssp. *ruficollis*), wintering adult (sex uncertain). Wing-formula intermediate between European and Crested while underwing barring fits neither species. This bird also has an orange iris. Oman, 12.12.2010 (DF)

78. Presumed hybrid/intergrade European x Crested Honey-buzzard *Pernis apivorus x ptilorhyncus*, adult male, wintering. Shows six fingers, but wingtip is too rounded and the arm is broader than the hand, as in European. Dark gorget and tail pattern similar to Crested, as well as lack of dark carpal patch, although dark primary coverts are odd for a pure Crested. Oman, 7.2.2013 (DF)

79. Presumed hybrid/intergrade European x Crested Honey-buzzard *Pernis apivorus x ptilorhyncus*, adult male, wintering. Possibly a Crested, but note European-like wing-shape with broad and bulging arm, tapering hand, and atypical underwing barring for a male Crested. Oman, 7.2.2013 (DF)

Identification of *Milvus* kites

The identification of *Milvus* kites can be a tough challenge, depending on where in the region you are. The Red Kite is always relatively easy to identify, given reasonable views, by its large white and finely barred underwing primary-window, reaching the trailing edge of the wing, its bright orange uppertail and whitish head, and its long-winged and long-tailed proportions.

It is the Black Kite group that poses the challenges. In Western Europe all birds should be typical Western Black Kites (see below), save for some vagrant Eastern-type birds visiting Europe during the summer and autumn. In the Middle East things get more complicated as several types may occur side-by-side. Some birds look like typical Western, while others clearly show influences from Black-eared Kite. For instance, among the tens of thousands of kites wintering in Israel, both Western and Black-eared-type birds occur, plus a vast range of intermediates. The true challenge here is to tell a possible Black-eared Kite from all the intergrades.

To separate the different forms of Black Kite, both structure and plumage are of importance. True Black-eared Kites are broad-winged, almost eagle-like birds, while more western populations are much slimmer in proportions, with narrower wings and tail. Also the wing-formula differs between populations, with Black-eared showing the broadest, squarest wingtip, while the African Yellow-billed and Western Black Kites have a clearly narrower and more pointed wingtip.

As for plumage characters, adults of the different forms are usually easily separated by plumage alone, while juveniles can be confusingly similar. One of the best separating characters is the pattern of the underwing primaries, with special emphasis on the amount of white and type of barring. Also the colour of the bare parts, such as feet, cere and iris colour, are of importance, particularly when separating Western and Black-eared Kites and their intergrades.

As always, good quality photographs are of great help, as they capture details that are easily overlooked even by the most experienced field observer.

BLACK KITE
Milvus migrans

VARIATION The Black Kite complex comprises several distinct forms across Eurasia, Africa and Australasia, some of which may be independent species. Occurs in Europe as nominate *migrans* (hereafter called Western Black Kite), and in E Asia as ssp. *lineatus* Black-eared Kite (see p. 91), regarded as a full species by some and occurring from C Siberia and Mongolia to China and Japan. The African form, the Yellow-billed Kite, is treated as a separate species in this work, *Milvus aegyptius*, with two subspecies: nominate *aegyptius* in extreme NE Africa and the more southerly *parasitus* (p. 99).

The migratory Black Kites from easternmost Europe (W Russia) to C Siberia and C Asia appear to represent a mixed population, a hybrid swarm, between *migrans* and *lineatus*, with birds showing mixed and/or intermediate characters (pers. obs.). These birds, which can be phenotypically almost identical to *lineatus*, winter in big numbers in the Middle East and E Africa, and many non-breeding immatures spend their first summers as far west as northern Europe (pers. obs.).

DISTRIBUTION Widely distributed over mainland S and C Europe, breeding up to the Arctic Circle in the north. Range extends east to E Siberia, China and Japan and south to Africa, India and Australasia.

BEHAVIOUR Mostly seen on the wing, either quartering above fields and wetlands at medium height, or following highways looking for road casualties. Skilfully swoops down to take prey from the ground, e.g. from a busy road without alighting. Perches upright in trees and on pylons. In winter congregates at communal roosts for the night, where thousands of birds may be involved.

SPECIES IDENTIFICATION Medium-sized, long-winged and long-tailed, largely dark brown raptor, with juveniles more distinctly marked. Wings of adults are typically parallel-edged, with arm and hand equally broad, at times giving an eagle-like impression (typically so in *lineatus*), while in juveniles the arm appears slightly broader than the hand, with a characteristic S-shaped trailing edge. Upperparts rather uniformly dark, except for diagnostic, diffuse pale area diagonally across upperwing-coverts, and with tail and mantle of same colour and darkness (cf. Red Kite). Rather uniformly dark also from below, with only marginally paler window on the hand. Old adult Western show distinctly pale grey heads, while fresh juveniles, as well as second plumage birds, show creamy or orange-buff heads respectively and a darker eye-mask. Feet small and unobtrusive, reaching only as far as the trailing edge of wings in flight (cf. Marsh Harrier and Booted Eagle). Some adults are rather richly rufous-brown below with rather rufous lesser upperwing-coverts and uppertail. In strongly underlit situations their underwings may appear brightly marked, and when strongly backlit the tail may appear orange and translucent, thus vaguely recalling Red Kite.

MOULT Complete annual moult starts in late spring, late Apr–May, with young birds starting ahead of the adults. Some 2nd cy Western Black Kites have already started their moult in the winter quarters and return to Europe in early May with missing or even freshly moulted inner primaries. Adult females start moulting with the onset of breeding, adult males considerably later. During autumn migration several outer primaries remain unmoulted. In adults the 3–5 outermost primaries (most commonly four) are retained, while in 2nd cy birds 0–3 outer primaries (mostly 1–2) remain unmoulted. Moult is resumed upon arrival in the winter quarters, where it is completed by Jan–Feb, in good time before the spring migration.

Western Black Kite *Milvus migrans migrans*

PLUMAGES Plumage development is slow and the process from juvenile to final adult plumage takes several years. Juvenile, second and adult plumage are rather distinctive, while later immature stages (third and fourth plumages) resemble adult, but colour of iris and head differs. *Feet and cere are bright yellow in all age-classes* (cf. Black-eared Kite), and the *contrast between the black bill and yellow cere is always striking* (cf. Yellow-billed and Red Kites).

Juvenile

Identified by neat plumage, with distinct creamy or ochre drops/streaks, lacking signs of flight-feather moult throughout autumn and winter. Head often pale creamy or ochre with a darker mask behind the eye; the eyes appear dark (greyish-brown) and the vent and undertail-coverts are pale cinnamon in many, often paler than the belly, but not in all. The contrast between brown belly and lighter vent and undertail-coverts is slight compared to juveniles of Black-eared Kite. The upperparts are equally adorned with light tips to most body feathers, including a distinct pale line along the tips of the greater coverts. Except for a pale carpal crescent the underwing appears rather uniformly dark and indistinctly marked, but some juveniles show a paler hand with clearly darker wingtips. Greater underwing-coverts show paler tips,

a good ageing character, which is retained until the first moult is completed, until at least late 2nd cy autumn. The amount of streaking and spotting varies individually and some birds may appear almost white-bodied, while others are poorly marked and may from a distance appear as uniform as adults. The *undertail is quite distinctly barred in most* (cf. juvenile Black-eared Kite) and the translucent tip of the tail may in certain light conditions be striking compared to adults.

During the winter the diagnostic pale streaking wears off and can be completely gone by the time of spring migration, making spring juveniles much harder to recognise. However, the extremely worn plumage is typical of spring juveniles, which also often show damaged primaries and tail feathers, while features like retained pale tips to the greater underwing-coverts and sharply pointed primary tips help to confirm the age. By Apr–May many birds have moulted into a fresh cream-coloured head with a darker eye-mask, while the iris colour has changed to a lighter brown in many.

Second plumage

This plumage resembles the juvenile, but the body plumage looks tatty and the flight feathers are moulting during 2nd cy summer and autumn. The head is still paler than the body, but a richer ochre colour compared to the pale creamy head of the juvenile, and birds retain the dark eye-mask, while the breast shows irregular juvenile-type pale streaking, but now with additional adult-type dark shaft-streaks. *In the autumn of their 2ⁿᵈ cy birds show a more advanced primary moult than adults*, with fewer retained outer primaries (see under Moult). Iris colour is a light greyish-brown in most, with the black pupil discernible.

By late second-winter (3ʳᵈ cy spring), after a complete moult, birds can be difficult to tell from first-winter juveniles on plumage, but the condition of the plumage is usually far better, the tips of the fingered primaries are rounded (sharply pointed in juveniles) and the outer primaries are black and in good condition compared to the worn and faded feathers of the juveniles.

Subadult plumages

The subsequent two or three age-classes resemble the adult in most respects, but the iris is still variably (light) brown (not yellow or whitish) and the head is brownish and concolorous with the rest of the plumage, not pale grey with fine dark streaking.

Definitive adult

Rather uniformly brown, varying from dark umber to quite vividly rufous in others. The head stands out as paler on most, being typically pale grey with fine black streaking. The underbody and particularly the breast shows dark shaft-streaks with light margins, broadest and most prominent across the upper breast, while the belly and vent are more uniform and can be rather rufous brown, even on duller birds. The uppertail is brown, sometimes even rufous-brown, with rather dense dark barring, but even on the brightest birds it is never the translucent bright orange of Red Kite. The iris colour is most conspicuous in full adults, very pale, light yellow or even whitish.

SEXING Birds are not possible to sex in the field.

CONFUSION RISKS Because of the generally dark plumage lacking distinct characters Black Kites are often mistaken for other raptor species, distant birds in particular. They are most often confused with female-type Marsh Harriers or dark morph Booted Eagles,

80. Western Black Kite *Milvus migrans migrans*, juvenile, migrant (with adult above). Note rather narrow and uniformly coloured wings, details of wing-formula and yellow feet; compare with Black-eared Kite. Spain, 4.9.2013 (DF)

which both show different silhouettes, the latter also has more distinctly marked upperparts. Autumn adults often show odd silhouettes, owing to heavy moult with missing tail feathers and disproportioned wings, and such birds account for numerous claims of odd eagles, vultures and even Bateleurs.

NOTE Identifying wintering Western Black Kites *M. m. migrans* in sub-Saharan Africa can be a tricky business because of the confusion with local immature Yellow-billed Kites. Western Black Kites are always best identified by *deep black bill, which contrasts sharply against the yellow cere*, and by wing-formula showing at least five long fingers and a shorter sixth (p5). African Yellow-billed Kites show a five-fingered wingtip (with a short and rounded p5) and already during the first winter the dark bill of the juvenile tends to turn grey from the base, which means that they lack the sharp contrast between black bill and bright yellow cere typical of Western Black Kite. Juveniles and young immatures are the hardest to tell apart, but some adults can also be challenging, as Yellow-billed Kites may sport a greyish and streaked head as adults, but they would never show the pale iris typical of full adult Black Kites, nor a black bill.

81. Black Kite *Milvus migrans*, adult, wintering/migrant. Birds with accidentally lost tails are not rare but always cause confusion. They have been reported as various eagles and even Bateleur. Israel, 2.10.2008 (DF)

82. Western Black Kite *Milvus migrans migrans*, juvenile, migrant. A rather average juvenile with primaries indistinctly barred in shades of grey, lacking the contrasting white and black markings of *lineatus*. Also note yellow feet and barred undertail. Spain, 4.9.2013 (DF)

83. Western Black Kite *Milvus migrans migrans*, juvenile, migrant. The head is always pale in juveniles, but the extent and clarity of the pale streaking/spotting of the breast and belly varies individually. Spain, 5.9.2013 (DF)

84. Western Black Kite *Milvus migrans migrans*, juvenile, migrant. A darker and less well marked individual. In autumn juveniles are always told from older birds by their intact wings, lacking signs of moult. Spain, 10.9.2012 (DF)

85. Western Black Kite *Milvus migrans migrans*, juvenile, migrant. Unusually light birds like this can be confusing to the unaware. Note how the tail-fork disappears completely when the tail is fully spread. Spain, 10.9.2012 (DF)

86. Western Black Kite *Milvus migrans migrans*, juvenile, migrant. Juveniles are beautifully spotted above, with the white line along the tips of the greater coverts a reliable ageing character. The diagonal light patch of the innerwing is a common feature of all Black Kites, but varies individually depending on age and wear of plumage. Spain, 3.9.2013 (DF)

87. Western Black Kite *Milvus migrans migrans*, juvenile, migrant. A lighter bird with a brighter upperwing patch. Spain, 10.9.2012 (DF)

88. Western Black Kite *Milvus migrans migrans*, 2nd cy, spring migrant. By spring juveniles are highly variable, depending on the extent of body moult in the previous winter, but the flight- and tail feathers are usually extremely worn. This bird has moulted very little, except for the new cream-coloured head, and appears as a very worn juvenile. Spain, 6.5.2008 (DF)

89. Western Black Kite *Milvus migrans migrans*, 2nd cy, spring migrant. Others are better preserved, even retaining the wing-band on the greater coverts. Spain, 5.5.2011 (DF)

90. Western Black Kite *Milvus migrans migrans*, 2nd cy, autumn migrant. 2nd cy birds differ from adults by their creamy head with a dark eye-mask. They have also moulted more primaries than adults. Spain, 3.9.2013 (DF)

91. Western Black Kite *Milvus migrans migrans*, 2nd cy, autumn migrant. Spain, 4.9.2013 (DF)

92. Western Black Kite *Milvus migrans migrans*, 2nd cy, autumn migrant. Note the narrow, pointed and extremely worn juvenile secondaries and outer primaries, good pointers for this age-class. Spain, 4.9.2013 (DF)

93. Western Black Kite *Milvus migrans migrans*, adult, spring migrant. Adults appear rather uniform (note primaries especially), except for a grey head with a light iris. Spain, 5.5.2011 (DF)

94. Western Black Kite *Milvus migrans migrans*, adult, spring migrant. A typical European adult, showing slightly rufous plumage with a grey, black-streaked head and black streaks across the upper breast. Spain, 9.3.2011 (DF)

95. Western Black Kite *Milvus migrans migrans*, adult, spring migrant. Another typical adult. Spain, 9.3.2011 (DF)

96. Western Black Kite *Milvus migrans migrans*, adult, autumn migrant. Adults have moulted fewer primaries compared to 2nd cy birds. Spain, 4.9.2013 (DF)

97. Western Black Kite *Milvus migrans migrans*, adult, wintering. A duller adult from the Middle East. Note grey head with light iris, yellow feet, poorly marked primaries and wing-formula, all diagnostic for nominate *migrans*; compare with Black-eared Kite. Israel, 5.10.2009 (DF)

98. Western Black Kite *Milvus migrans migrans*, adult, spring migrant. Brighter birds migrating through the Middle East get an extra boost to their colour from the stark desert light. Israel, 31.3.2008 (DF)

99. Western Black Kite *Milvus migrans migrans*, adult, spring migrant. Brightly coloured birds like this are no doubt behind many of the claims of Red Kite from the Middle East. Israel, 25.3.2013 (DF)

100. Western Black Kite *Milvus migrans migrans*, adult, spring migrant. The upperparts of adults are very uniform, except for the lighter patch diagonally across the arm. Note grey and streaked head, and light iris. The Gambia, 4.2.2008 (DF)

101. Western Black Kite *Milvus migrans migrans*, adult, spring migrant. Israel, 24.3.2010 (DF)

Black-eared Kite *Milvus migrans lineatus*

(See also Western Black Kite, p. 82)

VARIATION The Black-eared Kite is the Eastern Palearctic counterpart to Western Black Kite, with a debated taxonomic status; it is even regarded as a species in its own right by some. The numbers of Black Kites wintering in the Middle East have increased more than tenfold during the past two decades, and many of the tens of thousands now wintering in, for example, Israel resemble *lineatus*. However, since many of the birds are intermediates, showing a combination of both *lineatus* and *migrans* characters, claiming a 'pure' *lineatus* will remain difficult. The same applies to W Europe, Arabia and Africa, where all properly documented *lineatus*-type birds have been intergrades (pers. obs.).

IDENTIFICATION Differs from Western Black Kite *M. m. migrans* both in structure and plumage. The wings are broader with an almost square-looking wingtip, giving the bird an *eagle-like jizz*, which is explained by the wing-formula, showing *six prominently fingered primaries*, with the sixth (counting inwards) clearly longer and more pointed than in *migrans*.

One important feature separating the two forms is the colour of the bare parts. *In pure lineatus the cere and feet are always 'colourless', pale bluish-white or greenish-white, compared to bright yellow in all migrans.* This is a constant difference, and it is not age-related. The colourless cere of *lineatus* is further enhanced by pale lores and chin, giving the bird a pale-faced look, with a contrasting black and long, heavy-looking bill. Also the iris colour differs, remaining dark brown through all ages in *lineatus*, as opposed to *migrans*, where it gets paler with age to eventually become whitish in full adults.

PLUMAGES As for plumages, *lineatus* shows several differences compared to *migrans*. The *primaries are distinctly marked below*, with a variable white area at the base of the outer primaries, which in some spreads across all the primaries, with the inner distinctly barred in black, grey and white. This diagnostic feature is shared by all age-classes (and by intergrades).

Juvenile

Juvenile *lineatus* can approach some strongly marked juvenile *migrans*, but the different wing-formula, the *extensively pale belly and rear underbody* with unmarked creamy-white belly, vent and undertail-coverts, and the *colourless cere and feet* are good differentiating features.

The body plumage tends to be more heavily marked compared to juvenile Western, with a wide whitish tip to each of the upper median wing-coverts, while the breast shows bold whitish streaking. *The undertail looks uniformly pale* with a broad dusky tip and lacks distinct barring both above and below, contrary to most juvenile *migrans*.

Second plumage

Second plumage birds are still rather juvenile-like, heavily streaked below and the head is creamy to ochre-yellow with a dark eye-mask, and the rear underbody remains extensively pale. They differ from juveniles in showing rather irregular and ochre or tawny, rather than whitish, body markings and are either in active wing-moult or, after a completed moult, show flight feathers of different ages and wear, which, however, can be difficult to note in the field.

Adult

Rather similar to younger birds, but appear darker with a less distinctly marked body plumage. The plumage is a rich dark brown, including a *rufous-brown head* (darker than in juveniles and immatures) with a *dark eye-mask and a noticeably pale face*. The underbody is dark brown with rather indistinct rusty-ochre streaking across the upper and lower breast (black shaft-streaks typical of adult *migrans* are lacking), while the *rear underbody remains pale straw-coloured*. The main difference, compared to younger *lineatus*, is that adults are overall darker and more richly coloured, lacking the pale head and the whitish breast-streaking of younger birds. The cere and feet are whitish and the iris remains dark brown.

HYBRIDS Hybrids and intergrades between *lineatus* and *migrans* are highly variable, sometimes very similar to pure *lineatus*, but many of them show *migrans* characters, such as *pale iris, yellowish feet*, varying from pale to bright yellow, *the head may be greyish in adults with fine dark streaking* and the *pale breast-streaks show a prominent dark shaft-streak*. Confirming these characters requires close views, and the iris character can be used on older birds only, since both forms are dark-eyed as juveniles. Juvenile intergrades can be impossible to tell from pure *lineatus* under normal field conditions, but the *feet and cere of intergrades are yellowish* and the wing-formula is intermediate.

102. Black-eared Kite *Milvus migrans lineatus*, juvenile. Broad-winged and thickset, with dark secondaries and a variable white primary flash. Aged by the underwing band formed by pale tips to the greater coverts, the light, poorly marked undertail and extensively pale rear underbody. Japan, 9.2.2005 (DF)

103. Black-eared Kite *Milvus migrans lineatus*, juvenile. A lighter juvenile demonstrating the broad-winged, almost eagle-like silhouette typical of this subspecies. Note also the diagnostic pale blue feet and cere/base of bill. Japan, 9.2.2005 (DF)

104. Black-eared Kite *Milvus migrans lineatus*, second plumage. Still intensely streaked below but now with more clearly barred tail and secondaries. Differs from adult by lighter head and more brightly streaked underbody. Japan, 9.2.2005 (DF)

105. Black-eared Kite *Milvus migrans lineatus*, adult. More uniform than juveniles and immatures on breast and underwing-coverts but retains pale rear underbody, and blue feet and bill-base. Japan, 9.2.2005 (DF)

106. Black-eared Kite *Milvus migrans lineatus*, adult. Typical adult, brown with ochre streaking below. Note boldly marked primaries, pale vent and undertail-coverts, pale face contrasting with black bill and eye-mask, and pale feet and bill-base. Japan, 10.2.2005 (DF)

107. Black-eared Kite *Milvus migrans lineatus*, worn juvenile plumage of 2nd cy summer, nearly one year old. A brightly marked bird, which fulfils the criteria for *lineatus*. Mongolia, 4.6.2012 (DF)

108. Black-eared Kite *Milvus migrans lineatus*, worn juvenile plumage of 2nd cy summer, nearly one year old. The plumage features accord with *lineatus* but the structure feels somewhat slimmer. Mongolia, 10.6.2012 (DF)

109. Black-eared Kite *Milvus migrans lineatus*, worn juvenile plumage of 2nd cy summer, nearly one year old. Most plumage features are correct for *lineatus*, but barred tail, small bill and pale yellowish feet may indicate genetic influence from *migrans*. Mongolia, 10.6.2012 (DF)

110. Probable Black-eared Kite *Milvus migrans lineatus*, adult. Rather light grey head, dark breast streaks, lighter, more finely barred underwing, and pale yellow feet and cere indicate *migrans* influence. Mongolia, 10.6.2012 (DF)

111. Probable Black-eared Kite *Milvus migrans lineatus*, adult. Another bird with most features pointing towards *lineatus*, but grey head with small bill may indicate *migrans* influence. Mongolia, 10.6.2012 (DF)

112. Probable Black-eared Kite *Milvus migrans lineatus*, adult. Although looking superficially good for a *lineatus*, this bird features a grey head with dark streaking, yellowish feet, cere and gape-line and *migrans*-type breast-streaking, and could thus be an intergrade. Mongolia, 4.6.2012 (DF)

'Eastern Black Kite' – Identifying hybrids and intergrades between Western Black Kite and Black-eared Kite

The nominate subspecies of Black Kite *M. m. migrans*, here called Western Black Kite, has a westerly distribution within the Western Palearctic, largely confined to SW and W Europe. The Black-eared Kite *M. m. lineatus*, regarded by some as a separate species, is widespread in the Eastern Palearctic, particularly in China and Japan, reaching Mongolia and C Siberia in the west. However, bridging these two allopatric taxa is a vast zone of intergradation, thousands of kilometres wide, stretching from W Russia to C Siberia, Mongolia and C Asia. Within this area Black Kites are hugely variable with *lineatus* features increasing clinally from west to east. Many such *lineatus*-type birds will on closer inspection reveal characters related to *migrans*, indicating that they are of hybrid origin.

A typical feature of such a wide hybridisation zone is that the intergrades are highly variable (McCarthy 2006). Some of them may look more like Western, with the odd character (usually strongly marked underwing pattern) pointing towards Black-eared Kite, while others are very similar to Black-eared, save for a few minute details revealing genetic influence from Western (e.g. lighter iris and yellow legs and cere). However, in the field it may not be possible to separate between a pure *lineatus* and a *lineatus*-like intergrade, unless the bird is seen exceptionally well or preferably photographed at close range. Birds resembling *lineatus*, including *lineatus*-like intergrades, are hereafter called Eastern Black Kites (*M. migrans 'intermedius'*).

In Europe claims of Black-eared Kite have increased rapidly during the last two decades, no doubt boosted by its elevation to species level in some European countries. In recent years several claims have been reported annually all across Europe, mostly as autumn juveniles and again as first-summer birds in their 2nd cy. However, in many cases it has not been possible to decide whether the bird has been a true *lineatus*, rather than a more likely '*intermedius*'-type Eastern Black Kite intergrade, but in all well documented cases the birds have been identified as the latter.

Features to focus on when confronted with a suspected Black-eared Kite are, in order of importance:

- colour of feet and cere
- colour of iris (applicable to adults only)
- wing-formula
- primary pattern on underwing
- colour of head feathering (adults only)

If *lineatus* is regarded as a species, as it is by some authors, then the '*intermedius*' Eastern Black Kite creates a taxonomic problem, since it represents a huge, self-sustained hybrid population between the two species, thus by some definitions forming a species in its own right, albeit of hybrid origin.

References

McCarthy, E. M. 2006. *Handbook of Avian Hybrids of the World*. Oxford University Press.

113. Eastern Black Kite *Milvus migrans migrans x lineatus*, fresh juvenile, migrant. This brightly coloured bird combines features of *migrans* and *lineatus* in almost equal proportions, with shortish p5, distinctly barred undertail, yellowish feet and warm, sandy-brown coloration deriving from *migrans*, while largely pale rear underbody and extensive wing-flash may stem from *lineatus*. Egypt, 9.10.2010 (DF)

114. Eastern Black Kite *Milvus migrans migrans x lineatus*, fresh juvenile, migrant/wintering. Israel, 1.10.2008 (DF)

115. Eastern Black Kite *Milvus migrans migrans x lineatus*, juvenile, migrant/wintering. Rather dark vent and pale yellow feet differ from pure *lineatus*. Egypt, 20.3.2009 (DF)

116. Eastern Black Kite *Milvus migrans migrans x lineatus*, 2nd cy summer. Superficially similar to juvenile *lineatus*, but slimmer silhouette, less striking body plumage, finely barred undertail and lighter iris indicate *migrans* influence. Kazakhstan, 6.5.2010 (DF)

117. Eastern Black Kite *Milvus migrans migrans x lineatus*, juvenile, migrant/wintering. This striking bird is easily distinguished from local Yellow-billed Kites. Ethiopia, 18.11.2011 (DF)

118. Eastern Black Kite *Milvus migrans migrans x lineatus*, juvenile, migrant/wintering. Seen from above fresh juveniles do not differ from juvenile *migrans*. Egypt, 12.10.2010 (DF)

119. Eastern Black Kite *Milvus migrans migrans x lineatus*, 2ⁿᵈ cy summer in worn plumage, prior to first complete moult. By 2ⁿᵈ cy summer the distinctive juvenile markings have all worn off on the upperparts. Kazakhstan, 6.5.2010 (DF)

120. Eastern Black Kite *Milvus migrans migrans x lineatus*, adult summer. Wing-shape and -pattern rather similar to *lineatus*, but grey and streaked head and light iris indicate genetic introgression from *migrans*. Mongolia, 16.6.2012 (DF)

121. Eastern Black Kite *Milvus migrans migrans x lineatus*, adult summer. Superficially similar to *lineatus*, but note grey head and light iris indicating genetic influence from *migrans*. Mongolia, 7.6.2012 (DF)

YELLOW-BILLED KITE
Milvus aegyptius

VARIATION Until recently considered a subspecies of Black Kite *M. migrans*, but now commonly regarded as a full species. Two subspecies, nominate *aegyptius* in NE Africa (and S Arabia; see Note below), and ssp. *parasitus* in the rest of sub-Saharan Africa. The morphological differences between the subspecies are only slight, and intermediates possibly occur. Adult ssp. *aegyptius* are more brightly rufous with concolorous head, while adult ssp. *parasitus* are a duller greyish-brown, more clearly streaked on the breast and often with a slightly greyer head, thus approaching adults of Western Black Kite *M. m. migrans*, although adults of the former always sport a dark eye and a yellow bill.

DISTRIBUTION Widely distributed throughout sub-Saharan Africa, where it is the most widespread and numerous raptor, but also breeds in Egypt and S Arabia. Often lives in close proximity to humans, breeding in towns and villages, but also congregates around bush fires and swarming termites.

BEHAVIOUR Similar to Black Kite and mostly seen on the wing quartering open areas at moderate height looking for food; often seen patrolling major roads at first light. Urban populations are common in villages and towns throughout Africa. Freely mixes with wintering Black Kites at bush fires and termite emergences.

SPECIES IDENTIFICATION Similar to Western Black Kite and often difficult to separate, except for diagnostic adults. Differs structurally from Western Black Kite in showing *narrower wing and more pointed wingtip (only five fingers)* and slightly more deeply forked tail, but the slimmer proportions may be difficult to appreciate. Owing to narrower and more pointed wings and longer, slimmer and more deeply forked tail somewhat reminiscent of Red Kite in silhouette. Adults always identified by their all-yellow bill. Juveniles and immatures can be very similar to juvenile Western Black Kite *M. m. migrans* in plumage, but they normally show a different wing-formula and a more clearly forked tail. As a rule Yellow-billed Kites appear rather uniform, lacking any strong markings on the flight feathers of the underwing and the lighter window of the primaries is poorly defined. *Different timing of moult is also an important clue* when separating resident Yellow-billed from wintering Black Kites in Africa, as is colour of iris and bill.

MOULT Complete moult annually. Timing varies in different parts of Africa depending on local breeding season, and more information is needed. Adults normally start to moult at the onset of breeding, probably finishing some four months later. For instance, south of the Sahel, in both W and E Africa, Yellow-billed Kites commence primary moult in Jan–Feb, when migrant Black Kites are completing theirs (pers. obs.).

PLUMAGES Fresh juveniles and adults are easily told apart, but given good conditions a second plumage bird can also be separated. Easily confused with Black Kite, especially during the first two years of life, and many claims of Black Kites for sub-Saharan Africa actually refer to immature Yellow-billed Kites. The two species are best separated by different wing-formula, partly also by different timing of wing-moult, and adults also by colour of bill.

Juvenile

In fresh plumage very similar to juvenile Western Black Kite, with a largely cream- or tawny-spotted plumage, although many show less contrasting markings and a strong tawny colour to the tips of the upperwing-coverts, lacking in juvenile *migrans*. The pale markings soon bleach and wear off, and after a few months the birds look much more uniform in colour owing to the heavily abraded plumage. At this time many birds show a pale-mottled belly, unlike young Western Black Kites, while the head is turning ochre-yellow from freshly moulted feathers. The bill is black at first, with a yellow cere, but during the first year it turns a lighter grey from the base to eventually become yellow. Differs from juvenile Western Black Kite also by slimmer silhouette and different wing-formula, with just five fingered primaries (p6–10), and by slightly deeper tail-fork.

Second plumage

Similar to juvenile but lacks distinct streaking to underbody and pale spotting above. Moulting birds recognised by retained juvenile feathers and a pale greyish bill, lacking the strikingly sharp contrast between black bill and yellow cere shown by all Black Kites. Head is tawny yellow with a darker eye-mask in this plumage, while the body plumage appears rather uniform, often with lighter mottling on the lower breast and belly.

Adult

The all-yellow bill of adults is always a reliable feature, but this is obvious only from fairly close range. Note also that even Black Kites can appear yellow-billed in strong sunlight, caused by reflections from the sun. Adults of nominate *aegyptius*, from Egypt, Arabia and NE Africa, are rather colourful, brightly rufous and poorly streaked on the breast, and the head is the same colour as the rest of the bird. The bill and cere are yellow and the iris looks dark from a distance, being a deep chestnut-brown.

Adult *parasitus* is duller brown, more clearly streaked on the breast (although finer than in adult Black Kite) and the head is often more clearly streaked with a faint greyish tone, at times quite obvious. Despite a more similar plumage to adult Black Kite, the yellow bill and cere and the dark iris remain diagnostic.

SEXING Yellow-billed Kites are not possible to sex in the field.

CONFUSION RISKS Close views are always needed to separate Yellow-billed from migratory Western Black Kites, particularly younger birds, as winter ranges are overlapping. Distant birds could also be mistaken for Marsh Harrier or dark morph Booted Eagle, but behaviour and silhouettes differ, as well as details of plumage.

NOTE The birds of S Arabia, earlier referred to as *'arabicus'*, are quite distinct. The adults are bright rufous, often with a distinctly paler head, the iris is clearly lighter than in African *aegyptius* or *parasitus* and the bill may be partly grey also in old adults. Even the wing-formula appears to be slightly different, with a longer and more prominent sixth finger (p5) compared to the African counterparts, and the primaries are more distinctly barred. Could these be features from introgression of *migrans* genes?

122. Yellow-billed Kite *Milvus aegyptius*, juvenile ssp. *parasitus*. A rather light-coloured and well-marked juvenile. The Gambia, 24.11.2006 (DF)

123. Yellow-billed Kite *Milvus aegyptius*, juvenile ssp. *parasitus*. A darker and more worn juvenile. The Gambia, 24.11.2006 (DF)

124. Yellow-billed Kite *Milvus aegyptius*, juvenile ssp. *parasitus*. On this bird the bill has already started to turn yellow from the base, prior to its first moult. Note also the short 6th finger and compare with Black Kite. Ethiopia, 17.2.2012 (DF)

125. Yellow-billed Kite *Milvus aegyptius*, juvenile ssp. *parasitus*. The light and mottled belly appears to be a common feature among juvenile Yellow-billed Kites approaching their first moult. The Gambia, 29.11.2006 (DF)

126. Yellow-billed Kite *Milvus aegyptius*, juvenile ssp. *parasitus* in rather fresh plumage. Note short sixth finger (p5) and bill already turning yellow in its first year of life. The Gambia, 31.12.2011 (DF)

127. Yellow-billed Kite *Milvus aegyptius*, juvenile, presumed ssp. *parasitus*. By the time the first moult commences the plumage is very worn and lacks most of the light markings diagnostic of the juvenile plumage. Note cream-coloured head typical of second plumage and the bill is turning lighter from the base. Ethiopia, 22.2.2014 (DF)

128. Yellow-billed Kite *Milvus aegyptius*, second plumage of ssp. *aegyptius* (*arabicus*) from Arabia (Arabian birds were referred to as ssp. *arabicus* in the past). Note yellowish-buff head with hint of dark eye-mask and bill turning yellow, all typical of second plumage Yellow-billed Kite. Plumage is lighter and flight feathers more distinctly marked compared to ssp. *parasitus*. Oman, 16.2.2013 (DF)

129. Yellow-billed Kite *Milvus aegyptius*, second plumage of ssp. *aegyptius* (*arabicus*), same as **128**. Typical *aegyptius* also has a lighter brown iris compared to *parasitus*. Oman, 16.2.2013 (DF)

130. Yellow-billed Kite *Milvus aegyptius*, second plumage, possibly ssp. *aegyptius*. Note uniformly rufous-brown plumage indicating *aegyptius*, and cream-coloured head of this second-plumage bird. Ethiopia, 15.11.2011 (DF)

131. Yellow-billed Kite *Milvus aegyptius*, breeding adult of ssp. *aegyptius* (*arabicus*). Arabian birds differ from Yellow-billed Kites from further south in Africa by their brighter plumage colour, their lighter iris and tendency for p5 to be longer. Oman, 14.2.2013 (DF)

132. Yellow-billed Kite *Milvus aegyptius*, breeding adult of ssp. *aegyptius* (*arabicus*). Oman, 7.2.2014 (DF)

133. Yellow-billed Kite *Milvus aegyptius*, adult of ssp. *aegyptius* (*arabicus*). Oman, 25.11.2014 (DF)

134. Yellow-billed Kite *Milvus aegyptius*, adult of ssp. *aegyptius* (*arabicus*). A very brightly coloured, reddish-orange bird, typical of ssp. *aegyptius* from Arabia. Note also light-brown iris colour. Oman, 27.11.2014 (DF)

135. Yellow-billed Kite *Milvus aegyptius*, adult of ssp. *aegyptius* (*arabicus*), same as **133**. Lighter above than ssp. *parasitus*, with the patch across the upperwing-coverts brighter in particular. Oman, 25.11.2014 (DF)

136. Yellow-billed Kite *Milvus aegyptius*, adult. Many Ethiopian birds are more rufous than Yellow-billed Kites from further south in Africa (ssp. *parasitus*) and may be either wintering ssp. *aegyptius* or possibly intergrades between the two subspecies. Ethiopia, 3.11.2011 (DF)

137. Yellow-billed Kite *Milvus aegyptius*, adult. This bird from Ethiopia is displaying rather uniform upperparts. Ethiopia, 21.2.2006 (DF)

RED KITE
Milvus milvus

VARIATION Monotypic. The much-debated Cape Verde Kite *Milvus 'fasciicauda'*, from the Cape Verde islands, now believed to be extinct in the wild, is thought to have been a hybrid species, originating from the offspring of interbreeding Red and Black Kites marooned on the islands.

DISTRIBUTION The Red Kite is a typically W European species, with the bulk of the population occurring from S Sweden through W Europe to the Iberian Peninsula. Northern populations are migratory, spending the winter in SW Europe, young birds being more dispersive than adults. Some birds, mostly juveniles, cross the Strait of Gibraltar into Africa. Very rare visitor to the Middle East, where many claims refer to rufous-coloured Black Kites with bright primary windows.

BEHAVIOUR Highly aerial. Mostly seen on the wing, as it patrols open areas in a slow, buoyant searching flight. Prefers cultivated areas around villages and farms, and like Black Kite often follows roads scanning for road kills.

SPECIES IDENTIFICATION Normally not difficult to identify, although some more brightly coloured adult Black Kites may cause initial confusion. However, the shape of Red Kite is much more extreme, with long and narrow wings and a very long and slim, bright orange and deeply forked tail, giving the bird a much more delicate build compared to Black Kite. However, *juveniles are less extreme in silhouette* with a shorter, less deeply forked tail and with comparatively broader wings. The elegance of the bird is further enhanced by its unmatched, graceful flight. The entire body plumage is rufous and cinnamon in colour, the underwing shows a sharp contrast between dark secondaries and entirely white but finely barred inner primaries (cf. Black Kite), while the pale diagonal area across the upperwing-coverts is brighter and more contrastingly marked than in adult Black Kites. *In Black Kite the tail appears concolorous with the upperparts, while in Red Kite the tail is clearly lighter than the back and bright orange in colour.* The wing-formula differs from Black Kite in showing only five fingered primaries, the hand appearing narrower than the arm in full soar (Black Kite has more evenly broad wings), and the

tail is more deeply forked, although juveniles have a shallower fork than adults. The outer primaries show unbarred white bases below, which is often diagnostic compared to the various forms of Black Kite, which show more heavily barred primaries.

MOULT Juveniles undergo an individually variable partial body moult in their first winter, attaining the black breast streaking of the adults. The complete annual moult, comprising the flight feathers, starts in juveniles by Apr of the 2nd cy, in adults slightly later, and continues through the breeding season. In Sep–Oct, at the time of autumn migration, birds are still moulting actively and show unmoulted outer primaries, but in 2nd cy birds the moult has on average progressed further than in adults. The moult is completed in the winter quarters by Nov–Dec, possibly earlier in non-migratory residents.

PLUMAGES In favourable conditions three plumages can be recognised: juvenile, second plumage and adults. The second plumage can be recognised by retained juvenile flight feathers, when present. Important ageing characters are found in the details of head, breast, underwings, rear underbody and tail.

Juvenile

More compact in silhouette than the adult, with shorter and less deeply forked tail and comparatively broader-looking wings (fingers shorter) and much brighter and lighter colours. Upperwings more brightly marked than in adults, with pale upperwing-coverts patch wider, lighter and more contrasting, while the distinct pale tips to the greater secondary and primary coverts create a distinct wing-band. On approaching juveniles the light upperwing-coverts are visible as a bright patch near the carpal bend. The head is white and unstreaked and the black bill contrasts sharply against the yellow cere, while the iris is grey at first, turning white by early autumn. The underbody is lighter cinnamon compared to adults, with *broad creamy streaks on the breast*, while vent and undertail-coverts are uniformly pale. The underwing-coverts are clearly two-toned, with lighter *orange-cinnamon forewing contrasting with the darker band of the mid-wing*, while the greater underwing-coverts are pale-tipped and form a subtle wing-bar on the underwing, but depending on the light this can be difficult to discern. *The tail shows a complete albeit narrow dark subterminal band, which is lacking in adults.*

During their first autumn some juveniles can acquire new adult-type breast feathers with broad black streaks. The number of replaced feathers varies individually, as does the timing of this partial moult, but by spring most birds show at least some adult-type feathers on their underparts, while others may appear practically adult-like in this respect.

Second plumage

In most respects similar to adult, but some can be identified, most reliably by retained juvenile feathers. By early autumn of the 2nd cy birds still retain some juvenile feather-markers until moult is completed later in the season. The last juvenile feathers to be replaced are the outermost primaries, which are browner, shorter and more pointed compared to the new ones, while the retained juvenile greater underwing-coverts will show pale tips until they are replaced by adult-type feathers in late autumn. The last secondaries to be replaced are usually s4 and s8, which are sometimes retained until the next summer and can be recognised by being browner, as well as clearly narrower and more pointed compared to neighbouring new feathers. Birds of this age-class still show a *partial, dark subterminal tail-band* and the bill is largely dark, being yellow only just adjacent to the cere, and the vent and undertail-coverts are lighter than the breast (cf. adult). The iris is white as in adults.

Adult

Extreme in shape and proportions, with *very long and narrow wings and a long and slim, deeply forked tail.* Adults are overall darker rufous than juveniles, lacking the extensive and bright buffish areas to the upperwings and rear underbody. The head is whitish-grey with fine black streaking, the iris is whitish and the bill has turned yellow from the base, having thus lost the sharp contrast between black bill and yellow cere found in younger birds. In older birds still the bill may turn completely yellow, which is quite rare among raptors. The pale area in the upperwing-coverts is both reduced and subdued in colour compared to the juveniles, and the adults also lack the pale upperwing-band of the greater coverts. The more deeply forked and practically unbarred uppertail appears even more brightly coloured than in juveniles, partly because of the contrasting darker upperparts of the adults. The tail lacks the subterminal band of the juveniles, but shows finely barred outermost feathers and dark corners.

The underparts are darker rufous compared to juveniles and the breast shows broad black streaking. The underwing lacks the pale greater-covert tips of the juveniles and the forearm is a darker chestnut colour, making the dark band through the midwing less conspicuous compared to the juveniles.

Birds with a mostly black bill and a dark subterminal tail-band, but otherwise looking like adults, with no obvious retained juvenile feathers, could possibly be either birds in their second plumage having completed their moult, or birds in their third plumage.

SEXING is not possible in the field.

CONFUSION RISKS Not possible to mistake for anything else thanks to unique shape and diagnostic plumage, with rich cinnamon and rufous tones and a striking white window to the outer underwing. Distant birds in poor light could be mistaken for a more colourful Black Kite, but the rich cinnamon upperparts, in particular the brightly coloured tail, are diagnostic.

References

Johnson, J. A., Watson, R. T., & Mindell, D. P. 2005. Prioritizing species conservation: does the Cape Verde Kite exist? *Proc. Royal Soc. London* B 272: 1365–1371.

138. Red Kite *Milvus milvus*, juvenile. Juveniles are streaked pale on the breast and the contrast between orange forewing and dark mid-wing band is striking. Note also the distinct dark subterminal tail-band of juveniles. Sweden, 5.9.2010 (DF)

139. Red Kite *Milvus milvus*, juvenile. Seen at close range, the light feather-tips of the greater coverts form a diagnostic wing-band in juveniles. Sweden, 5.10.2014 (DF)

140. Red Kite *Milvus milvus*, juvenile. Juveniles are clearly shorter-tailed than adults. Spain, 30.10.2009 (DF)

141. Red Kite *Milvus milvus*, juvenile. The first adult-type dark breast-streaks start to appear in the first autumn. Spain, 29.10.2009 (DF)

142. Red Kite *Milvus milvus*, juvenile (same as **139**). The broad, light fringes of the upperwing-coverts create bright upperwing patches. Sweden, 5.10.2014 (DF)

143. Red Kite *Milvus milvus*, second plumage. Similar to adult, but recognised by some narrow and faded retained juvenile secondaries and some pale-tipped juvenile greater underwing-coverts. Note also narrow subterminal tail-band, and light vent and undertail coverts. Spain, 23.11.2010 (DF)

144. Red Kite *Milvus milvus*, adult plumage. Overall darker than juvenile, with black breast-streaks. Note translucent and largely unbarred, uniform tail with long and sharply pointed corners. Spain, 23.11.2010 (DF)

145. Red Kite *Milvus milvus*, adult plumage. Spain, 23.11.2010 (DF)

146. Red Kite *Milvus milvus*, adult plumage. Adults appear longer- and slimmer-winged and longer-tailed than juveniles. Spain, 23.11.2010 (DF)

147. Red Kite *Milvus milvus*, adult plumage. Old adults may show practically unbarred tails, while the bill turns all-yellow. Spain, 21.11.2010 (DF)

148. Red Kite *Milvus milvus*, adult plumage. Adults look much darker above than juveniles and the far less contrasting light upperwing patch is dotted with wide, black feather-centres. Sweden, 4.10.2013 (DF)

149. Red Kite *Milvus milvus*, adult plumage. Note the practically unmarked uppertail of the adult with long and pointed corners. Sweden, 7.10.2014 (DF)

150. Red Kite *Milvus milvus*, adult plumage. Adults show strong dark markings to upperwings and mantle, with upperparts showing stark contrast with largely uniform, deep orange-cinnamon tail. Sweden, 4.10.2013 (DF)

Identifying hybrids between Black Kite and Red Kite

Hybridisation between Black Kite and Red Kite has been documented in Sweden and Germany, usually in situations where both species occur, but with one being much rarer than the other. It is also known that the hybrids are fertile and capable of reproducing, as shown by a case in Sweden, where a hybrid paired with a Red Kite and successfully raised chicks. Thus, one has to take into account not only hybrids, but also later generation back-crosses.

Hybrid and intergrade Black x Red Kites can be highly variable. First generation hybrids are often clearly intermediate, featuring characters from both parental species, while later generation back-crosses may approach one or other parental species, being thus much more difficult to identify.

Typical features would include intermediate structure and coloration, with a wing-formula tending more towards Black Kite (longer p5), a wide and whitish underwing primary-window extending across all primaries, similar to Red Kite, but often barred throughout, as in Black Kite. The body plumage often shows a rufous or cinnamon colour, sometimes including the upper tail, which has a shallower fork compared to Red Kite. The lesser underwing-coverts may be clearly cinnamon, with a dark greater-covert band, similar to Red Kite.

The wing-formula of Red Kite shows a short p5 followed by a long gap before the tip of p6 (counting outwards), while in Western Black Kite the tips of primaries p5, p6 and p7 are spaced more evenly. The longest primaries (p10–p6) are white without barring inside the fingers in Red Kite, while they are barred in Western Black Kite.

References

Forsman, D. 1998. *The Raptors of Europe and the Middle East – A Handbook of Field Identification.* T. & A. D. Poyser.

Forsman, D. & Nye, D. 2007. A hybrid Red Kite x Black Kite in Cyprus. *Birding World* 20: 480–481.

BLACK-SHOULDERED KITE
Elanus caeruleus

Other names: Black-winged Kite

VARIATION Depending on source, up to four subspecies are recognised. Breeds in our region in Iberia and N Africa, as nominate *caeruleus*, but recent records from Israel (now breeding) and Turkey refer to ssp. *vociferus* of continental S Asia (Pakistan and India), showing recent westward range expansion. Vagrants to the Middle East need to be scrutinised carefully, as either subspecies may be involved. In ssp. *vociferus* the primaries of the underwing are darker than in *caeruleus* and the dark area bleeds onto the secondaries, leaving a clearly paler trailing edge to the wing. In ssp. *caeruleus* the dark pigmentation is confined to the primaries, while the secondaries are pale and lack the white trailing edge of *vociferus*.

DISTRIBUTION Africa, SW Europe and Asia to Australia. In the Western Palearctic confined to the Iberian Peninsula, S France, NW Africa and Egypt, with the western population increasing and expanding northwards. An increasing number of vagrants have reached as far north as C Europe and S Scandinavia, while birds of Asian origin have bred in Israel and been recorded as far west as Turkey and Georgia.

BEHAVIOUR A bird of open and semi-open areas. Often perches prominently on telegraph poles or dead tree-tops. When hunting flies at medium height over open ground, with frequent stops to hover, and when stooping down on prey lifts wings up in a deep V. Also soars with wings lifted in a diagnostic marked dihedral. Black-shouldered Kites are largely crepuscular in summer, but active throughout the day in winter.

SPECIES IDENTIFICATION Unmistakable owing to typical behaviour, silhouette and plumage. In flight big-headed and short-tailed, with relatively broad and long wings with pointed triangular tips. Colour very pale, appearing white from a distance, with darker wingtips below and a black patch on upper forearm. When hunting, hovers frequently like a kestrel, and glides and dives on wings held in a rather deep V.

The two subspecies occurring in the Western Palearctic differ in the underwing pattern. In African *caeruleus*, occurring in SW Europe, only the primaries are dark below, while in Asian *vociferus*, turning up in the Middle East, the primaries are blacker below, and also the secondaries are smoky grey with white tips, creating a white trailing edge to the wing.

MOULT May breed at any time of year and may raise several broods per year, and since juveniles start to moult their body plumage within weeks after fledging, ageing becomes very difficult. Complete moult including flight feathers starts at 3–4 months and takes about three months to complete.

Adults undergo a complete moult annually, but timing varies individually depending on breeding schedule.

PLUMAGES Normally two plumages can be distinguished, juvenile and adult. Immatures in transition can be aged as long as juvenile wing and tail feathers are retained.

Juvenile

In fresh plumage easily recognised by brown streaking to crown and upper breast, brownish mantle with creamy tips and distinct pale tips to flight feathers and greater upperwing-coverts. Tail is brownish-grey with a faint, darker grey subterminal band. The dark upperwing patch is less contrasting, dark brown rather than black as in adults, and the iris is orange-brown to start with, later changing gradually from orange to red. The body moult starts within weeks after fledging, and the last remaining signs of immaturity should be looked for among the tail feathers and the greater upperwing primary coverts.

Adult

Plumage is beautifully white, grey and black, lacking any markings to particular feathers; overall cleaner-looking and more contrasting, compared to slightly duskier juveniles. The mantle is uniformly grey and the upperwing patch is black, while the tail feathers are clean and look white from a distance, lacking the dark subterminal suffusion of the juveniles. The iris is deep ruby-red and looks dark in normal field conditions.

SEXING is not possible in the field.

CONFUSION RISKS May be mistaken for a male harrier in brief views, but diagnostic silhouette, plumage and behaviour make it rather unmistakable.

151. Black-shouldered Kite *Elanus caeruleus,* juvenile ssp. *caeruleus.* Aged by conspicuous white tips to brownish scapulars and greater upperwing-coverts as well as orange, not ruby-red, iris. Identified as ssp. *caeruleus* by dark primaries contrasting with whitish secondaries on underwing. The Gambia, 19.4.2007 (DF)

152. Black-shouldered Kite *Elanus caeruleus,* juvenile ssp. *caeruleus* (same as **151**). Note retained dirty brownish-grey juvenile upperparts and dusky juvenile tail, although the first primary moult has already commenced. The Gambia, 19.4.2007 (DF)

153. Black-shouldered Kite *Elanus caeruleus,* juvenile ssp. *vociferus.* Note smoky greyish secondaries of ssp. *vociferus* contrasting with white underwing-coverts (cf. *caeruleus*). India, 15.12.2007 (DF)

154. Black-shouldered Kite *Elanus caeruleus*, fresh juvenile ssp. *vociferus*, showing features of juvenile upperwing. Note also orange, not red, iris. India, 15.12.2007 (DF)

155. Black-shouldered Kite *Elanus caeruleus*, fresh juvenile ssp. *vociferus* (same as **154**). The dark upperwing patch is not as black as in adults. India, 15.12.2007 (DF)

156. Black-shouldered Kite *Elanus caeruleus*, immature ssp. *caeruleus* in advanced transitional plumage. Note pale tips to retained juvenile upperwing primary coverts. Spain, 19.2.2011 (DF)

157. Black-shouldered Kite *Elanus caeruleus*, immature ssp. *caeruleus* in advanced transitional plumage (same as **156**). About half of tail feathers are retained juvenile feathers, which are longer, more pointed and greyer with dusky tips. Note also the juvenile outer primaries. Spain, 19.2.2011 (DF)

117

158. Black-shouldered Kite *Elanus caeruleus*, adult ssp. *caeruleus*, demonstrating translucent white tail (cf. juv) and contrast between dark primaries and light secondaries (cf. *vociferus*). Spain, 30.10.2009 (DF)

159. Black-shouldered Kite *Elanus caeruleus*, adult ssp. *caeruleus* (same as **158**). Lacks pale feather tips of coverts and scapulars of juvenile and the iris is bright red. Spain, 30.10.2009 (DF)

160. Black-shouldered Kite *Elanus caeruleus*, adult ssp. *vociferus* with dusky secondaries contrasting with white wing-linings, typical of this subspecies. India, 13.12.2007 (DF)

161. Black-shouldered Kite *Elanus caeruleus*, adult ssp. *vociferus* (same as **160**). India, 13.12.2007 (DF)

SWALLOW-TAILED KITE
Elanoides forficatus

Other names: American Swallow-tailed Kite

VARIATION Two subspecies, migratory *forficatus* in USA and resident *yetapa* in Mexico and the northern half of South America.

DISTRIBUTION Rare vagrant to our area, with three confirmed records. An adult was seen and photographed on 19–23 Mar 1993 at Costa Calma, Fuerteventura, Canary Islands, Spain, another bird was seen on Flores, Azores, Portugal on 17 Mar 2005 and a 2nd cy was photographed at Vigia das Feteiras, São Miguel, Azores, Portugal on 24 Aug–7 Sep 2008.

In North America, the main autumn migration is in Jul-Sep. Breeders return in Feb–Mar, yearlings in Apr–May.

BEHAVIOUR Highly aerial and a masterful, elegant flier. Feeds on insects and small vertebrates.

SPECIES IDENTIFICATION Unmistakable, owing to unique proportions and plumage.

MOULT Complete moult begins May–Jun in yearlings and females, in Jun–Jul for males. Moult completed by Aug–Sep.

PLUMAGES Adult and juvenile superficially similar, but separable by fine plumage details; adult male and female inseparable.

Juvenile
Head, underbody and underwing-coverts white, flight feathers, tail and upperparts black with purplish-blue iridescence. Head and neck with a tawny wash in fresh plumage, wearing soon off after fledging. Gradation of tail fork regular, with outermost sixth tail feather not considerably longer than fifth, clearly shorter than adult's and resulting in a shorter tail and a shallower fork.

Adult
Similar to juvenile, but outer tail feathers much elongated and tail obviously longer with a deeper fork. Dark upperparts are iridescent blue, often appearing silvery-grey from a distance, while forewing and anterior part of mantle contrast as matt black.

SEXING Adult male and female similar.

CONFUSION RISKS None.

162. Swallow-tailed Kite *Elanoides forficatus*, presumed 2nd cy immature, based on worn and pointed remiges and rectrices. Unmistakable owing to diagnostic plumage and structure. USA, 16.5.2013 (Richard Crossley)

163. Swallow-tailed Kite *Elanoides forficatus*, adult. An unmistakable species, but difficult to age as plumage of young and adults is practically identical. USA, 8.4.2010 (Richard Crossley)

164. Swallow-tailed Kite *Elanoides forficatus*, adult. Upperparts are largely iridescent, with colour changing depending on angle of light, but tail, forewing and mantle are permanently darker. USA, 8.4.2010 (Richard Crossley)

WHITE-TAILED EAGLE
Haliaeetus albicilla

Other names: White-tailed Sea Eagle

VARIATION Occurs as nominate *albicilla* throughout our region. The somewhat larger *groenlandicus* is found in S Greenland only.

DISTRIBUTION Patchily across C, E and N Europe along coasts, major rivers, lakes and wetlands. Population density increasing from W to E and from S to N. Recently reintroduced in Scotland, where population now self-sustaining. Mostly resident, but northern inland populations are migratory, with immatures moving further than adults.

BEHAVIOUR A sluggish raptor spending most of the time perched, keeping an eye out for hunting opportunities. Active flight looks heavy and laboured, with long series of shallow wing-beats interspersed by short glides. Soars freely on rather flat and level wings, sometimes to considerable heights. Unbelievably agile and fast when pursuing gulls, cormorants or Ospreys to steal their catch, manoeuvring skilfully even through the tightest corners, like a giant skua. Also descends to shores at low tide, where it walks around looking for carcasses or trapped fish.

SPECIES IDENTIFICATION A large eagle with long and broad, rectangular wings, a shortish, wedged tail and a strong and well protruding head. Flight feathers lack barring. Juveniles and adults differ considerably in silhouette due to different proportions of wings and tail. Juveniles and immatures appear blotchy below with pale axillaries and light underwing bands in most, while tail feathers are variably white with dark edges. Adults are more uniform, with head and neck being palest and contrasting against darker brown body and underwing-coverts, tail feathers are white and translucent. Even after first-adult plumage is reached head, neck and breast, and iris continue to turn increasingly pale with age. Feet yellow with bare, unfeathered tarsi.

Powered flight consists of long series of up to 15–30 flaps interspersed by rather short glides, the flapping sequences being clearly longer than in other large eagles.

MOULT As with other large raptors first moult-cycle takes several years to complete. Moults mainly during breeding season, between Apr and Oct. First three age-classes can be safely aged by the number of retained juvenile remiges, but later plumages not safe to age owing to extensive individual plumage variation and geographical variation in the progress of moult. Northern migratory populations moult more slowly than birds from further south, which should be taken into account when ageing birds by moult pattern.

PLUMAGES Five plumage-types can be recognised, and although they are linked to age, they do not necessarily convey the exact age of the bird. This is particularly true for second to fourth plumages, in which advanced birds may look as if one year older than their actual age, while slowly moulting birds may appear one year younger than their actual age. Use of a suite of multiple characters usually guarantees correct ageing.

Juveniles have much longer flight and tail feathers compared to adults, and thus a different silhouette, with broader wings and longer tail. The second generation of feathers, grown in the first moult, are considerably shorter than the juvenile feathers, but the feather length continues to decrease also in the following moults, although to a lesser extent (pers. obs.).

Juvenile

Plumage colours are a mixture of mustard and dark browns, with little or no visible white. No signs of moult. Head and neck are usually darkest, with a contrasting pale grey loral patch, grey cere and dark bill. The breast is regularly streaked, while the pale bases to the underwing-coverts form variable pale lines and bands; the axillaries are also largely pale, even whitish, which is a good species character. The trailing edge is serrated, formed by the spiky tips of juvenile secondaries, while the tail varies from largely white to almost dark, but when fanned always shows dark margins to each feather. The upperwing-coverts are nicely patterned, with regular lines of dark feather-tips set against a mustard ground colour, with the entire coverts area gradually becoming paler towards the inner wing. The general impression of juveniles varies considerably, partly reflecting geographic origin. Birds from more southerly populations tend to be darker and more uniform, while some northern birds can look strikingly pale and distinctly marked, but individual variation is still considerable within each region. Regardless of colour form, by the following spring the plumage will have bleached notably, and by this time also the cere will have turned yellow, making the bill look clearly bi-coloured. Iris colour remains darkish throughout the first year of life, but the dark pupil may start to show by late spring.

Second plumage

Differs from juvenile by extensive white mottling, especially to lower breast, belly and mantle. From a distance the *underbody is variably mottled and contrasts sharply against the dark head and neck and the darker thighs*. The upperwing lacks the clear-cut contrast of the juvenile, and the paler coverts appear untidy, lacking the regular pattern of the juveniles, while the underwing looks still largely juvenile, save for moulted inner primaries and the odd secondary. The bill is mostly dark, contrasting with the now yellow cere, but the base of the bill and the brown iris both turn lighter towards spring in the 3rd cy.

Third plumage

Some birds are rather similar to the second plumage, but most are usually more uniformly brown, with far less white mottling below. The head is brown matching the rest of the body in colour, the bill appears all-yellow from a distance, but looks clearly greyer in close views, and the eye has turned a light brown in most. The underwings still show white bands on the coverts and the tail varies individually from nearly all-dark to mostly white with darker margins and tips to the individual feathers. Most northern birds still show a few retained juvenile remiges, mostly p9–10 and s9 which are extremely faded and worn by now, and with retained juvenile secondaries well protruding from the trailing edge of the wing, while birds from southern populations may already have lost their last juvenile feathers.

Fourth plumage

Similar to adult, and distant birds are not necessarily separable. Compared to a full adult the head is still more or less the colour of the body, the white tail feathers often show some narrow dark tips and the underwing-coverts and axillaries hold the odd pale feather. Bill and iris are yellow as in adults, the bill often with a visible grey smudge distally. No retained juvenile remiges.

Adult

By the fifth plumage the birds have acquired all the features of the adult, with yellow bill, white tail and uniformly brown underwings and axillaries, yellow iris and deep yellow bill. Even after reaching this first adult stage, the plumage will continue to get paler still. Notably the head, neck, upper breast and mantle can be strikingly pale in old birds, while the iris turns almost white and the bill becomes a bleached yellow.

SEXING Although females are on average larger, heavier and larger-billed than males, as a rule single birds are not possible to sex with certainty. Sometimes sexing is difficult even when directly comparing the birds in a breeding pair.

CONFUSION RISKS Adults are rather unmistakable, although it is worth remembering that the white tail is not always that obvious, mostly depending on the light. Juveniles and immatures could be mistaken for other mostly dark, large eagles, but the combination of pale axillaries and underwing bands and the white mottling of immatures below is diagnostic.

165. White-tailed Eagle *Haliaeetus albicilla*, recently fledged juvenile. Juveniles in fresh plumage appear dark, but the streaking below and the upperwing markings are still easy to see. Finland, 12.7.2013 (DF)

166. White-tailed Eagle *Haliaeetus albicilla*, recently fledged juvenile (same as 165). Finland, 12.7.2013 (DF)

167. White-tailed Eagle *Haliaeetus albicilla*, juvenile in 2ⁿᵈ cy spring. Some young birds are rather juvenile-like and still dark in 2ⁿᵈ cy spring. Finland, 8.4.2011 (DF)

168. White-tailed Eagle *Haliaeetus albicilla*, juvenile in 2ⁿᵈ cy spring. Others are more brightly marked and may recall the next immature plumage-stage. Finland, 14.3.2012 (DF)

169. White-tailed Eagle *Haliaeetus albicilla*, juvenile in 2ⁿᵈ cy spring. An even more white-speckled juvenile, but note the colour and markings of the belly and compare with the next plumage. Finland, 3.3.2012 (DF)

170. White-tailed Eagle *Haliaeetus albicilla*, juvenile in 2nd cy spring. Some juveniles and immatures show white-speckled flight feathers. Japan, 12.2.2005 (DF)

171. White-tailed Eagle *Haliaeetus albicilla*, autumn juvenile. Juveniles are easy to age by their regular rows of markings on their upperwing-coverts. Finland, 11.10.2013 (DF)

172. White-tailed Eagle *Haliaeetus albicilla*, 2nd cy summer. This juvenile has just started its first moult and dropped its innermost primary, but most of the plumage is still juvenile. Norway, 1.6.2014 (DF)

173. White-tailed Eagle *Haliaeetus albicilla*, 2nd cy summer. The first moult is proceeding slowly in northern birds, with only two inner primaries and part of the upperwing-coverts replaced by late July. Norway, 20.7.2014 (DF)

174. White-tailed Eagle *Haliaeetus albicilla*, second plumage in 3rd cy spring. The second plumage is characterised by largely white breast with contrasting dark neck and head, and dark trousers. The new tail is shorter than the juvenile tail, giving the bird a more broad-winged appearance. Finland, 31.3.2011 (DF)

175. White-tailed Eagle *Haliaeetus albicilla*, second plumage in 3rd cy spring. This bird in its second plumage has just started its second moult, and can be aged by the combination of plumage characters, like white-speckled breast, and the location of the moult front with growing p4. Finland, 20.4.2014 (DF)

176. White-tailed Eagle *Haliaeetus albicilla*, second plumage in 3rd cy spring/summer. This bird from arctic Norway is lagging one month behind in its moult compared to its relatives from S Finland (cf. **175**). The upperparts of the second plumage are characterised by a largely white mantle, while most of the secondaries and more than half of the primaries are still juvenile feathers. Norway, 26.5.2014 (DF)

177. White-tailed Eagle *Haliaeetus albicilla*, third plumage in 4ᵗʰ cy spring. The third plumage is rather variable but has, as a rule, less white on the underparts compared to the previous plumage. The bill is turning lighter, looking yellow from a distance in many. Only a few juvenile remiges remain, mostly 1–2 outer primaries and, like here, the odd median, long and pointed secondary. Japan, 12.2.2005 (DF)

178. White-tailed Eagle *Haliaeetus albicilla*, third plumage in 4ᵗʰ cy spring. Similar in plumage to 177 this bird already has a yellow bill. All juvenile secondaries have been moulted, but the outer two juvenile primaries still remain. Finland, 27.3.2014 (DF)

179. White-tailed Eagle *Haliaeetus albicilla*, third plumage in 4th cy spring. Note rather uniform general impression, save for white markings to axillaries and underwings, and immature-type tail. Outer three primaries are still retained juvenile feathers. Finland, 31.3.2011 (DF)

180. White-tailed Eagle *Haliaeetus albicilla*, third plumage moulting to fourth plumage in 4th cy summer. The head, bill and iris are gradually getting lighter and the plumage loses its white speckling. The active primary moult has now reached the outermost primaries. Finland, 26.7.2011 (DF)

181. White-tailed Eagle *Haliaeetus albicilla*, third plumage moulting to fourth plumage in 4th cy summer (same as **180**). The upperparts are becoming greyish-brown, but the new tail still retains its immature look. Finland, 26.7.2011 (DF)

182. White-tailed Eagle *Haliaeetus albicilla*, fourth plumage in 5ᵗʰ cy spring. This plumage can be quite similar to the third plumage, but has on average a paler head, brighter yellow bill, a lighter iris and fewer white markings on the underbody and underwings. The tail varies individually and can be anything from a typical immature's to almost white with some dark markings towards the tip. Note that the outermost primaries are fresh, having been recently moulted. Japan, 10.2.2005 (DF)

183. White-tailed Eagle *Haliaeetus albicilla*, (first?) adult plumage. This bird from arctic Norway looks like a full adult, except for the pale marbling visible on some of its retained remiges from an earlier immature plumage. Birds from the far north tend to be lighter in colour compared to birds of similar age from more southerly populations. Norway, 7.6.2011 (DF)

184. White-tailed Eagle *Haliaeetus albicilla*, adult plumage. Despite looking fully adult at first glance, this bird also shows some pale markings on some of its underwing-coverts, indicating a younger age. Finland, 14.3.2012 (DF)

185. White-tailed Eagle *Haliaeetus albicilla*, adult male, breeding. Uniform underwings and body, pale head and iris, and all-white tail indicate a fully adult plumage. Finland, 25.5.2012 (DF)

186. White-tailed Eagle *Haliaeetus albicilla*, adult, wintering. Finland, 1.4.2013 (DF)

187. White-tailed Eagle *Haliaeetus albicilla*, adult. Norway, 7.6.2011 (DF)

188. White-tailed Eagle *Haliaeetus albicilla*, adult male, breeding (same as 185). Fully adult birds are rather pale greyish above with wings darkening gradually towards the trailing edge. Finland, 25.5.2012 (DF)

189. White-tailed Eagle *Haliaeetus albicilla*, adult female, breeding (paired to 185/188). Finland, 25.5.2012 (DF)

PALLAS'S FISH EAGLE
Haliaeetus leucoryphus

Other names: Pallas's Sea Eagle

VARIATION Monotypic.

DISTRIBUTION Breeds from C Asia, S Siberia, Mongolia and China to the Indian subcontinent. Formerly also in easternmost Europe, in the Volga delta and elsewhere around the Caspian Sea, but recently populations have decreased rapidly over most of the range. There is a scattering of records, mostly from N and E Europe from the early 20th century, but in the light of the current situation further sightings seem unlikely.

BEHAVIOUR Hunts live birds and fish, but is also a scavenger and a food pirate. Birds spend a lot of time perched, either on the open shore or on a prominent tree, usually next to the hunting grounds.

SPECIES IDENTIFICATION A large eagle, but clearly slimmer in proportions and more lightly built than White-tailed Eagle, the two sometimes occurring together. Wings are rather narrow but parallel-edged and the tail is fairly long with a squarish tip, giving the bird a rather unique, slim silhouette in comparison to other sympatric large raptors. Adults are easy to recognise by all-dark wings and body, with contrasting pale head, breast and mantle and a white tail with a broad black tip. Juveniles are more uniform, but recognised by the diagnostic underwing markings, although immature White-tailed Eagles may occasionally show hints of similar wing markings. Feet rather colourless, pale grey to pale flesh, but never bright yellow as in White-tailed.

MOULT Immatures moult only a part of their flight feathers annually, similar to other large raptors, and birds can be aged by the number of retained juvenile primaries. The onset of the first moult is described by three 2nd cy birds: one had dropped its first primary in Mar, another had replaced four inner primaries by May and yet another showed three new primaries on 5 Jun.

Adults are probably capable of replacing the entire set of flight feathers in one moulting season.

PLUMAGES Normally 3–4 plumages can be recognised, although the first two are rather similar and best separated by comparing details of moult.

Juvenile

A rather uniform dark raptor, except for striking underwing pattern. Crown and hindneck are a lighter brown and contrast with the dark face and neck-sides, the latter showing notably long lanceolate feathers, often visible even in flight. Upperwings and tail look uniformly dark, but upperwing-coverts are browner with paler tips and fringes forming variably distinct upperwing-bands. The underbody is somewhat paler greyish-brown and often contrasts with the darker head and tail. Underwings distinctly marked, with largely white inner primaries and a broad underwing-band, formed by buffish axillaries and median coverts, sandwiched between a dark leading edge and greyer greater coverts, the latter with pale tips. Iris is dark, while the dark grey bill and cere emphasise the fleshy yellow gape-flange.

Second plumage

Similar to juvenile and best identified by moulting inner primaries, which show a similar pattern to the juvenile feathers, largely white with dark tips. The new tail feathers show an emerging light-mottled tail-band half way out on the feather, but this can be difficult to see in field conditions.

Subadult plumages

The adult plumage is achieved gradually after another 1–2 subadult plumages, in which the primary flash is reduced, the tail-band gets brighter and the head, neck and iris get paler.

Adult

Except for pale sandy-coloured head, breast and mantle the adults are overall uniformly dark, including upper- and underwing surfaces. The rather long and narrow tail is white with a broad black terminal band. Bill and feet are pale greyish, iris pale.

SEXING Females are larger than males, but size difference is usually not obvious in the field, unless both birds of a pair are seen together.

CONFUSION RISKS Adults could be mistaken for a Golden Eagle, based on similar tail pattern, but the narrower and parallel-edged wings differ. Juveniles could be mistaken for immature White-tailed Eagle, but wings are slimmer, the tail is longer and the striking underwing markings are diagnostic.

190. Pallas's Fish Eagle *Haliaeetus leucoryphus*, juvenile. Easily identified by rather narrow, rectangular wings and striking underwing pattern. Mongolia, 7.6.2012 (DF)

191. Pallas's Fish Eagle *Haliaeetus leucoryphus*, juvenile. Unmistakable thanks to striking plumage. Mongolia, 7.6.2012 (DF)

192. Pallas's Fish Eagle *Haliaeetus leucoryphus*, juvenile (left same as 190, right same as 191). Upperparts are comparatively uniform, with coverts paler greyish-brown compared to dark flight feathers and tail. The pale margins to the greater coverts and uppertail-coverts form faint bands when seen from a distance. Mongolia, 7.6.2012 (DF)

193. Pallas's Fish Eagle *Haliaeetus leucoryphus*, 2nd cy. The new inner primaries are largely white as in juveniles, while the new tail feathers start to show a faint lighter central band. Mongolia, 6.6.2012 (DF)

194. Pallas's Fish Eagle *Haliaeetus leucoryphus*, 2ⁿᵈ cy (same as **193**). Plumage still rather similar to juvenile, but told by ongoing moult of inner primaries. Mongolia, 6.6.2012 (DF)

195. Pallas's Fish Eagle *Haliaeetus leucoryphus*, adult. Note rather narrow, rectangular wings, creamy head and neck and conspicuous tail pattern. India, 1.12.2007 (Annika Forsten)

196. Pallas's Fish Eagle *Haliaeetus leucoryphus*, adult. India, 1.12.2007 (Annika Forsten)

BALD EAGLE
Haliaeetus leucocephalus

VARIATION Two subspecies, with nominate *leuco-cephalus* and northwestern *alascanus* differing mainly in size.

DISTRIBUTION A very rare transatlantic vagrant from North America, with two accepted records of juveniles from Ireland, one shot in Jan 1973 and another taken into care in Nov 1987.

BEHAVIOUR Similar to White-tailed Eagle.

SPECIES IDENTIFICATION Adults are unmistakable, but juveniles and immatures could easily be mistaken for White-tailed Eagles. The different age-classes of Bald Eagle correspond to the plumages of White-tailed Eagle. *The Bald Eagle has narrower wings than the White-tailed, with only six clearly fingered primaries in the wingtip* (seven in White-tailed) and the tail is longer, with a squarer tip. Because of the narrower wings the tail looks longer compared to the compact and broad-winged silhouette of the White-tailed Eagle.

MOULT Similar to White-tailed Eagle.

PLUMAGES Plumages can be divided into juvenile plumage, second and third plumages, fourth plumage-type and adult plumage, thus corresponding to the plumage succession in White-tailed Eagle. As in White-tailed Eagle, the extensive plumage variation found within each of the subadult age-classes makes an accurate ageing difficult or even impossible, and often one has to rely on plumage-types instead of trying to age the bird precisely.

Juvenile

Juveniles have longer, narrower and more pointed secondaries and rectrices compared to older birds, giving them a broader wing and a longer tail, just as in White-tailed Eagle. Always aged by their uniform plumage, lacking signs of moult. By and large, the plumage is rather similar to that of a juvenile White-tailed Eagle, but Bald Eagles come in dark and lighter forms, which are quite different from each other. Dark birds are on average more uniform and darker and less clearly marked on the underbody, mantle and upperwings, compared to White-tailed Eagles, while pale birds can be strikingly coloured, with extensive white markings to their underwing remiges, particularly on the inner primaries and inner secondaries. The tail is longer than in White-tailed Eagle and has a rather square, not wedge-shaped tip; it looks dark when folded but exposes white feather-centres when fanned, just as in White-tailed.

Second plumage

Best aged by moult pattern comparing freshly moulted and retained juvenile flight feathers, as for White-tailed Eagle: fresh inner primaries, while the outer primaries and the majority of the secondaries are retained, long and spiky juvenile feathers. The plumage is extensively white-mottled in many, with darker head and neck, thighs and upperwings. Bill mostly grey, with a yellow cere and base, and the iris colour is light brown.

Third plumage

Differs from the previous plumage in retaining just a few outermost juvenile primaries and a few, if any at all, median juvenile secondaries. The individual plumage variation in this age-class is considerable, and some can still be very similar to the previous plumage with extensive white mottling, while others are quite dark and more uniform, looking distinctly 'older'. Bill and iris are both turning lighter, greyish-yellow and light brown respectively, but the individual variation in colour is considerable.

Fourth plumage

More advanced birds may appear adult-like, being dark with a whitish head and tail, but they always retain signs of immaturity in their body plumage in the form of visible pale feather-bases, particularly in the axillaries and underwing-coverts, and by showing darker markings on their white body parts. More slowly developing birds are still mottled and thus difficult to distinguish from advanced birds of the preceding plumage.

Adult

Unmistakable. All dark, except for white head and neck, rump, undertail-coverts and tail. Bill and feet bright yellow, iris pale to nearly whitish. Some birds may attain the adult plumage in their fifth plumage, but most birds a year later.

SEXING Although females are on average larger than males, birds can rarely be sexed in the field, unless both birds of a pair are seen together.

CONFUSION RISKS Adults are unmistakable, but immatures could easily be overlooked as immature White-tailed Eagles. Slimmer proportions with narrower wings is the main difference compared to White-tailed Eagle.

197. Bald Eagle *Haliaeetus leucocephalus*, juvenile. Similar to juvenile White-tailed Eagle, but differs by narrower, more rectangular wings, with six, not seven fingers. Plumage is also less streaked, and remiges and underwing-coverts show more white, although amount varies individually. USA (Jerry Liguori)

198. Bald Eagle *Haliaeetus leucocephalus*, juvenile. Note narrower wings and comparatively longer (and darker) tail. Upperwing-coverts, mantle and scapulars are more uniform compared to juvenile White-tailed Eagle. USA (Jerry Liguori)

199. Bald Eagle *Haliaeetus leucocephalus*, second plumage. Immatures are very similar to corresponding plumages of White-tailed Eagle, but notice diagnostic wingtip, with six fingers only. Immature Bald Eagles also show on average more white in their primaries and secondaries. USA (Jerry Liguori)

200. Bald Eagle *Haliaeetus leucocephalus*, second plumage. Largely similar to White-tailed Eagle. Note how much shorter the moulted secondaries are compared to the retained brown juvenile feathers, explaining the even narrower wing of an older Bald Eagle compared to a White-tailed of corresponding age. USA (Jerry Liguori)

AFRICAN FISH EAGLE
Haliaeetus vocifer

VARIATION Monotypic

DISTRIBUTION Widely distributed across sub-Saharan Africa. Rare vagrant to S Egypt.

BEHAVIOUR Most of its time is spent on a prominent perch near water. Snatches prey from the surface, often after impressive swoops; also feeds by kleptoparasitising kites, storks and herons. Social behaviour includes high soaring flights with diagnostic loud calling.

SPECIES IDENTIFICATION Adults are unmistakable owing to their unique shape and plumage. Immatures are less extreme in shape and the plumage is more varied. In a Western Palearctic context juveniles and immatures could be mistaken for immature White-tailed Eagle or Pallas's Fish Eagle, but proportions and plumage details are diagnostic.

MOULT Timing of annual moult varies locally, depending on the timing of the breeding cycle. Replacing the entire plumage takes more than one year, although moult appears to progress rapidly without long stops.

PLUMAGES Plumage development is rather quick for the size of the bird, probably owing to continuous moult, with possibly no more than two immature stages between juvenile and full adult plumages.

Juvenile

A rather varied and striking plumage, bearing some resemblance to the adult plumage owing to partly white head and broad white breast-band, while upper breast is dark and heavily streaked and rear underbody mottled. The undersides of the wings are marked with a pale ochre leading edge to the arm, pale axillaries and individually variable white windows on the inner primaries and inner secondaries. Tail feathers are white with dark tips. Bill, cere and facial skin dark grey, feet pale flesh, iris greyish-brown. Upperwings are dark with browner coverts, paler brown along the inner parts of the leading edge.

The silhouette in flight is far less extreme compared to the adults, with narrower and more parallel-edged wings and a slightly longer tail. Secondaries are pointed, creating a serrated trailing edge to the wing.

Immature

The exact ageing of immatures and subadults is not possible, partly owing to individual variation, but also due to great variation in timing of breeding cycle, even locally.

Moulting out of the juvenile plumage, the immatures first lose the dark streaking of the head and forebody, which turn white, while the underwings still retain the pale axillaries, forewing and pale windows. In the next stage the new tail feathers lack the dark tips, the new remiges are blacker and the body turns dark. Gradually the immature features give way to the adult plumage, but the axillaries still remain pale, as do some underwing-coverts. The cere and facial skin are off-white and colourless, the feet pale flesh.

Subadults look superficially like adults but are darker, lacking the adult's vivid chestnut colour. They may still show some dark streaking across the breast, while upperwing- and underwing-coverts and axillaries show pale feather-bases. The last signs of immaturity are usually some dark markings on the tail and some pale markings on the underwing-coverts. By now the cere and face have turned yellow.

Adult

Easy to recognise by striking plumage with white head, neck, upper breast, mantle and tail, which contrast with the rich chestnut body and wing-coverts and black flight feathers and greater coverts. Bill is black, iris medium brown; cere, base of bill and facial skin bright yellow, feet yellow.

SEXING Sexes similar.

CONFUSION RISKS Adults are unmistakable, while juveniles and immatures could be mistaken for other large, broad-winged and short-tailed raptors, although silhouette and plumage, particularly of the underwings, are unique.

201. African Fish Eagle *Haliaeetus vocifer*, fresh juvenile. Plumage fresh and uniform, lacking signs of moult. Note grey iris, cere and bill, and darkish head and neck. Ethiopia, 7.2.2012 (DF)

202. African Fish Eagle *Haliaeetus vocifer*, fresh juvenile, not yet independent. Upperwing-coverts with regular scaly pattern; whiter-headed than 201 of similar age. Ethiopia, 24.2.2006 (DF)

203. African Fish Eagle *Haliaeetus vocifer*, immature with first moult initiated. Innermost primary missing/growing and outer tail feathers moulted but largely juvenile-like. Ethiopia, 5.11.2011 (DF)

204. African Fish Eagle *Haliaeetus vocifer*, immature of same age as **203** with new inner primary and half of tail, belly becoming darker and more uniform. Ethiopia, 5.11.2011 (DF)

205. African Fish Eagle *Haliaeetus vocifer*, immature of similar age to **203** and **204**. Ethiopia, 11.1.2010 (DF)

206. African Fish Eagle *Haliaeetus vocifer*, immature, possibly around 18 months of age. Roughly half of primaries and secondaries moulted, and tail turning white. Ethiopia, 7.2.2012 (DF)

207. African Fish Eagle *Haliaeetus vocifer*, immature, approximately same age as **206**. This bird has some retained juvenile primaries and secondaries. Ethiopia, 5.11.2011 (DF)

208. African Fish Eagle *Haliaeetus vocifer*, older immature, probably about two years of age. All flight feathers have by now been replaced at least once, gradually turning uniformly dark as moult progresses. Iris is getting lighter while cere and facial skin are turning yellow. Ethiopia, 5.11.2011 (DF)

209. African Fish Eagle *Haliaeetus vocifer*, adult. Diagnostic shape and plumage render adults unmistakable. Note that wings are more rounded and tail shorter compared to juveniles, making the flight silhouette even more extreme. Ethiopia, 16.1.2010 (DF)

210. African Fish Eagle *Haliaeetus vocifer*, adult. Ethiopia, 6.11.2011 (DF)

211. African Fish Eagle *Haliaeetus vocifer*, adult. Ethiopia, 7.1.2010 (DF)

212. African Fish Eagle *Haliaeetus vocifer*, adult. Ethiopia, 11.1.2010 (DF)

EGYPTIAN VULTURE
Neophron percnopterus

VARIATION Two subspecies: nominate *percnopterus* over most of its European, Asian and African range, and ssp. *ginginianus* on the Indian subcontinent (pale rather than dark bill). Birds from the Canary Islands are sometimes separated as ssp. *majorensis*.

DISTRIBUTION Mostly confined to rocky areas or mountains, but extends foraging tours down to lowlands. In our region, main population is in Spain and S France, with smaller numbers scattered in SE and E Europe. As a whole, the European populations are threatened and numbers are declining. Winters in the Sahel, straddling the southern fringe of the Saharan desert, where partly commensal with man.

BEHAVIOUR Mostly seen on the wing, usually either when moving between breeding, roosting and foraging sites, or on migration. Normally flies at medium to high altitudes, descending only at the nest or roost site or to feed. Also spends time perched and walking around on the ground.

SPECIES IDENTIFICATION Adults are unmistakable, but immatures are often mistaken for other large brown raptors, such as different *Aquila* eagles, despite their diagnostic silhouette. Told from similar-sized brown eagles by narrower wings (only five long fingers and one shorter, compared to 6–7 long fingers in eagles), with hand narrower than arm, by clearly wedge-shaped brown tail with translucent edges, and by diagnostic small head with long and narrow bill. Superficially similar-looking Hooded Vulture has broader and more rectangular wings, a short square tail and always shows a wide pale central area on remiges of underwing.

MOULT As in other large raptors the moult of the flight feathers is a prolonged process, and the first moult cycle takes more than a year to complete. Moults partly in breeding areas, partly on the wintering grounds. Immatures can be safely aged by details of moult. Birds wintering in Arabia and India seem to moult more feathers in a season than birds migrating to Africa.

PLUMAGES Four plumage-types can be separated, with the first two being rather similar, largely dark and juvenile-like. The third plumage is variable, but in the field gives the impression of a milky coffee-coloured somewhat adult-like bird, while the fourth plumage-type is the clean-looking, black-and-white adult-type plumage. The adult plumage can be split into two, with younger adults still showing some odd dark feathers, especially among the underwing-coverts, while definitive adults only show uniformly white underbody feathers.

It appears that eastern populations, wintering in Arabia and S Asia, have a more rapid moult, replacing more feathers per moulting season, and that these birds acquire the adult plumage up to one year sooner than their European counterparts (pers. obs.; for details see below).

Juvenile

Plumage is rather variable, with upperparts ranging from rather uniformly dark brown and featureless to more mottled and lighter-coloured, with distinct pale areas on the upperwings, mantle and rump. Some juveniles may even be mistaken for adults owing to their largely whitish upperparts. Best told from older age-classes by neat and uniform plumage, lack of wing moult, but also by the serrated trailing edge of the wing due to the pointed juvenile secondaries. The juveniles also show a pale vent and undertail-coverts. On the upperwing the pale outer edge of the secondaries is restricted to the proximal parts of the feather and the pale secondary panel is at best narrow and poorly defined (cf. later plumages). *The bare skin of the face is bluish or whitish and the fluffy ruff of the upper neck is dark.*

An unknown (major?) proportion of European juveniles stay on the African wintering grounds throughout the following summer, but some return to European breeding areas.

Second plumage

Safely aged by details of wing moult, *showing at least a few retained juvenile outer primaries*, often also some secondaries, all by now looking faded brown and very pointed compared to the new, moulted primaries and secondaries. As with the juvenile plumage, this plumage is also highly variable depending on the extent of the body moult, and while some still look rather dark and juvenile-like, others are more advanced with a high percentage of light-coloured body feathers throughout the plumage. The dark birds are often not possible to tell from juveniles at a distance, while more advanced birds approach the next, more variegated third plumage, but t*hey still retain a dark ruff and neck.* Dark second plumage birds usually appear even darker and more uniform compared to juveniles, as the upperparts have lost the pale tips and fringes due to wear and the belly and vent have turned darker. The new secondaries show broader white margins on the upperwing contrasting sharply with the retained juvenile feathers, which are faded brown with barely noticeable lighter margins. *The facial skin is still colourless, whitish or pale blue in most,* but some eastern birds in particular may already have developed a yellowish face.

European birds are long-distance migrants and appear to moult more slowly compared to birds wintering in Arabia. European birds thus show fewer moulted feathers compared to Asian birds at any one time. Most second-plumage European birds still appear dark, while birds from further east can be either darkish, or they may look like third-plumage birds, although still with a few retained juvenile outer primaries.

Third plumage

This plumage looks pale buffish from a distance, but consists of a mixture of white, greyish-buff and darker feathers, in individually varying proportions. Compared to the two previous plumages *the ruff and neck are the same colour as the rest of the body,* often a mixture of pale and darker feathers, *the facial skin is gradually turning yellow* and the outer primaries are fresh, black adult-type feathers. The tail shows a mixture of buffish and white feathers and the collective impression is of a buffish tail with paler edges. By and large, the bird looks like a dirty, mottled adult. The margins of the new secondaries above are white and broad, as in adults, adding to the contrasting pattern of the upperwing.

Fourth plumage

These birds look like adults in all respects, but at least some still show a few dark feathers, especially on the underwing-coverts.

Adult

Unmistakable thanks to unique silhouette and plumage. All white from below, including wedge-shaped and translucent tail, but with black remiges. The bare

face is a bright yellow and the stained ruff stands out as being yellowish-buff, while the pale pinkish feet are often visible below the tail. Upperparts more variegated, as remiges are proximally partly white and the greater coverts partly black, the latter forming a black rectangle on the inner wing.

Note that birds may vary a lot in colour, depending on the degree of staining from dirt at rubbish dumps etc., which will particularly affect the ruff and the underbody.

SEXING Adults can tentatively be sexed, but more material is needed: males have a brighter yellow-orange facial skin than females, which in turn shows a dark 'eye-shadow' below the eye, but these differences are best used when directly comparing the birds of a breeding pair. Immatures are not known to differ.

CONFUSION RISKS Immatures could easily be mistaken for a brown eagle, such as a Lesser Spotted Eagle, with which they mingle on migration, but notice narrower wings and longer, clearly wedge-shaped tail and the slim and pointed head. Lacks the pale central underwing of Hooded Vulture, which also has broader, more rectangular wings and a short, rounded tail.

213. Egyptian Vulture *Neophron percnopterus*, second plumage (right) with adult Greater Spotted Eagle. Note narrower wings and clearly wedge-shaped tail of the vulture. Oman, 2.11.2004 (DF)

214. Egyptian Vulture *Neophron percnopterus*, juvenile, migrant in fresh plumage. Uniform condition of plumage identifies this bird as a juvenile. Note translucent edges of partly opened tail. Egypt, 9.10.2010 (DF)

215. Egyptian Vulture *Neophron percnopterus*, juvenile, wintering. Differs from various eagles by light tail, naked face and short p5 (6th finger). India, 9.12.2007 (DF)

216. Egyptian Vulture *Neophron percnopterus*, juvenile, migrant in fresh plumage. Juveniles can be variably light-mottled, but are always told by lacking signs of wing-moult. Egypt, 11.10.2010 (DF)

217. Egyptian Vulture *Neophron percnopterus*, juvenile, migrant in fresh plumage. This bird is at the lighter end of variation and could easily be mistaken for an older immature. Egypt, 11.10.2010 (DF)

218. Egyptian Vulture *Neophron percnopterus*, juvenile, migrant. The most brightly marked juveniles can even be mistaken for adults. Spain, 15.9.2005 (DF)

219. Egyptian Vulture *Neophron percnopterus*, juvenile, wintering. Bluish face and poorly defined light edges to secondaries are typical of juveniles, as is the uniform condition of the plumage. Oman, 6.11.2004 (DF)

220. Egyptian Vulture *Neophron percnopterus*, juvenile, wintering. A lighter juvenile, superficially resembling the second plumage, but notice how the pale feather-tips of the upperwing form regular bands in juveniles. India, 7.12.2007 (DF)

221. Egyptian Vulture *Neophron percnopterus*, second plumage, wintering. This plumage is highly variable, but is usually identified by retained pointed and worn juvenile primaries and secondaries, a generally dark collar and a pale face, while the body plumage varies from mostly dark to as pale as in this advanced bird. Oman, 22.11.2014 (DF)

222. Egyptian Vulture *Neophron percnopterus*, second plumage, wintering (same as **221**). Note faded juvenile remiges and a mix of light and brown upperwing-coverts. Oman, 22.11.2014 (DF)

223. Egyptian Vulture *Neophron percnopterus*, second plumage, wintering. A more average-looking second plumage, but note advanced remex moult, with all juvenile feathers replaced, except for a few outer primaries. Oman, 2.11.2004 (DF)

224. Egyptian Vulture *Neophron percnopterus*, second plumage, wintering. Similar to **223**, but with less advanced moult. Oman, 2.11.2004 (DF)

225. Egyptian Vulture *Neophron percnopterus*, second plumage, wintering. Moult pattern similar to **224**, but underwing-coverts largely white. Oman, 2.11.2004 (DF)

226. Egyptian Vulture *Neophron percnopterus*, second plumage, wintering. A rather typical, darkish second-plumage bird, reliably aged by its good number of retained juvenile secondaries and primaries. India, 17.12.2007 (DF)

227. Egyptian Vulture *Neophron percnopterus*, second plumage, wintering. This bird has only one retained juvenile secondary and two primaries, and the upperparts are speckled with the first white feathers. India, 17.12.2007 (DF)

228. Egyptian Vulture *Neophron percnopterus*, third plumage, migrant. This Spanish bird shows how dark European birds still may look despite their age. In this case the moult pattern and the yellow face tell it from a second-plumage bird. Spain, 4.9.2013 (DF)

229. Egyptian Vulture *Neophron percnopterus*, third plumage, wintering. Plumage not that different from many second-plumage birds, but note different moult pattern, with fresh outer primaries and a second wave in the inner. Naked face is turning yellow. Oman, 6.11.2004 (DF)

230. Egyptian Vulture *Neophron percnopterus*, third plumage, wintering. A lighter bird showing two moult fronts in the primaries, with fresh outermost and inner primaries, while the median primaries are the most abraded. Also note lighter collar and yellowish face. Oman, 2.11.2004 (DF)

231. Egyptian Vulture *Neophron percnopterus*, third plumage, wintering. Upperparts resemble adult, but with a lot of greyish-brown feathers admixed. Ruff is still partly dark and face is yellow. Oman, 6.11.2004 (DF)

232. Egyptian Vulture *Neophron percnopterus*, adult, wintering. Adults are distinctly black-and-white with a yellow face, but underbody and ruff are often stained from dirt. Oman, 3.11.2004 (DF)

233. Egyptian Vulture *Neophron percnopterus*, adult, wintering. Adults are unmistakable, even from a distance. Oman, 2.11.2004 (DF)

151

234. Egyptian Vulture *Neophron percnopterus*, adult, wintering. Oman, 6.11.2004 (DF)

235. Egyptian Vulture *Neophron percnopterus*, adult, wintering. Oman, 6.11.2004 (DF)

HOODED VULTURE
Necrosyrtes monachus

VARIATION Two subspecies recognised, but slight size difference negligible in the field: ssp. *monachus* in W Africa, reaching W Sudan, is marginally larger than *pileatus* from N Uganda and Ethiopia through E and S Africa.

DISTRIBUTION Restricted to sub-Saharan Africa, from the dry Sahel in the north to southern parts of Africa. Prefers open and semi-open areas, often in association with humans, and at times in big concentrations around villages and towns. Vagrant to Morocco and Mauritania.

BEHAVIOUR Often found roosting in trees around villages, but covers huge areas of countryside during foraging flights. Joins other vultures at carcasses and frequently visits rubbish tips.

SPECIES IDENTIFICATION An all-dark largish raptor, with rectangular and parallel-edged wings, a small pointed head with a long, narrow bill and a shortish, square-cut tail. The wingtip is broad with six prominent fingers. Dark underwing-coverts contrast with a large and ill-defined pale area across the mid-wing, which is the single most diagnostic identification feature. Due to lack of striking plumage features, distant birds can easily be mistaken for other large, dark raptor species, such as immatures of Egyptian or Lappet-faced Vultures or all-brown eagles. Peculiar head shape rules out all but Egyptian, which has clearly narrower wings and a longer, strongly wedge-shaped tail.

MOULT The replacement of the flight feathers is a lengthy process, taking several years to complete, and moult pattern can be used for ageing during the first years of life.

PLUMAGES Usually four plumages can be identified given good views of the moulting pattern, while adults and juveniles can also be separated by plumage differences from a greater distance.

Juvenile

Dark brown above, with a noticeable contrast between brown coverts and darker flight feathers. Secondaries are typically pointed, creating a distinctly serrated trailing edge to the wing. Underparts are more uniform than in the adult, with a barely noticeable paler area to the mid-wing, while the thighs conform with the rest of the underbody. The bare face is pale bluish and the crown, nape and hindneck are covered in dark down. Feet are pale grey and the iris is dark.

Second plumage

Similar to juvenile, but identified by moulted inner primaries, while secondaries are still juvenile. Plumage overall is more faded brown compared to the juvenile.

Third plumage

Similar to second plumage, but moult has reached the outer primaries and some 50% of the secondaries have been replaced. Juvenile primaries and secondaries stand out as very worn and faded brown. The head is still covered in dark down, as in juveniles, but the colour is somewhat sun-bleached.

Adult

More uniformly coloured above than juvenile, lacking the sharp upperwing contrast between browner coverts and darker remiges. The trailing edge of the wing lacks the serration found in juveniles and immatures. From below the underwing is characterised by a diagnostic large silvery area across the bases of the remiges, while the thighs show woolly, whitish inner parts. The naked face is pale flesh-coloured, but may become deep red when blushing, and the crown, nape and hindneck are covered in short, dense 'lambs-wool' (sometimes likened to the wig of an English judge).

SEXING is not possible in the field.

CONFUSION RISKS In a Western Palearctic context the main confusion risk is with immature Egyptian Vulture, which is never as uniformly dark brown above as Hooded, and also shows a narrower wing (only five fingers) and a clearly longer, wedge-shaped tail, with pale and translucent edges and tip. Distant birds could also be mistaken for a spotted eagle, for example, but the diagnostic head shape and lack of strong plumage contrasts, such as pale uppertail-coverts or wing-flashes, should soon become apparent, while the pale central underwing of Hooded is a diagnostic feature.

236. Hooded Vulture *Necrosyrtes monachus*, juvenile. Juveniles have pointed flight feathers, creating a serrated trailing edge to the wing, and the pale area of the mid-wing is poorly defined compared to the adults. The colour of the bare face varies depending on the bird's mood, but the crown and nape are covered with short dark 'hair' in juveniles and young immatures. The Gambia, 29.11.2006 (DF)

237. Hooded Vulture *Necrosyrtes monachus*, juvenile. Note the uniform condition of the plumage. Juveniles are also slimmer-winged and longer-tailed than adults. The Gambia, 29.11.2006 (DF)

238. Hooded Vulture *Necrosyrtes monachus*, moulting immature. This bird, approximately two years of age, has replaced most of its juvenile primaries and about half of its juvenile secondaries. Note how the retained juvenile secondaries are shorter and more pointed compared to the new ones, explaining the difference in shape between juveniles and adults. Ethiopia, 16.2.2012 (DF)

239. Hooded Vulture *Necrosyrtes monachus*, moulting immature. This immature, at a similar stage of moult as **238**, has retained its juvenile-type dark head. The Gambia, 29.11.2006 (DF)

240. Hooded Vulture *Necrosyrtes monachus*, adult. Compare the wider and brighter underwing patch, the broadly rounded secondary-tips and the colour of the 'hair' of this adult with the younger birds in the previous images. Note also the white facial skin of this less excited bird. The Gambia, 29.11.2006 (DF)

241. Hooded Vulture *Necrosyrtes monachus*, adult. Note the bright underwing flash and grey 'lamb's-wool wig'. The Gambia, 18.4.2007 (DF)

242. Hooded Vulture *Necrosyrtes monachus*, adult with Fan-tailed Raven *Corvus rhipidurus*. Adults are rather uniformly dark brown above, while immatures show a clearer distinction between browner coverts and darker flight feathers. Ethiopia, 5.2.2012 (DF)

EURASIAN GRIFFON VULTURE
Gyps fulvus

Other names: Eurasian Griffon, Griffon Vulture

VARIATION Nominate *fulvus* in Europe, N Africa, Middle East and C Asia and ssp. *fulvescens* further east from Pakistan and N India through the lower Himalayas to Nepal and Bhutan. Most *fulvescens* are easily recognised as Eurasian Griffons by their rich mustard-coloured plumage, particularly obvious in juvenile and younger immature plumages, but some are more greyish-brown and more distinctly streaked in juvenile plumage, thus superficially resembling subadult/adult Himalayan Griffon *G. himalayensis. In fulvescens most birds, both juveniles and older birds, show largely pale and unmarked median and greater underwing-coverts*, a trait that rarely occurs in juveniles of western populations and a feature to further enhance the similarity with Himalayan Griffon Vulture (which see).

DISTRIBUTION Across S Europe, with strong populations in Spain and S France, smaller populations in the Alps, Sardinia, Cyprus, Crete and the Balkans. Discontinuous range extends from Asia Minor and north of Black Sea eastwards to mountains of C Asia and the Himalayas. Middle Eastern populations are mostly small and dwindling.

BEHAVIOUR Mostly seen on the wing, either thermalling over mountain ridges or gliding high across the sky, often several together. Congregates at carcasses, sometimes in hundreds. Roosts on steep cliffs or rocks, also on high-voltage pylons.

SPECIES IDENTIFICATION Easy to recognise as a vulture by huge wings, short tail and small whitish head. Easily told from other vultures *by strong contrast to upper- and underwings, with pale mustard- or milk coffee-coloured coverts contrasting sharply with dark remiges.* However, some adults are darker and the contrast above may not differ markedly from that of adult Black Vulture or immature Rüppell's. Even the underwing contrast may in some adults be hard to discern, particularly in dull lighting conditions. Soars with wings held in a marked dihedral (cf. Black Vulture).

MOULT Replacing the juvenile plumage takes several years. From the first moult onwards the plumage always shows a mixture of fresh and old feathers. Knowledge about moult is essential for the correct ageing of immatures.

Juveniles start to moult their primaries in Mar–Jun of their 2nd cy (occasionally as early as February), when about one year old, replacing only the inner 2–4 primaries. The subsequent moults start earlier, in Jan–Feb and continue until Nov. Adults replace their primaries simultaneously at several different points, and are thus capable of replacing their entire set of remiges in a much shorter period of time compared to immatures.

PLUMAGES Plumage changes are only slight and ageing older immatures must be considered rather difficult. Birds in the first three plumages are best aged by details of moult, while later age-classes should be seen as plumage types, rather than precisely aged plumages. Apart from details of moult, colour of bill, structure and colour of ruff and colour of iris are also important to note.

Juvenile

Plumage is uniform and fresh lacking signs of moult. Body is rather deep mustard-coloured, with variable light streaking, while wing- and tail feathers are uniformly blackish, creating a striking contrast between remiges and body-plumage. *Secondaries pointed, producing a serrated trailing edge to the arm.* Greater upperwing-coverts are brown and pointed, with narrow and poorly defined pale fringes. Head pale grey, bill and cere dark and iris dark. Ruff consists of brownish, long and narrow lanceolate feathers.

Second plumage

Active moult in the summer of 2nd cy includes only inner primaries, occasionally the odd secondary. Once moult is suspended in late autumn birds are still very similar to juveniles, except for 2–4 moulted inner primaries, which can be difficult to detect (the darker colour of the fresh inner primaries is best seen on the upperwing). The plumage is more worn and somewhat faded in colour compared to fresh juveniles. Some already start to show a pale tip to the bill, and iris is still dark.

Third plumage

By and large still resembles the juvenile, but signs of wing moult are apparent. The second moult continues in the 3rd cy spring from where the first moult was suspended the previous autumn, and actively moulting birds are now replacing their median primaries (p3/4–p6/7). Secondary moult is also extensive, with feathers moulted at several active points. From now on the new greater upperwing-coverts start to show the adult-type pattern, with dark centres and distinct broad pale fringes. When moult is suspended in late autumn two to four juvenile primaries still remain (usually p8–p10), and about half of the secondaries are still the faded brown and pointed juvenile feathers, with fresh ones being broader and more rounded at the tip. Head, ruff and iris as in juvenile, but bill can be largely pale.

Fourth plumage

Many birds of this age group can still be identified early in the year by their retained outermost juvenile primaries (p8–10), which are being replaced during the summer (4th cy). A small percentage of birds still retain their juvenile p10 even after this moult, and can thus still be identified in the 5th cy spring, until the last remaining juvenile feathers are shed early in the year. The bill is by now largely pale, while iris and ruff are still mostly similar to the juvenile's, although individual differences are great.

Subadult

Although the plumage is largely adult, birds still carry some traits of immaturity, but the combinations vary individually. Colour of bill and iris change slowly as the bird matures, as does the structure of the ruff. These features develop independently and some characters may look adult-like, while others still look immature.

Birds probably appear in a subadult plumage for 1–2 years before acquiring the final adult plumage.

Adult

Body plumage is usually paler and greyer than in younger birds, the head is white, the bill and iris are pale, and the ruff is white and fluffy (at times soiled with blood). Seen from below the remiges are clearly greyish with darker margins and the upperwing shows a band of distinctly marked rounded greater coverts with black centres and wide pale margins.

SEXING Eurasian Griffons cannot be sexed in the field.

CONFUSION RISKS Proportions in flight are very similar to other *Gyps* vultures, and even to Lappet-faced Vulture, with a broad, bulging arm, a narrower hand and a small tail. In a European context, told from other large vultures by striking upper- and underwing contrasts, but beware of clearly smaller African White-backed Vulture and the *erlangeri* subspecies of Rüppell's Griffon Vulture, which can both be very similar in this respect. Darker Eurasian Griffons occur, adults in particular, and great care should be taken not to misidentify them as Rüppell's Griffons, which frequently happens in Spain (for further details see under Rüppell's Griffon Vulture). In the easternmost part of its range the Eurasian Griffon, represented by ssp. *fulvescens*, is most likely to be confused with the larger Himalayan Griffon Vulture. Himalayan Griffons are not only larger and broader-winged, but also darker greyish-brown and distinctly streaked as juveniles and immatures, while adults have a very contrasting plumage with whitish wing-coverts and dark flight feathers.

243. Eurasian Griffon Vulture *Gyps fulvus*, juvenile. Uniformly black and pointed flight feathers and a rich mustard-coloured body characterise a juvenile, and the underwing-coverts are on average lighter and plainer compared to adults. Spain, 25.10.2009 (DF)

244. Eurasian Griffon Vulture *Gyps fulvus*, juvenile. Juveniles have a dark bill and the ruff is the same colour as the body, and is made from long, lanceolated feathers. Spain, 16.2.2011 (DF)

245. Eurasian Griffon Vulture *Gyps fulvus*, juvenile. In juveniles all flight feathers are uniform and of the same age, and the pattern of the greater coverts is rather indistinct and differs clearly from adult-type feathers. Note also the narrow and pointed shape of all coverts. Spain, 14.2.2011 (DF)

246. Eurasian Griffon Vulture *Gyps fulvus*, 2nd cy spring, migrant. By spring birds can be very worn but are still recognised by their largely juvenile flight feathers. Some new median upperwing-coverts and new growing inner primaries can be seen on this bird returning from Africa. Bill and eye are still predominantly dark. Spain, 4.5.2011 (DF)

247. Eurasian Griffon Vulture *Gyps fulvus*, 2nd cy autumn. Aged by three moulted inner primaries and p4 recently dropped, while rest of remiges are juvenile (note pointed secondaries). Spain, 15.9.2012 (DF)

248. Eurasian Griffon Vulture *Gyps fulvus*, 3rd cy immature, approaching two years of age. This bird has replaced its four inner primaries and 1–2 secondaries, but the remaining remiges are still juvenile feathers. The median and lesser upperwing-coverts are largely new, while half of the greater coverts are still juvenile (note difference in shape). Head is still juvenile-like with dark bill and eye, and brown ruff. Spain, 14.2.2011 (DF)

249. Eurasian Griffon Vulture *Gyps fulvus*, presumed 4th cy immature, approaching three years of age. Only 2–3 outer juvenile primaries and about half of the juvenile secondaries remain, by now faded and worn. Bill and iris are getting lighter but ruff is still brown and made of feathers. Spain, 17.2.2011 (DF)

250. Eurasian Griffon Vulture *Gyps fulvus*, presumed 4th cy immature, approaching three years of age. This bird shows a moult pattern comparable to **249**, with distinctive, worn and pointed juvenile secondaries. By now most of the greater coverts are of adult-type, while head characters are still largely juvenile, but the bill is becoming lighter. Spain, 14.2.2011 (DF)

251. Eurasian Griffon Vulture *Gyps fulvus*, adult. Adults are on average lighter and greyer, but body colour varies individually. This bird, superficially looking like a full adult, still has a ruff of feathers rather than the white fluff of older birds. Note the bicoloured flight feathers of adults, with lighter centres and darker edges and tips. Spain, 23.11.2010 (DF)

252. Eurasian Griffon Vulture *Gyps fulvus*, adult. The light horn-coloured bill, light iris and a fluffy white ruff are characters of an adult. Spain, 23.11.2010 (DF)

253. Eurasian Griffon Vulture *Gyps fulvus*, adult. Adults can be very pale milky coffee-coloured above. Note pale bill and iris, and fluffy ruff, all features of an adult, as are 'frosty' secondaries above. Spain, 9.1.2007 (DF)

254. Eurasian Griffon Vulture *Gyps fulvus*, adult. Note the distinctly marked greater coverts in adults. Spain, 18.4.2010 (DF)

163

255. Eurasian Griffon Vulture *Gyps fulvus*, juvenile ssp. *fulvescens*. Similar to juvenile nominate, but the underwing-coverts show a wide light area and the body is more clearly streaked; however, similar juvenile *fulvus* do occur. India, 17.12.2007 (DF)

256. Eurasian Griffon Vulture *Gyps fulvus*, immature ssp. *fulvescens* after second moult, presumed 3rd cy (approximately 2.5 years old). Aged like ssp. *fulvus* by number of retained juvenile remiges. Note wide pale area on underwing-coverts and distinctly streaked body, indicative of *fulvescens*. India, 17.12.2007 (DF)

257. Eurasian Griffon Vulture *Gyps fulvus*, subadult/young adult ssp. *fulvescens*. No signs left of juvenile plumage, but iris is still darkish, bill greyish and the ruff consists of feathers, indicating a younger age. Note the largely pale underwing-coverts and the streaky body, both typical of *fulvescens*. India, 9.12.2007 (DF)

258. Eurasian Griffon Vulture *Gyps fulvus*, adult ssp. *fulvescens*. Pale bill and iris indicate full adult. The largely pale underwing-coverts could lead to confusion with adult Himalayan Vulture, but such underwings would not be expected among adults of nominate *fulvus* in the west. India, 7.12.2007 (DF)

RÜPPELL'S GRIFFON VULTURE
Gyps rueppellii

Other names: Rüppell's Griffon, Rüppell's Vulture

VARIATION Two subspecies are recognised, with nominate *rueppellii* from W Africa to Sudan, Uganda, Kenya and Tanzania and ssp. *erlangeri* in Ethiopia, Eritrea and Somalia. The nominate adults of W Africa are very dark birds, even looking black from a distance, with rather fine whitish scaling and a gleaming white patagial band. Further to the east adult birds get gradually paler, as the black centre of each contour feather shrinks while the light margin gets wider. The palest and least scalloped birds are found in Ethiopia, Eritrea and Somalia (ssp. *erlangeri*) and some of the *extremes can be practically identical to Eurasian Griffon Vulture*. In East Africa Rüppell's Griffon has a distinct pale morph, which is fawn-coloured even in juvenile plumage and looks confusingly similar to Eurasian Griffon.

The subspecific affinity of the East African Rüppell's Griffons, away from Ethiopia, has been widely debated. Morphologically they are much closer to Ethiopian *erlangeri* than to the dark nominate birds of W Africa, and are therefore perhaps best included in *erlangeri*. However, since the occurrence of pale and almost uniform birds seems to be a largely Ethiopian phenomenon, the core area of *erlangeri* probably lies in the Ethiopian highlands, from where the genes have spread, becoming gradually diluted by nominate genes from the west. Genetic studies are needed to clarify whether *erlangeri* in itself is of hybrid origin, dating back to times when Eurasian Griffons were wintering in good numbers in NE Africa, as suggested in older literature (see e.g. Mundy *et al.* 1992). The temperate climate of the Ethiopian highlands corresponds to the climatic conditions of, for example, the C Asian mountains, from where many of the ancient migrants may have originated.

DISTRIBUTION Breeds in the dry northern parts of sub-Saharan Africa, reaching further south in East Africa. Birds entering SW Europe belong to the darker nominate subspecies, probably originating in W Africa where Rüppell's join troops of wintering Eurasian Griffons.

BEHAVIOUR In a European context Rüppell's Griffons are mostly found together with Eurasian Griffons, often mingling with the large flocks of non-breeding birds. The birds turn up in Spain in late spring, joining migrating parties of Eurasian Griffons on their return flight from wintering grounds in W Africa. They spend the summer, sometimes a full year, mostly in S Spain, but eventually they return back to Africa together with the Eurasian Griffons, usually during Oct–Nov. Most of the birds seen in Europe are about 12–24 months old, but some lingering birds may show a more advanced plumage, approaching the adult's. Birds in juvenile plumage are rare, but have been recorded.

SPECIES IDENTIFICATION Similar in shape to Eurasian Griffon, with which it normally associates in Europe, but slightly smaller in direct comparison. Plumage of immatures is a dull greyish-brown, from a distance rather dark and quite similar to adult Black Vulture, lacking the strong contrasts and the mustard colours typical of Eurasian Griffon. The single most diagnostic feature, apart from the uniformly dark and greyish-brown impression, is the gleaming white line along the forewing. In closer views white scalloping and anchor-marks first appear among the undertail-coverts and axillaries, later spreading to all body and wing-coverts.

MOULT Similar to other large vultures, taking approximately three moulting seasons to replace the entire set of juvenile flight feathers. Moult stage can thus be used for ageing immatures as long as juvenile feathers are retained. Owing to the stepwise moult, the adults are capable of replacing their flight feathers in a considerably shorter time-span, probably within two moulting seasons.

PLUMAGES Given good views at least four different plumage-types can be separated, but to identify the second and third plumages (sometimes even a fourth) good views and an understanding of the primary moult is required. Details of head, bill, iris colour and collar are also important. The descriptions refer to the darker, western form of Rüppell's Griffon, which is the one reaching SW Europe. The much paler and hence more Eurasian Griffon-like Rüppell's Griffons of East Africa, which have recently turned up in the Middle East, are described opposite under 'Note'.

Juvenile

Darker and duller greyish-brown compared to any Eurasian Griffon, but rather similar to African White-backed, with barely discernible contrast between browner coverts and dark remiges of upperwing. All body feathers and all upperwing- and underwing-coverts have a distinct and contrasting buffish shaft-streak, and the patagium shows one broad white line. In fresh plumage similar to only slightly smaller African White-backed Vulture, which see. Bill dark horn, cere grey, iris dark and feet grey.

Second plumage

This plumage and the next are the ones in which birds are normally seen in Europe. Most of the juvenile body plumage has been replaced once and the first white anchor-marks start to appear among the upperwing-coverts, underbody and, in particular, among the axillaries and undertail-coverts. The best ageing character is the stage of primary moult, with active moult in the inner 1–4 primaries and also the first few secondaries are now being replaced. Birds seen in Sep–Oct in S Spain have all been moulting actively.

Third plumage

Similar to second plumage, but the primary moult is active in p4–6 and up to a third of the secondaries have been replaced. The body plumage has turned rather dark by now with increasing white scalloping.

Adult

From a distance may be mistaken for a Eurasian Griffon, although always darker, duller and overall more uniform, lacking the strong upperwing contrast of Eurasian Griffon. Closer views will reveal the scaly appearance of the upperwing- and underwing-coverts, as well as scaly underbody and scapulars. The bill and iris are pale, as in adult Eurasian Griffon, while the white ruff is even more contrasting than in Eurasian Griffon.

SEXING As far as is known, Rüppell's Griffons cannot be sexed in the field.

CONFUSION RISKS In Spain Eurasian Griffons are frequently misidentified as Rüppell's by overly keen birders. Most Eurasians show a sharp contrast both above and below, between pale coverts and darker remiges, which is not seen in nominate Rüppell's. However, some Eurasians, both juveniles as well as adults, are darker and browner than average, and the lack of wing contrast is not enough to claim a Rüppell's.

NOTE Birds belonging to the subspecies *erlangeri* could turn up in Arabia, or elsewhere in the Middle East. In fact, during 2014 two separate birds were indentified in N Israel. These birds can be *confusingly similar to Eurasian Griffon* and would easily be overlooked as such. Ethiopian birds of the ssp. *erlangeri* are much paler than birds from further west in Africa – identical to many Eurasian Griffons in colour – and the scaling may be almost completely missing. In particular, some immatures, lacking the scaling completely, can be virtually identical to Eurasian Griffons of similar age, but early on they acquire the diagnostic anchor-shaped markings to the axillaries and undertail-coverts. Since Eurasian Griffons are becoming increasingly rare as migrants along the eastern flyway, any griffons encountered in Arabia or NE Africa should be checked with utmost care, and if possible photographed.

Hybridisation is another problem that has arisen from those Rüppell's that have stayed in Spain and Portugal and which, as adults, have taken up residence in Eurasian Griffon colonies. Birds with intermediate plumage characters are already known from the Iberian Peninsula, indicating hybridisation, and this could become an increasing problem in the future.

References

Forsman, D. 2005: Rüppell's Vultures in Spain. *Birding World* 18 (10): 435–438.

Mundy, P., Butchart, D., Ledger, J. and Piper, S. 1992: *The Vultures of Africa*. Academic Press.

259. Rüppell's Griffon Vulture *Gyps rueppellii*, immature second from left, with Eurasian Griffons, the most common situation in a European context. Note the uniform darkness compared to the bicoloured griffons. Spain, 27.10.2009 (DF)

260. Rüppell's Griffon Vulture *Gyps rueppellii*, juvenile. The rather dark greyish-brown ground colour with striking streaks makes a juvenile stand out in a European context, although birds in juvenile plumage are quite rare in Europe. Spain, 11.9.2005 (DF)

261. Rüppell's Griffon Vulture *Gyps rueppellii*, juvenile commencing its first moult, approximately one year old. Note the rather dark ground colour with distinct pale streaking and the bright white patagial mark on the forewing. Differs from similar juvenile African White-backed by lighter face and feet, and by clean white forewing band. Bill and eyes are dark in juveniles. Ethiopia, 3.1.2010 (DF)

262. Rüppell's Griffon Vulture *Gyps rueppellii*, juvenile starting its first moult, approximately one year old (same as **261**). Still in juvenile plumage save for some new darker upperwing-coverts. Very similar to juvenile African White-backed but note pale head, which is also longer and slimmer. Ethiopia, 3.1.2010 (DF)

263. Rüppell's Griffon Vulture *Gyps rueppellii*, immature, about two years old. The inner four primaries have been replaced and s1 is new. The body plumage shows diagnostic light anchor-shaped tips, and the bill is turning paler. Overall dark, with striking white patagial line. Spain, 10.9.2005 (DF)

169

264. Rüppell's Griffon Vulture *Gyps rueppellii*, immature, approaching two years of age. Note overall dark plumage with light anchor-markings and white forewing band. Spain, 12.9.2005 (DF)

265. Rüppell's Griffon Vulture *Gyps rueppellii*, immature, approaching two years of age (same as **264**). The dark upperside, with barely noticeable contrast between coverts and flight feathers, shows scattered diagnostic pale scaling, which will increase with age. Spain, 12.9.2005 (DF)

266. Rüppell's Griffon Vulture *Gyps rueppellii*, immature, approximately two years of age. Despite more advanced moult and plumage bleached from wear and sun, this bird does not differ markedly from the previous individual. Note the great asymmetry in the moult of the secondaries. Ethiopia, 8.2.2012 (DF)

267. Rüppell's Griffon Vulture *Gyps rueppellii*, adult. Note overall black impression of 'western' adults with striking patagial band and white scalloping. Cameroon, 24.4.2006 (Ralph Buij)

268. Rüppell's Griffon Vulture *Gyps rueppellii*, adult. 'Western' adult showing dark plumage with distinct white covert-tips. Cameroon, 4.4.2006 (Ralph Buij)

269. Rüppell's Griffon Vulture *Gyps rueppellii*, juvenile, commencing its first moult at approximately one year of age. These pale birds from East Africa look very different from the ordinary type of Rüppell's (cf. **261**). They have not been properly described before, and are therefore frequently misidentified as Eurasian Griffons. Note the few new darker feathers on the body and underwing-coverts with diagnostic anchor-shaped markings. Ethiopia, 5.2.2012 (DF)

171

270. Rüppell's Griffon Vulture *Gyps rueppellii*, immature/subadult. This bird still retains juvenile outer primaries and several secondaries. It looks almost identical to a Eurasian Griffon, but the markings of the axillaries are diagnostic. Ethiopia, 3.1.2010 (DF)

271. Rüppell's Griffon Vulture *Gyps rueppellii*, immature/subadult (same as **270**). Very similar to Eurasian Griffon and only told by a small number of scapulars with anchor-shaped markings. Ethiopia, 3.1.2010 (DF)

272. Rüppell's Griffon Vulture *Gyps rueppellii*, adult, breeding. Similar to Eurasian Griffon owing to uniform body, but identified by diagnostically marked axillaries and underwing- and undertail-coverts. Note the full, dark crop of this bird returning to its nest. Ethiopia, 3.11.2011 (DF)

273. Rüppell's Griffon Vulture *Gyps rueppellii*, adult. This moulting adult has lost most of its underwing-coverts and hence looks more like an adult African White-backed Vulture. Note, however, the long-winged appearance and the diagnostic axillaries and undertail-coverts, as well as the pale face and grey feet. Ethiopia, 8.2.2012 (DF)

274. Rüppell's Griffon Vulture *Gyps rueppellii*, adult. This is a more normal-looking East African Rüppell's Griffon Vulture, pale compared to West African birds, but with distinct scaling all over. Note pale bill and light iris typical of adults. Ethiopia, 16.1.2010 (DF)

275. Rüppell's Griffon Vulture *Gyps rueppellii*, adult, breeding. A far less scalloped individual, but dark feather-bases are visible through the wing-coverts. However, from a distance this bird would look like a perfect Eurasian Griffon. Ethiopia, 3.11.2011 (DF)

AFRICAN WHITE-BACKED VULTURE
Gyps africanus

VARIATION Monotypic.

DISTRIBUTION Widely distributed across sub-Saharan Africa, only avoiding continuous forest. Immatures recently recorded as rare vagrants to Spain and Morocco, possibly joining Eurasian Griffons (and Rüppell's Griffons) returning from their wintering grounds; may have been overlooked in the past.

BEHAVIOUR Spends a lot of time on the wing when foraging. Roosts communally on tops of trees, often together with other vulture species.

SPECIES IDENTIFICATION Adults are rather unmistakable, but juveniles and immatures are very similar to only marginally larger Rüppell's Griffons in corresponding plumage. In direct comparison told from Rüppell's as well as Eurasian Griffon by clearly smaller size, but when size is not obvious blackish face and bill and dark legs remain diagnostic. Silhouette is slightly shorter-winged than in either Rüppell's or Eurasian Griffon, with comparatively broader arm and narrower hand, thus producing a more pronounced S-curve to the trailing edge of the wing. The diagnostic white lower back of older birds can be difficult to see as it is often covered by the large scapulars, even in flight, and in any case it is not developed until an age of approximately three years. The white back is best seen when birds are actively flapping or flying away, while soaring birds with wings held in a dihedral may sometimes appear to lack it completely.

MOULT Starts about 10 months after leaving the nest and probably progresses as in other large vultures, but details are not well known. The statement by Houston (1975) that the moult is a continuous process needs confirmation, as immatures and adults with suspended moult do occur. By the time the last juvenile primaries are shed, the birds are still in a juvenile-like, streaky brown immature plumage.

PLUMAGES Adults and juveniles are easy to separate by plumage. Later immature plumages are superficially juvenile-like, but signs of moult as well as bicoloured remiges with rounded tips reveal the birds as non-juveniles. The plumage development from juvenile to full adult probably takes some 5–6 years, but studies with marked known-age birds are needed to confirm this.

Juvenile

Overall rather dark, greyish-brown, with distinct buffish streaks to every body feather. Remiges and tail very dark, both above and below, contrasting slightly against the browner coverts, with the trailing edge of the wing sharply serrated. The patagium shows a downy white line, similar to juvenile Rüppell's Griffon, and this is the only obvious plumage feature. The neck and crown are covered in pale down, but the eye and bill stand out as being truly dark, almost blackish.

Older immatures

These can be recognised from juveniles by moult in the remiges, with new feathers clearly bicoloured from below, grey with dark margins, and showing a rounded not pointed tip (in immature Rüppell's Griffon the remiges remain dark below until the birds reach adult plumage). The light patagial band varies in length and width, depending on the stage of moult, and additional light areas sometimes appear in the underwing-coverts, only to disappear again when the feathers are regrown. Birds carry this type of streaked immature plumage possibly for two years, and towards the end of this period the white back starts to appear.

Adult

Best recognised by partially or entirely white underwing-coverts (save for a narrow dark leading edge), which contrast strongly with the darker remiges and body. Adults are, however, highly variable, and it probably takes several more years before the plumage reaches the final adult stage. Younger adults show a brownish, heavily streaked body, and while the underwing-coverts are already largely whitish the upperwing-coverts are still dull greyish-brown. The final adult plumage is very smart, with a uniform, pale sandy or even milky coffee-coloured body plumage contrasting with dark flight feathers and shining white wing-linings. The upperparts show a strong contrast between pale wing-coverts and dark flight feathers and large scapulars, while the back is sparkling white, but often hidden by the scapulars. The head, neck and feet stand out as very dark and seen from below the flight feathers are silvery-grey with distinct dark margins.

SEXING is not possible in the field.

CONFUSION RISKS Juveniles and immatures are very difficult to tell from juvenile Rüppell's Griffon, which shares much the same plumage features. However, during their first year of life juvenile White-backed develop a dark face in addition to the black bill and cere, and the flight silhouette is slightly more compact, with proportionately shorter and broader wings, just enough to make a slight difference to the experienced eye.

References

Houston, D. C. 1975: The moult of the White-backed and Rüppell's Griffon Vultures *Gyps africanus* and *G. rueppellii. Ibis* 117: 474–488.

276. African White-backed Vulture *Gyps africanus*, juvenile. This bird, approximately one year old, is still in a completely juvenile plumage, but has just started its first moult by dropping its innermost primaries (p1) in both wings. Note uniform condition of flight feathers, with pointed tips, and the regular streaking of the body plumage. Ethiopia, 16.1.2010 (DF)

277. African White-backed Vulture *Gyps africanus*, juvenile (same as **276**). Similar in plumage to a juvenile Rüppell's Griffon Vulture, the young African White-backed is smaller and shorter-winged, and its face and feet are darker. Ethiopia, 16.1.2010 (DF)

278. African White-backed Vulture *Gyps africanus*, juvenile (same as **276**). Upperparts appear uniform with a marked contrast between lighter coverts and darker flight feathers and tail. A few new darker feathers with white shaft-streaks have emerged on the mantle and upperwings. Ethiopia, 16.1.2010 (DF)

279. African White-backed Vulture *Gyps africanus*, young immature. Slightly more advanced in its moult compared to **278**, but still in mostly juvenile plumage. Note dark face and fairly large white patches on underwing-coverts. Ethiopia, 8.2.2012 (DF)

280. African White-backed Vulture *Gyps africanus*, young immature. Similar stage of moult to **279**, but ongoing moult of underwing-coverts is revealing large white areas. Note dark face and compare with immature Rüppell's Griffon. Ethiopia, 6.3.2006 (DF)

281. African White-backed Vulture *Gyps africanus*, immature, probably close to two years of age. About half of the flight feathers are moulted, with new secondaries showing typically lighter centres and dark edges. Body and underwing-coverts are by now largely moulted but appearance is still streaked and juvenile-like, while the dark face is becoming more obvious. Ethiopia, 6.3.2006 (DF)

282. African White-backed Vulture *Gyps africanus*, immature plumage. Upperparts rather uniformly brown on this bird, at a similar moult stage to 281, with no sign of a white back yet. Ethiopia, 16.1.2010 (DF)

283. African White-backed Vulture *Gyps africanus*, subadult. Body plumage still all-streaked but juvenile flight feathers have all been replaced (perhaps save for right p10). Told from similar immature Rüppell's Griffon by bicoloured remiges and black face. Namibia, 7.4.2006 (DF)

284. African White-backed Vulture *Gyps africanus*, young adult. Body still dark and partly streaked and upperwing-coverts dull brown, but underwing patch and white back fully developed. Ethiopia, 1.3.2006 (DF)

285. African White-backed Vulture *Gyps africanus*, adult. Body uniformly light brown, lacking streaking of previous plumages, and largely white underwing-coverts contrast with black face and feet and dark flight feathers. Ethiopia, 13.2.2012 (DF)

286. African White-backed Vulture *Gyps africanus*, old adult. Old birds become almost entirely greyish-white with no discernible contrast between body and underwing-coverts. Namibia, 4.4.2006 (DF)

287. African White-backed Vulture *Gyps africanus*, old adult. The upperwing-coverts also become very pale with increasing age, but note diagnostic black face and bill. Ethiopia, 12.1.2010 (DF)

178

HIMALAYAN GRIFFON VULTURE
Gyps himalayensis

Other names: Himalayan Griffon

VARIATION Monotypic.

DISTRIBUTION Breeds in the high mountains of interior Asia, but immatures in particular roam widely, reaching further south on the Indian subcontinent during post-breeding dispersal. The Himalayan Griffon Vulture is one of the latest additions to the region's avifauna, with an immature recorded in Dubai, United Arab Emirates, on 13 October 2012, during an influx of Eurasian Griffons.

BEHAVIOUR Similar to other large vultures.

SPECIES IDENTIFICATION Clearly a *Gyps* vulture, with long and rectangular wings, a short tail and a small head. Differs from other *Gyps* species by larger size (although overlaps with Eurasian Griffon) and *broader and more rectangular wings* with very long fingered primaries and a broader and more square wingtip, with up to *eight fingered primaries*, compared to seven in the smaller species. The 8th finger (p3) is shorter, yet pointed, in juveniles, but longer and often clearly fingered in adults. Owing to the huge wings and body size the *head appears notably small in flight* by comparison. Adults are unmistakable with very pale body plumage and practically white underwing-coverts (not that dissimilar to adult African White-backed Vulture), while immatures are more similar to other immature *Gyps* vultures, but clearly darker and duller brown than Eurasian Griffon, with contrasting broad white streaking to the entire body plumage.

MOULT Slow and prolonged, as in other large vultures, and the state of moult can be used for ageing, following the principles explained under Eurasian Griffon.

PLUMAGES Immatures and adults are easily told apart, but ageing immatures requires a closer study of the primary moult. Adult plumage is acquired in about 6–7 years and is preceded by up to four streaked immature plumages, followed by another 1–2 subadult stages.

Juvenile

Easily identified as a juvenile by the intact plumage lacking signs of moult to flight feathers. Plumage greyish-brown with prominent white streaks to

practically every contour feather and a short but distinct light band in the patagium and another less well-defined but wider light band in the median underwing-coverts. The remiges appear uniformly dark from below, with spiky tips to the secondaries and tail feathers. Unlike in Eurasian Griffon the bill turns pale horn at an earlier age, from the first autumn. Iris is dark.

Older immatures

Similar to juvenile, but the white streaking is even brighter and bolder, with individual markings more drop-shaped. Best told from juvenile by irregularities in plumage owing to ongoing moult of body and flight feathers. Immatures can be roughly aged based on the number of moulted primaries and secondaries, similar to Eurasian Griffon.

Adult

Unmistakable owing to huge size and contrasting plumage, with almost whitish body plumage contrasting with dark flight feathers and tail. Underwing-coverts are white, save for a narrow stained carpal-patch along the leading edge of the wing. The upperwing shows a marked contrast between nearly white coverts and dark remiges, while the darker greater coverts show poorly defined lighter tips only, a pattern very different from the distinctly marked greater coverts of adult Eurasian Griffons. Younger adults and subadults are less clean-looking compared to old birds, with a sandy-brown and diffusely streaked body and a smaller white area on the underwing-coverts.

SEXING It is not possible to sex birds in the field.

CONFUSION RISKS Easily confused with other *Gyps* vultures, particularly Eurasian Griffon, with its eastern subspecies *fulvescens* being partially sympatric with Himalayan. Adult Himalayan told by striking plumage, while juveniles and immatures are more similar to immatures of other species. Often joins Eurasian Griffon, both species venturing far south of their breeding ranges in winter. In direct comparison immature Himalayans are told from Eurasian Griffons by their larger size, broader and more rectangular wings and their duller and darker brown plumage with distinct white streaking.

288. Himalayan Griffon Vulture *Gyps himalayensis*, one-year-old immature gliding together with juvenile Eurasian Griffon. Himalayan Griffons are massive, broad-winged birds dwarfing even birds, as large as Eurasian Griffons. India, 17.12.2007 (DF)

289. Himalayan Griffon Vulture *Gyps himalayensis*, juvenile with juvenile Eurasian Griffon. Note the broader wings and the heavier body of Himalayan Griffon. India, 13.12.2007 (DF)

290. Himalayan Griffon Vulture *Gyps himalayensis*, juvenile. From this angle the wings look more rectangular than they actually are. India, 17.12.2007 (DF)

291. Himalayan Griffon Vulture *Gyps himalayensis*, juvenile (same as **290**). Juveniles are greyish-brown and boldly streaked with a narrow white patagial band and a wider pale area across the rear wing-coverts. India, 17.12.2007 (DF)

292. Himalayan Griffon Vulture *Gyps himalayensis*, juvenile. The size of the head is about the same as in Eurasian Griffon, but because the Himalayan Griffon is overall clearly larger, it looks conspicuously small-headed for its size. India, 17.12.2007 (DF)

293. Himalayan Griffon Vulture *Gyps himalayensis*, juvenile. Note conspicuous streaking on the upperside compared to any other large vulture of the region. India, 12.12.2007 (DF)

294. Himalayan Griffon Vulture *Gyps himalayensis*, immature, about 18 months of age. Similar to juvenile, but inner two primaries are new and the plumage is faded. India, 17.12.2007 (DF)

295. Himalayan Griffon Vulture *Gyps himalayensis*, adult. Note very broad wings and tiny head compared to, for example, Eurasian Griffon. Nepal, 12.11.2013 (Martti Siponen)

296. Himalayan Griffon Vulture *Gyps himalayensis*, adult. Adults have white underwing-coverts, whereas the body may be streaked or plain, depending on age, much as in African White-backed Vulture. Nepal, 11.11.2013 (Martti Siponen)

297. Himalayan Griffon Vulture *Gyps himalayensis*, adult. Adults are very pale above, but some old Eurasian Griffon Vultures may approach them in paleness. Note diagnostic white back and uppertail-coverts and differently marked greater coverts compared to adult Eurasian Griffon. Nepal, 12.11.2013 (Martti Siponen)

BEARDED VULTURE
Gypaetus barbatus

Other names: Lammergeier

VARIATION Two subspecies, with nominate *barbatus* in Europe, NW Africa and Asia from Asia Minor eastwards, and African ssp. *meridionalis* in S Egypt, Sudan, Yemen, Horn of Africa and patchily further south through E and S Africa. African *meridionalis* differs from nominate in showing partly bare, unfeathered tarsi, which can be difficult to see in flight. It also lacks both the partial breast-band and the dark ear-speck of nominate *barbatus*, and the crown and cheeks are whiter and cleaner-looking, lacking the dark streaking of *barbatus*. More information is needed on the morphology of the population around the Red Sea.

DISTRIBUTION Confined to high mountain ranges, where it prefers alpine meadows at altitudes above 2000 m. The population is increasing slowly in the Pyrenees and introduction schemes in the Alps have been successful. Some pairs still remain on Corsica and Crete. Widely distributed through the mountains of Asia, from the Middle East and Central Asia to China and Mongolia, and south to the Himalayas. Discontinuous range in E and S Africa with an isolated population in NW Africa.

BEHAVIOUR Mostly seen singly or at the most a few together. Spends most of the time on the wing, patrolling mountain slopes in slow gliding flight. Circles at times to considerable heights. Joins other vultures at carcasses, but keeps to the periphery until the worst of the feeding frenzy is over.

SPECIES IDENTIFICATION Adults are truly unmistakable owing to unique silhouette, with long tapering wings and long wedge-shaped tail. However, younger immatures are more compact in their outline compared to adults, showing clearly broader and more rounded wings and a comparatively shorter tail. Immatures could be mistaken for immature Egyptian Vulture, but enormous size, revealed by majestic slow movements, and unique shape soon become apparent. In active flight, somewhat like huge Montagu's Harrier, with deep but softly sweeping wing-beats, almost as if in slow motion. Glides on wings characteristically pressed down, with clearly lowered arm but flat hand.

MOULT of the remiges is a lengthy process, normally taking four years from fledging before the last juvenile flight feathers have been replaced. Thanks to the stepwise moult, which develops with age, adults are probably capable of replacing their entire plumage in two moulting seasons.

PLUMAGES Some six different plumage types can be separated. Although the first three plumages appear very similar from a distance, birds can be reliably aged by details of primary moult. Later immature stages can be identified by changing proportions of light and dark feathers in their body plumages. Because of great individual variation in plumage development later immature stages are best referred to as plumage types.

Juvenile

No signs of moult. Inner primaries and all secondaries are uniformly dark in colour and pointed. Head and neck black, rest of upperparts dark greyish-brown, with variable amount of white speckling on upperwing-coverts and mantle, and with a marked contrast between lighter coverts and dark flight feathers and tail. From a distance the underparts appear greyish with contrasting black head, tail and remiges, but on closer views the underbody is variable and irregularly speckled with light grey and some rusty-orange, often creating an impression of ongoing body moult. Tail feathers are dark with individually variable white wedges at the tips, and the underwing-coverts are brownish-grey, with the lighter grey greater coverts forming a visible wing-band.

Silhouette rather eagle-like, owing to broad and bulging arm (with long spiky secondaries), making the tail look comparatively short and broad. Iris is dark to begin with, but may turn light during the first winter.

Second plumage

Very similar to juvenile, both in silhouette as well as in plumage, best identified by moulted black inner primaries with rounded tips. The new primaries are darker than the remaining juvenile primaries, and the contrast is best seen on the upperwing. Secondaries are still juvenile, or at most one or two have been moulted (most commonly s1 and s5); the tail is also juvenile in most birds, sometimes with new central or outer feathers, not markedly longer than the juvenile feathers. Underwing-coverts are mostly retained juvenile feathers. The iris colour has changed to whitish.

Third plumage

Superficially still resembles the juvenile, although individual variation is extensive, but the wings now appear narrower and more adult-like. Typically, birds *retain a group of long spiky juvenile secondaries in the mid-wing, by now faded brown in colour.* Primary moult has reached p7–p8, but the outer 2–3 primaries are still juvenile, worn and faded brown. White feathers start to appear around the face, but the neck-boa is still black. Seen from a distance, the impression may be contradictory, with long tail and narrow wings creating the shape of an adult, but with the plumage colours of a juvenile, although rusty yellowish-brown feathers may have appeared on the breast and belly in numbers.

Fourth plumage

The silhouette is by now that of an adult, with narrow wings and a long, slim tail. The head is largely white, but still retains a dark neck-band. Underbody is a mottled mix of '*café au lait*' and tawny-yellow. One or two juvenile secondaries, by now extremely worn and faded, may still remain, sticking out beyond the trailing edge, but all the juvenile primaries will have been replaced at least once.

Fifth plumage

Much like the adult, but the underwing-coverts are still darkish grey-brown, the underwing thus lacking the marked contrast of the adults. Most birds still show some dark feathers on the underbody, while some of the secondaries are still all-dark, contrasting with the silvery secondaries of adults.

Adult

The entire body is pale below, some being pure white, but most with a more or less intense rusty-yellow stain. The lesser and median underwing-coverts are blackish, with fine white shaft-streaks, contrasting with greyer greater coverts and remiges. Remiges are glossy grey below, with distinct dark tips. Upperparts appear an extraordinary silvery-grey from a distance, changing in shade depending on the angle of light, while the mantle appears dark, contrasting strongly with the white head.

SEXING Adult males appear slim and long-tailed compared to adult females, owing to narrower wings, which explains the proportionally longer tail. In males, the bill is smaller and the forehead steeper, giving them a more rounded head profile compared to females, which have longer and heavier bills and a more sloping forehead, but this feature requires some experience and can be difficult to assess on flying birds.

CONFUSION RISKS Adults are unmistakable thanks to the distinct shape and plumage, but younger immatures in particular can be mistaken for some other vultures or large eagles, despite the unique silhouette.

298. Bearded Vulture *Gypaetus barbatus*, adult (above) and immature. Spain, 23.11.2010 (DF)

299. Bearded Vulture *Gypaetus barbatus*, fresh juvenile. Note broad wings and comparatively shorter tail of juvenile compared to adults and older immatures. Rather uniformly grey, with distinct black hood. Spain, 23.11.2010 (DF)

300. Bearded Vulture *Gypaetus barbatus*, fresh juvenile. Note how all remiges are sharply pointed and of similar age. Spain, 23.11.2010 (DF)

301. Bearded Vulture *Gypaetus barbatus*, 2nd cy, second plumage. Still in largely juvenile plumage, but notice inner four freshly moulted primaries. Spain, 23.11.2010 (DF)

302. Bearded Vulture *Gypaetus barbatus*, 2nd cy, second plumage. Spain, 22.11.2010 (DF)

303. Bearded Vulture *Gypaetus barbatus*, 2nd cy, second plumage. Rather uniform and juvenile-like above but moulted inner four primaries stand out as clearly darker. Spain, 22.11.2010 (DF)

304. Bearded Vulture *Gypaetus barbatus*, 3rd cy, third plumage, approximately 2.5 years old. Differs from previous plumage by more advanced moult, with just a few retained outer juvenile primaries, while up to half of the juvenile secondaries are still left, standing out as longer, pointed and faded brown compared to the freshly moulted feathers. Plumage colour still juvenile-like, but new tail feathers are longer, giving the bird a more adult shape. Spain, 23.11.2010 (DF)

305. Bearded Vulture *Gypaetus barbatus*, 3rd cy, third plumage. A more slowly moulting bird with just a few moulted secondaries and four retained juvenile primaries. Spain, 23.11.2010 (DF)

306. Bearded Vulture *Gypaetus barbatus*, 3ʳᵈ cy, third plumage. A more advanced bird with retained juvenile secondaries showing clearly. Spain, 23.11.2010 (DF)

307. Bearded Vulture *Gypaetus barbatus*, 3ʳᵈ cy, third plumage. Upperwing-coverts are becoming more variegated and white feathers appear around the face. Retained juvenile flight feathers stand out as very faded brown by now. Spain, 22.11.2010 (DF)

308. Bearded Vulture *Gypaetus barbatus*, fourth plumage, approximately 3.5 years old. A few juvenile remiges still remain and the inner primaries have been moulted for a second time. Spain, 23.11.2010 (DF)

309. Bearded Vulture *Gypaetus barbatus*, subadult plumage, approximately 4.5 years old. The odd juvenile secondary remains and the first dark underwing-coverts and axillaries of adult plumage start to appear. Head and underbody is turning whiter. Spain, 23.11.2010 (DF)

310. Bearded Vulture *Gypaetus barbatus*, adult. A striking plumage in black and white, stained below in rusty-yellow and orange. The active flight is reminiscent of a massive harrier in slow motion, while the spectacular proportions render it unmistakable. Note dark necklace and the small dark streak behind the eye, both features lacking in the African ssp. *meridionalis*. Spain, 23.11.2010 (DF)

311. Bearded Vulture *Gypaetus barbatus*, adult. Reflection from the snow is lightening the underside, exposing the subtle shades and details of the underwing. In less dramatic light the underwings look simply dark and the body white. Spain, 23.11.2010 (DF)

312. Bearded Vulture *Gypaetus barbatus*, adult, presumed female. Spain, 18.11.2004 (DF)

313. Bearded Vulture *Gypaetus barbatus*, adult, presumed female. Depending on the light the upperparts may appear dark or they may shine as silver. Spain, 18.11.2004 (DF)

314. Bearded Vulture *Gypaetus barbatus*, adult of African ssp. *meridionalis*. African birds appear darker on the underwing while the head is a cleaner white. They lack the dark necklace and the ear-covert streak and their legs are less feathered compared to adults of the nominate subspecies. Ethiopia, 3.11.2011 (DF)

315. Bearded Vulture *Gypaetus barbatus*, adult of African ssp. *meridionalis*. Although African birds look darker above, their feathers are reflective and may suddenly flash silver. Ethiopia, 3.11.2011 (DF)

BLACK VULTURE
Aegypius monachus

Other names: Cinereous Vulture, Monk Vulture

VARIATION Monotypic, but slight increase in size from west to east.

DISTRIBUTION In Europe healthy populations are found in SW and C Spain, with some remaining on Mallorca. Local, isolated populations are also found in the Balkans (E Greece) and Crimea. Extensive Asian distribution ranges from Turkey and the Caucasus to E China. Stragglers reach the Middle East, Africa, the Indian subcontinent and SE Asia in winter.

BEHAVIOUR Seen mostly on the wing, often joining other vultures. At carcasses dominates other vultures. In Europe breeds in trees and is thus less dependent on cliffs than, for example, Eurasian Griffon.

SPECIES IDENTIFICATION Huge dark raptor, with massive rectangular wings with parallel edges (cf. *Gyps* vultures and Lappet-faced) and fairly noticeable, wedge-shaped tail. The head looks small in relation to the impressive flight silhouette. Easy to tell from other vultures by uniformly dark underparts, but depending on light often shows a contrast between darker forewing and greyer flight feathers. The pale feet are usually obvious against the dark plumage. Upperparts equally dark, with only slight contrast between browner coverts and darker remiges, the contrast more sharply defined in juveniles. Trailing edge of wing notably serrated. In head-on views, wing position diagnostic when gliding, with arms level and hands drooping, with a notable angle at the carpal.

MOULT Adults moult their flight feathers between Jan and Oct/Nov, while 2nd cy birds moulting for the first time do not start until May–Jun. Only some of the flight feathers are moulted annually, and the last juvenile remiges are not replaced until the bird is more than 3 (4) years old. The number and location of retained juvenile remiges is the only reliable way of ageing immatures. Thanks to the stepwise moult the adults are capable of renewing their plumage at a much faster rate than the immatures, moulting their entire plumage probably in two moulting seasons.

PLUMAGES Given good views at least four, sometimes five, age-classes can be separated. Juveniles and adult-type birds can be identified by plumage characters, while the second and third plumages (and sometimes a fourth plumage) are identified by details of primary moult.

Juvenile

This is the darkest of all the plumages and fresh birds appear almost black. On the upperwing a slight but sharp contrast can be seen between the black remiges and slightly browner coverts. From below the forewing is black, while the glossier remiges and greater coverts are somewhat greyer, the latter standing out as an even paler wing-band. The small-looking head is covered in dense black down, while the bare areas, including the feet, are pale pink or violet-pink and the bill is black. The trailing edge of the wing is deeply serrated owing to sharply pointed secondary tips.

Second plumage

Very similar to juvenile in every respect, but told by moulted, black innermost 2–3 primaries, which can be difficult to discern. The rest of the plumage is only slightly browner (more faded) than in juveniles, including the down on the nape, which has become browner (and is no longer black).

Third plumage

This is still very similar to previous plumages, but identified on close views by more advanced moult of body plumage and flight feathers. The primary moult has now reached the median fingered primaries (p6–7) and a considerable part of the secondaries (30–50%) have also been moulted. The new secondaries show more rounded tips than the retained, by now rather worn, juvenile secondaries. The upperparts are becoming more uniformly dark brown, approaching the colour of the adults, and the head starts to show a hint of a paler supercilium and a paler crown. The dark bill is turning paler at the base in some.

Fourth plumage

This is still possible to recognise, as long as the bird is moulting. The primary moult has now reached the outer primaries, while the last juvenile secondaries are also about to disappear. Even after this moult some individuals can still be recognised by the odd extremely worn, retained juvenile remex, which should be looked for among the outermost primaries and the median secondaries. The plumage has become overall brown and the head turns increasingly pale, with the pale nape now extending down the neck to meet the dark throat, leaving only a dark face-mask around the eyes, while the pale area of the inner bill expands. However, changes in plumage, as well as colours of soft parts, vary individually, as does progression of moult, and therefore the above should only be seen as average traits subject to individual variation.

Subadult

After the last juvenile remiges are shed the key to ageing is lost. The subsequent subadult plumages are characterised by a darker head and a darker bill compared to the definitive adult plumage, but in most other respects they are similar to adults. The hitherto pinkish bare skin of the head turns pale bluish and the feet become deeper pink in colour, while the plumage of the underbody becomes more streaked.

Adult

Overall more uniformly brown than younger immatures, both above and below, but the easiest way to tell full adults from younger birds is by looking at the head. The head is largely pale grey, with darker markings confined to the area around the eye, gape and throat. Also the bill becomes paler with age, starting from the base of the upper mandible at 2–3 years and extending outwards from there, leaving only a dark culmen in the oldest birds.

SEXING is not possible in the field.

CONFUSION RISKS Most likely to be mistaken for an all-dark *Aquila* eagle, such as adults of either Steppe or Greater Spotted Eagle, owing to uniformly dark plumage and eagle-like silhouette with parallel-edged wings, but Black Vulture is usually revealed by its huge size and slow-motion movements in flight. Both Lappet-faced Vulture and Rüppell's Griffon, which are also generally rather dark, can be told from Black by their different silhouette, with bulging arm, a clearly narrower hand than arm and a comparatively shorter tail, apart from apparent differences in plumage.

316. Black Vulture *Aegypius monachus,* immature (left) with Eurasian Griffon *Gyps fulvus*. Note the overall dark impression and the more rectangular wings of the Black Vulture. India, 13.12.2007 (DF)

317. Black Vulture *Aegypius monachus*, immature. Note the typical drooping wing-posture of gliding birds. India, 17.12.2007 (DF)

318. Black Vulture *Aegypius monachus*, juvenile (1st cy), wintering. Juveniles show a marked underwing contrast between dark coverts and lighter remiges. Note the uniform condition of the remiges and the black head, both typical features of fresh juveniles. India, 17.12.2007 (DF)

319. Black Vulture *Aegypius monachus*, juvenile (1ˢᵗ cy), wintering. The juvenile plumage shows a rather marked contrast between browner coverts/mantle and blackish flight feathers. India, 17.12.2007 (DF)

320. Black Vulture *Aegypius monachus*, second plumage (3ʳᵈ cy), after the first moult. The second plumage is still rather juvenile-like, including dark head, although more faded brown in colour and lacking the marked upperwing contrast of the juvenile. The freshly moulted inner primaries, black in colour, are the single most reliable ageing feature and are best seen on the upperwing. Spain, 17.2.2011 (DF)

321. Black Vulture *Aegypius monachus*, third plumage (3ʳᵈ cy) after second moult, wintering. By and large approaching adult plumage, but outer primaries and most secondaries are still juvenile feathers. India, 17.12.2007 (DF)

322. Black Vulture *Aegypius monachus*, third plumage (3rd cy) after second moult, wintering. This bird, hatched in 2005, can be aged by its retained juvenile pp7–10. Pp4–6 were moulted in summer of 2007, pp1–3 in 2006. India, 17.12.2007 (DF)

323. Black Vulture *Aegypius monachus*, third plumage (4th cy), commencing its third moult. Outer primaries and nearly half of the secondaries are still juvenile feathers, which could possibly be shed in this moult. Mongolia, 5.6.2012 (DF)

324. Black Vulture *Aegypius monachus*, adult, wintering. No juvenile flight feathers left. Note advanced stepwise moult, with moult-fronts in primaries and secondaries running just a few feathers apart, and streaked underbody, typical of adults. India, 17.12.2007 (DF)

325. Black Vulture *Aegypius monachus*, adult. Note streaked underbody of this adult, just about to complete its annual moult. Spain, 29.10.2009 (DF)

326. Black Vulture *Aegypius monachus*, adult. Heavily moulting birds with exposed downy tracts may resemble Lappet-faced Vulture, but note dark throat and thighs of Black Vulture. Mongolia, 10.6.2012 (DF)

327. Black Vulture *Aegypius monachus,* adult. Adult-type birds are rather uniform and brown above, with coverts barely lighter than flight feathers. Spain 14.2.2011 (DF)

LAPPET-FACED VULTURE
Torgos tracheliotos

VARIATION Two, sometimes three subspecies recognised in Africa and the Middle East, with ssp. *negevensis* breeding in Arabia, nominate *tracheliotos* over most of Africa and intermediate ssp. *nubicus* (when recognised) in NE Africa (Egypt, Sudan, Ethiopia). Status of latter disputed, and often seen as an intermediate in a cline between *tracheliotos* and *negevensis*. Nominate *tracheliotos* is the darkest form; adults are practically black, with a strongly contrasting white patagial underwing-band and white thighs and flanks, more deeply red head, the most prominent lappets and the brightest bill colour. Ssp. *negevensis* is much browner and more uniform; it lacks a prominent underwing-band, the head is pale, appearing colourless and the lappets are less conspicuous.

DISTRIBUTION In our region, confined to hot and dry areas of the Arabian peninsula and S Egypt. Breeds in acacia-strewn desert and semi-desert, but ranges far on foraging flights. Vagrants are likely to be non-breeding immatures.

BEHAVIOUR Mostly seen in flight, covering huge areas, often at considerable height. Where common, congregates at carcasses and animal markets, where it is much the dominant vulture species.

SPECIES IDENTIFICATION (ssp. *negevensis*) Easy to identify as a vulture by huge wings, with comparatively large, pale head and short tail. Wings not as rectangular as in Black Vulture, shaped more like a griffon's, with clearly bulging secondaries and narrower hand, and with the *diagnostic pale flanks and abdomen of adults visible from afar.* Tail shorter than in Black Vulture, wedged at first when fresh, but later becoming more rounded when worn. The pale feet do not stand out as in Black Vulture, because of the light belly and vent. Contrast of upperwings rather obvious, with coverts lighter brown than the remiges. The massive head and bill, the biggest of all our vultures, are pale and conspicuous and stand out as an additional identification feature.

MOULT First moult starts at an age of about 12 months, during which only the inner 3–4 primaries are replaced. The next moult, one year later, comprises the median primaries and some secondaries. During the third moult, at the age of approximately three years, the primary moult reaches the outer juvenile primaries, but many individuals still keep one or two outer feathers until the fourth moult. The same applies also to juvenile secondaries, which are largely replaced during the third moult, but most birds retain some until the fourth moult. Adults are capable of replacing most of their flight feathers in two successive moulting seasons.

PLUMAGES (ssp. *negevensis*) Plumage differences are only slight between various age-classes, and details of progression of moult are important for ageing. The first three age-classes are safely aged by the moult pattern of the primaries. Given good views juveniles are also easy to identify by plumage features.

Juvenile

Juveniles have a uniform plumage with no signs of moult, and the uniformly dark secondaries are very pointed, with almost spiky tips. The woolly thighs are dark brown and often stand out as the darkest part of the underbody. The patagial band is greyish-brown, woolly and poorly defined. Head is pale grey, with prominent dark eye, and bill is uniformly horn-grey, somewhat paler along the cutting edge. The upperwing-coverts are clearly browner than the darker flight feathers and the contrast is quite sharp compared to older birds.

Second plumage

Much like the juvenile, but inner 3–4 primaries are freshly moulted and stand out as darker than the rest. Rest of plumage more faded and worn, compared to fresh juveniles.

Third plumage

Largely similar to previous plumage, but primary moult has now reached the median primaries, making the outer juvenile primaries stand out as a group of very faded brown, very worn and pointed feathers contrasting with the fresh and dark median primaries. Also the secondaries have been extensively replaced, a major difference compared to the previous plumage. The plumage has become overall more faded and the thighs are no longer as distinctly dark as in juveniles, the lanceolate feathers of the breast appear more contrasting and the diagnostic light flank-patch starts to appear.

Fourth plumage

Some birds (probably not all) can still be aged by one or two retained outermost juvenile primaries, which by now are very worn and faded.

Adult

No juvenile remiges remain, and all secondaries have narrow but rounded, not spiky tips. The rear underbody and flanks, including thighs, are contrastingly pale and fluffy, while the breast is whitish but prominently streaked with long lanceolated feathers. The adult-type remiges are paler below with darker margins compared to the all-dark juvenile-type feathers. Head is more colourful than in juvenile, pale pinkish, and the bill is distinctly bicoloured, dark horn-grey with an extensive pale cutting edge.

SEXING It is not possible to sex birds in the field.

CONFUSION RISKS Other vultures, especially Black, but also Rüppell's Griffon or even Eurasian Griffon, but large, pale head with heavy bill and dark eye and pale rear underbody are diagnostic.

328. Lappet-faced Vulture *Torgos tracheliotos*, juvenile (1st cy) of ssp. *negevensis*. Silhouette differs from other large vultures by large head, rather short tail and by broader inner arm than outer wing; note also diagnostic light rear underbody. Oman, 2.11.2004 (DF)

329. Lappet-faced Vulture *Torgos tracheliotos*, juvenile (1st cy) of ssp. *negevensis*. Aged as a juvenile by uniform condition of spiky-tipped flight feathers. Note the dark thighs, typical of juveniles, and that the underwing-coverts look rather mottled and the body is already streaked. Oman, 3.11.2004 (DF)

330. Lappet-faced Vulture *Torgos tracheliotos*, juvenile (1ˢᵗ cy) of ssp. *negevensis*. Large, pale head with dark eye differs from other vultures; note also exposed pale legs. Oman, 6.11.2004 (DF)

331. Lappet-faced Vulture *Torgos tracheliotos*, immature *'nubicus'* in first moult, probably just over 12 months of age. Plumage still largely juvenile, except for a few new median upperwing-coverts, mantle feathers and scapulars, some of which are white, an unnatural condition but previously known from African Lappet-faced Vultures. Ethiopia, 13.2.2012 (DF)

332. Lappet-faced Vulture *Torgos tracheliotos*, immature *negevensis* towards the end of its first moult, around two years of age. Plumage is worn and faded, still partly juvenile, but note new inner three primaries and half-grown p4 (hidden). Oman, 19.11.2014 (DF)

333. Lappet-faced Vulture *Torgos tracheliotos*, immature (2ⁿᵈ cy) *negevensis* towards the end of its first moult, around two years of age (same as **332**). Oman, 19.11.2014 (DF)

334. Lappet-faced Vulture *Torgos tracheliotos*, immature *negevensis,* in the final stages of its second moult (presumed third plumage/3rd cy). The outer three primaries and about half of the secondaries are still juvenile. Note the lighter flanks and trousers approaching the colour of the adults. Oman, 2.11.2004 (DF)

335. Lappet-faced Vulture *Torgos tracheliotos*, immature *negevensis,* in the final stages of its second moult (presumed third plumage/3rd cy). The upperwing contrast may be quite distinct and approach that of *Gyps* vultures. Note stage of primary moult and number of retained juvenile secondaries and compare with **334**. Oman, 10.11.2013 (DF)

336. Lappet-faced Vulture *Torgos tracheliotos,* adult *negevensis*. Adults are typically lighter on the body, with whitish flanks and rear underbody visible from afar and acting as good species characters. Note the rather uniform underwing of this subspecies, even in adult plumage, and compare with birds from further south in Africa. Oman, 3.11.2004 (DF)

337. Lappet-faced Vulture *Torgos tracheliotos,* adult *'nubicus'*. The upperwing contrast is usually clearer than in Black Vultures, for example. Huge naked head with peering dark eyes is diagnostic among vultures. Ethiopia, 8.2.2012 (DF)

338. Lappet-faced Vulture *Torgos tracheliotos*, adult *'nubicus'*. The plumage is much more contrasting compared to ssp. *negevensis*, with prominent white patagial band, flanks and thighs. Ethiopia, 13.2.2012 (DF)

339. Lappet-faced Vulture *Torgos tracheliotos*, adult *'nubicus'*. Rather similar to nominate birds owing to its more striking plumage and deeper red head. Ethiopia, 16.1.2010 (DF)

SHORT-TOED SNAKE EAGLE
Circaetus gallicus

Other names: Short-toed Eagle

VARIATION Monotypic.

DISTRIBUTION Breeding migrant occurring across S and E Europe, with strongholds in the Iberian peninsula and France. Distribution continues to C Asia and Indian subcontinent. Vagrants reach N Europe annually. Winters mainly in the Sahel, along the southern fringe of the Sahara, occasionally also in S Europe. For breeding, prefers open hilly country with scattered woods but hunts over all kinds of open areas.

BEHAVIOUR Mostly seen on the wing, either circling low or hovering against the wind, looking for prey. Perches prominently on tree-tops, and typically also roosts on high-voltage pylons.

SPECIES IDENTIFICATION Large eagle, with long and broad wings and full tail. Mostly easy to identify, even from afar, by uniformly pale underparts, with or without contrasting darker hood. From a distance, the underwings look completely pale, lacking the striking dark plumage patterns of buzzards or Osprey, while the upperparts show a marked contrast between lighter brown coverts and dark remiges. The greyish uppertail is paler towards the edges and shows 2–3 sparse bands. Flapping flight is impressive, with slow and deep wing-beats, almost as if in slow-motion. Often seen hunting over open slopes, where it frequently stops to hover in mid-air with typically slow and twisting wing-beats.

MOULT Autumn juveniles show no signs of moult, while older birds always show feathers of different age and wear, often with growing feathers in the wings and tail, even on migration. Older birds moult actively both in summer and on the winter quarters, yet most are not capable of replacing all the flight feathers in the course of one year. Many juveniles start their first wing-moult while still in Africa, returning in spring with a few fresh inner primaries. Knowledge of the details of primary moult is a prerequisite for ageing second and third plumage immatures.

PLUMAGES Three age-classes can be readily separated: juveniles, second plumage and adult, and sometimes a third plumage can also be identified. Important ageing features to look for are moult contrasts in the primaries, type of underwing barring and type of head coloration. Juveniles and adults are quite diagnostic, while second and third plumage birds are usually white-headed and very similar-looking; the latter are reliably separated only by details of wing moult.

Juvenile

No signs of moult in wings and tail, with white trailing edge uniform and complete, while dark subterminal band on wing is lacking or very faint. Hood, when present, clearly rufous-brown, not grey, and underwing remiges only finely and poorly barred, or not barred at all. Underparts may be almost completely white, while more patterned (hooded) birds may show intensely marked underwing-coverts and body. Upperparts show a marked contrast between uniform, pale greyish-brown coverts and darker remiges. Greater upperwing-coverts have pale tips, forming a uniform thin white line on the mid-wing.

During winter the juvenile plumage fades due to bleaching and wear. Some birds start to moult their innermost primaries before the spring migration, while others return in their juvenile plumage, which is by then extremely worn and faded.

Second plumage

Superficially similar to juvenile, but always told by moulting inner primaries, while the outer primaries and most of the secondaries are still retained juvenile feathers in the 2nd cy autumn, narrow and pointed and clearly worn. Head often very white, catching the eye from afar, and upperwing-coverts are very worn and faded, creating a striking contrast to the upperwing, similar to third plumage. During the following winter the moult progresses, but birds can still be identified in 3rd cy spring by their overall pale plumage and juvenile outermost primaries, which are by now extremely faded and worn.

Third plumage

Very similar to second plumage, with pale head and blonde overall impression; the two are not normally separable in the field, but can be distinguished in close views by details of moult: the primaries show two moult fronts, the outer being active in the outermost primaries with another active moult front in the inner primaries; the latter are now being replaced for the second time. Advanced birds may already have replaced their last outermost juvenile primaries by the autumn, while more slowly moulting birds still retain 1–2 juvenile outermost primaries, but regardless of moulting progress the most abraded feathers are the median primaries, usually p4–p6. Third plumage birds lack a full hood and many are truly white-headed (crown, face and throat), similar to second plumage, with many showing a contrasting necklace of distinct dark streaks across the upper breast. The dark trailing edge, typical of adults, is still largely missing and the whole underwing is thus very similar to second plumage birds. The upperwing-coverts are still very pale, but by now are largely replaced and fresh, which is another difference compared to second plumage birds.

Adult

Adults show completely and rather heavily barred flight feathers, with a clearly darker, grey subterminal band along the trailing edge of the wing. The upperwing contrast is duller with darker coverts compared to younger birds, while the hood is often more conspicuous and grey in colour, differing from the rufous hood of the juveniles. The wing moult appears irregular as a result of several simultaneous moult fronts, often progressing just a few feathers apart. The body plumage is quite variable, with some birds showing a solid dark hood and heavily marked belly and underwing-coverts, while others show a streaked hood and fewer and fainter bars or spots on the body and underwing-coverts.

SEXING The sexes are said to differ by type of hood, with adult females showing a darker and more complete hood, while the males are paler and more streaked on the throat and upper breast. This, however, appears not to be a reliable feature, since females with streaked hoods are known from breeding pairs. More information is needed.

CONFUSION RISKS Often wrongly regarded as difficult to identify from other largely pale raptors, such as pale buzzards, Ospreys or adult Bonelli's Eagles, but the rather uniform appearance of the pale underwing is diagnostic. Whether the truly featureless underwing of a juvenile, or a more prominently marked underwing of an adult, the underwings of distant Short-toed Snake-Eagles always look featureless below, lacking the solid dark areas shown by all other raptors of similar size and structure.

340. Short-toed Snake Eagle *Circaetus gallicus*, juvenile. Juveniles have a rufous hood and the underwing-coverts are dotted with the same colour. Remiges are poorly marked and show no signs of moult. Spain, 16.9.2011 (DF)

341. Short-toed Snake Eagle *Circaetus gallicus*, juvenile. A more strongly marked juvenile, but note colour of body markings and fine underwing barring. Spain, 19.9.2012 (DF)

342. Short-toed Snake Eagle *Circaetus gallicus*, juvenile. Juveniles marked this strongly are rare. Israel, 2.10.2009 (DF)

343. Short-toed Snake Eagle *Circaetus gallicus*, juvenile. Note the clean look of the juvenile underwing. Oman, 11.12.2010 (DF)

344. Short-toed Snake Eagle *Circaetus gallicus*, juvenile. Fresh juveniles show a marked contrast between lighter coverts and darker remiges, and the pale tips of the greater coverts form a distinct wing-band. Israel, 8.10.2009 (DF)

345. Short-toed Snake Eagle *Circaetus gallicus*, juvenile, 2ⁿᵈ cy spring migrant. Returning juveniles in spring are worn and faded above. This bird arriving from Africa has already replaced two inner primaries in the winter quarters. Spain, 4.5.2011 (DF)

346. Short-toed Snake Eagle *Circaetus gallicus*, second plumage, 2ⁿᵈ cy autumn migrant. Birds of this age have typically replaced half of their primaries while most secondaries are still juvenile feathers, worn and faded. Note juvenile underwing-coverts. Spain, 10.9.2012 (DF)

347. Short-toed Snake Eagle *Circaetus gallicus*, second plumage, 2ⁿᵈ cy autumn migrant. Easily identified by typical moult pattern and by generally pale plumage. Israel, 2.10.2009 (DF)

348. Short-toed Snake Eagle *Circaetus gallicus*, second plumage, 2nd cy autumn migrant/wintering. This late autumn bird shows suspended moult retaining juvenile pp7–10 and ss3–4 and ss6–9 (ss6–10 in right wing). Note very pale head of this and next plumage. Oman, 9.12.2012 (DF)

349. Short-toed Snake Eagle *Circaetus gallicus*, third plumage, 3rd cy autumn migrant. Birds in third plumage differ from similarly pale second plumage by showing two simultaneous moult fronts in the primaries, one in the outer primaries and one in the inner primaries. As a rule, all secondaries have been moulted at least once, and typically birds show a necklace of dark spots across the crop-area. Spain, 16.9.2011 (DF)

350. Short-toed Snake Eagle *Circaetus gallicus*, third plumage, 3rd cy autumn migrant. Note fresh outer and inner four primaries and dark spotting on crop. The new flight feathers are still sparsely marked compared to adults. Spain, 15.9.2011 (DF)

351. Short-toed Snake Eagle *Circaetus gallicus*, third plumage, 3rd cy autumn migrant. Another typical, very white bird with bold spots on the upper breast. This bird had still retained its outermost juvenile primary. Spain, 5.9.2013 (DF)

352. Short-toed Snake Eagle *Circaetus gallicus*, third plumage, autumn migrant (same as **351**). Second and third plumage birds appear practically white-headed with contrasting light upperwing-coverts. Spain, 5.9.2013 (DF)

353. Short-toed Snake Eagle *Circaetus gallicus*, adult, spring migrant. Adults are distinctly barred below with a sharply defined darker trailing edge to the wings. Note sparse tail-banding, typical of the species. Israel, 19.3.2010 (DF)

354. Short-toed Snake Eagle *Circaetus gallicus*, young adult, autumn migrant. Some of the secondaries are still the thinly barred immature type, while the most recently replaced feathers show bold bars and a more distinct dark tip. Spain, 5.9.2013 (DF)

355. Short-toed Snake Eagle *Circaetus gallicus*, young adult, autumn migrant. This bird also retains some immature-type median secondaries, which differ from the more heavily barred fresh feathers. Spain, 10.9.2012 (DF)

356. Short-toed Snake Eagle *Circaetus gallicus*, adult, autumn migrant. Heavily marked adult with boldly barred remiges and distinct dusky trailing edge. Spain, 10.9.2012 (DF)

357. Short-toed Snake Eagle *Circaetus gallicus*, adult, autumn migrant. A well-marked adult. Spain, 16.9.2011 (DF)

358. Short-toed Snake Eagle *Circaetus gallicus*, adult, wintering. Birds from the eastern part of the range are often more finely marked and neater looking than western birds. India, 13.12.2007 (DF)

359. Short-toed Snake Eagle *Circaetus gallicus*, adult, autumn migrant. Adults lack the neat pale fringes above of autumn juveniles. Spain, 9.9.2013 (DF)

360. Short-toed Snake Eagle *Circaetus gallicus*, adult, spring migrant. Although distinct, the upperwing contrast of adults is far less striking compared to juveniles and immatures. Spain, 28.4.2011 (DF)

BATELEUR
Terathopius ecaudatus

VARIATION Monotypic.

DISTRIBUTION Widely distributed across open and semi-open areas of sub-Saharan Africa and SW Arabia. Rare vagrant away from breeding areas, mostly immatures, including several records from Israel and single recent records from Turkey and Spain.

BEHAVIOUR Flight is highly typical and unmistakable, a fast rolling glide with body rocking from side to side and wings lifted to form a deep V-shape. Covers huge distances in this peculiar flight, which is only occasionally interspersed by brief sequences of rapid flaps.

SPECIES IDENTIFICATION Adults are unmistakable, owing to unique colouration, silhouette and flight. Juveniles and young immatures are less extreme, with a more normal wing-shape and a longer tail, and could potentially be mistaken for some other medium-sized dark raptor having lost its tail accidentally, but unique flight remains diagnostic.

MOULT Replacing the entire juvenile plumage takes several years. Adults are capable of replacing their flight feathers in a much shorter period of time, possibly in just over one year.

PLUMAGES The plumage development is a protracted process, taking several years to complete. The adult plumage is acquired gradually, although the estimated 7–8 years is probably an overestimate. During this period the juvenile plumage is followed by at least another two entirely brown plumages, which in turn are replaced by subadult stages, during which the adult features are gradually acquired.

Juvenile

All brown with more evenly broad wings and a longer tail compared to adults (legs not protruding beyond tip of tail). The inner hand is broader and the outer arm is narrower than in adults, the wingtip appears more rounded and the tail extends beyond the tips of the toes. The silhouette is, in fact, very different from the adult's, although most field guides show them as being identical. The body plumage is uniformly brown, with pale tips to most feathers, and the wings lack signs of moult. The remiges are uniformly brown above and below, and the trailing edge of the wing is moderately serrated. Bill is grey with a light bluish cere and facial skin, and the feet are pale flesh.

Second plumage

Similar to juvenile, but the inner primaries have been replaced and stand out as a group of darker feathers, contrasting with the retained outer juvenile primaries, which by now have faded to a lighter brown colour. Wing-formula as in juvenile.

Third plumage

Still resembles juvenile, but primary moult has reached the outer primaries and some secondaries have also been replaced. The tail has already been moulted and is now short, with toes projecting beyond the tip of the tail. Bill is still pale, feet are yellowish (but can be red in some), and the facial skin is yellow or red.

Subadult

During the subsequent subadult plumages the birds gradually acquire adult features over a period of 2–3 years, but details await further studies. During this period the remiges get their final coloration and the underwing-coverts become mottled at first, later turning plain white.

Adult

Adult plumage is very colourful and unique among raptors. The large head and the underbody are black, while the underwing-coverts are white. The remiges are white with black tips in females, all black in males, except for paler outer primaries in both. Upperparts are more variegated with black scapulars, chestnut (or buff in some – from wear?) mantle, back, rump and tail, while upperwing-coverts are largely silvery-grey. Base of bill is yellow with a dark tip, while cere, facial skin and feet are bright red.

SEXING Sexing adults and older immatures is easy by checking the underwing: in males the secondaries are all dark, both above and below, while in females they are pale with dark tips.

CONFUSION RISKS Adults are unmistakable. In immatures the only confusion may arise from all-dark raptors, such as Black Kite, Marsh Harrier or *Aquila* eagles, which have lost their tails. The wing-shape and behaviour in flight are, however, unique and should be enough to correct any mistake.

NOTE The great difference in flight silhouette between immatures and adults has largely been overlooked in the literature. The wing-formula of juveniles shows a rounded wingtip with shorter fingers compared to the extreme wingtip shape of the adults, with long fingers forming the wingtip but with the inner primaries rapidly decreasing in length. In juveniles the outer five primaries show equally long fingered sections of each feather, with the emarginated part making up to about 50% of the visible part of the feather. In adults the outer fingers are notably narrower, spikier and longer, with the fingered section clearly longer, making up for more than 50% of the visible part of the feather and with inner primaries rapidly decreasing in length while the secondaries are more bulging than in juveniles.

361. Bateleur *Terathopius ecaudatus*, juvenile. Juveniles are longer-tailed than adults and the wings are more rectangular with a broader tip. Ghana, 22.2.2010 (DF)

362. Bateleur *Terathopius ecaudatus*, juvenile growing innermost primaries. In juveniles the inner primaries are longer and the secondaries shorter than in adults, and the feet do not project beyond the tip of the tail. Compare wing-formula with adult. Uganda, 15.12.2009 (DF)

363. Bateleur *Terathopius ecaudatus*, immature during early stages of first moult, approximately one year old. This bird has replaced half of its primaries, most of its body plumage and some underwing-coverts, but looks still largely juvenile. Note shorter tail compared to juvenile. Ghana, 23.2.2010 (DF)

216

364. Bateleur *Terathopius ecaudatus*, moulting from first to second plumage, near completion of first moult. Aged by retained juvenile outer two primaries. Juvenile body plumage replaced by similar-looking second plumage but silhouette approaching that of adult. Uganda, 7.12.2008 (DF)

365. Bateleur *Terathopius ecaudatus*, immature towards end of first moult, about two years old. Primaries and upperparts mostly moulted but half of juvenile secondaries still retained. Plumage still resembles juvenile, but wing-shape is adult-like and tail is short. Uganda, 15.12.2009 (DF)

366. Bateleur *Terathopius ecaudatus*, immature, 2–3 years old. Has replaced all juvenile feathers and primary moult now on second round. Body plumage still juvenile-like, but bill is turning yellow and feet red. Uganda, 16.12.2009 (DF)

367. Bateleur *Terathopius ecaudatus*, older immature. Has replaced all juvenile feathers, underwing-coverts showing a mixture of brown and blackish feathers. Uganda, 16.12.2009 (DF)

368. Bateleur *Terathopius ecaudatus*, subadult male. Underwing-coverts a mixture of black and white feathers while part of secondaries are all black and of adult male type. Uganda, 15.12.2009 (DF)

369. Bateleur *Terathopius ecaudatus*, near-adult female. Some underwing-coverts are still dark on this near-adult female. Adult females are easy to tell from adult males by their whitish underwing with a black trailing edge. Uganda, 7.12.2008 (DF)

370. Bateleur *Terathopius ecaudatus*, adult male (with adult Tawny Eagle). Adult males are easily sexed by all-black secondaries and inner primaries. Ethiopia, 12.2.2012 (DF)

371. Bateleur *Terathopius ecaudatus*, adult male (same as **370**). Unmistakable. Note how narrow and pointed the fingered primaries are compared to juveniles and immatures. Ethiopia, 12.2.2012 (DF)

372. Bateleur *Terathopius ecaudatus*, adult female. Remiges are white in females save for a narrow black trailing edge to the wings. Uganda, 18.12.2009 (DF)

373. Bateleur *Terathopius ecaudatus*, adult female. Uganda, 16.12.2009 (DF)

374. Bateleur *Terathopius ecaudatus*, adult female (same as **373**). Note silvery secondaries; in adult males secondaries are also dark above. Uganda, 16.12.2009 (DF)

WESTERN MARSH HARRIER
Circus aeruginosus aeruginosus

Other names: Eurasian Marsh Harrier, Marsh Harrier

VARIATION Nominate *aeruginosus* in Western Palearctic, with eastern subspecies *spilonotus*, sometimes regarded as a full species: Eastern Marsh Harrier *Circus spilonotus*, replacing the nominate subspecies from Lake Baikal, C Mongolia and China eastwards; nominate ssp. is then called Western Marsh Harrier. The ranges of the two populations are overlapping with hybridisation reported from Lake Baikal (Fefelov 2001) and W Mongolia (pers. obs.), but the extent of hybridisation is not well known and more information is needed.

DISTRIBUTION Widely distributed around wetlands across Europe and Asia. Migrating birds can be found in all kinds of open country. Winters from W and S Europe to S Asia and sub-Saharan Africa

BEHAVIOUR Mostly seen on the wing when hunting. Quarters swamps, meadows and fields in a low searching flight, progressing in a slow and leisurely manner. Perches from time to time on the ground, but also roosts in trees during the hottest part of the day. Outside the breeding season congregates, like other harriers, in communal roosts for the night.

SPECIES IDENTIFICATION A highly variable species with a wide spectrum of plumages. Plumages can be roughly grouped into two, the brown female-type plumage representing juveniles and older females, and the rather striking plumage of the adult males. Note, however, that plumages vary considerably, both as a result of individual variation, but also depending on sex and age.

Adult males are as a rule rather easily identified, while the dark, female-plumaged birds are often mistaken for other species, except when hunting low over reedbeds or fields in typical harrier fashion. Particularly when migrating high in the sky, female-type birds can be mistaken for Black Kite or dark morph Booted Eagle, which both share the overall dark and rather featureless plumage of the female-type Marsh Harrier. Note the less rectangular wing-shape of Marsh, with a broader and bulging arm and a narrower hand with a more rounded wingtip, while both Black Kite and Booted Eagle have a squarer wingtip with

more deeply splayed fingers and more parallel-edged wings. Adult males on migration look surprisingly slim and slender when high up in the sky and could easily be mistaken for one of the smaller harriers.

MOULT The annual complete moult in adults commences for females in May–Jun, at the start of breeding, while breeding males may start up to several weeks later. Yearlings start their complete moult slightly earlier than adult females, and the moult continues throughout summer for both age-groups. By Aug–Sep, at the time of autumn migration, adult birds are usually less than halfway through their primary moult, females usually being far more advanced than males, while 2nd cy birds moulting for the first time are clearly more advanced. At this time of year the difference in the location of the moult front in the primaries can be, at least tentatively, used for ageing (see below under Plumages). The moult is completed in the winter quarters during Oct–Dec, with first-timers completing their moult up to several weeks before older birds.

Juveniles undergo an individually variable body moult during their first winter comprising feathers of mantle, upperwing- and uppertail-coverts, head and breast. The central tail feathers may also be moulted, in which case the birds can be sexed: the new tail feathers are brown in females, greyish in males. Juveniles with an extensive body moult can be difficult to tell from older females in spring, but the narrower hand and dark underwings and tail differ from older females.

PLUMAGES A highly variable species with a wide selection of plumages. Adult males are usually easy to single out as a group, and given good views first-adult males can be told from older birds; sometimes, even the second-adult plumage is separable from older males. Females are equally variable, with the very dark-coloured first-adults greatly resembling juveniles, except for their broader wings, while older birds tend to get browner and less uniform in plumage.

This rather logical general picture is badly disturbed by the fact that some adult males retain a female-like brown plumage for life. On the other hand, some (old?) females also develop a male-like grey bloom to their upperwings and tail, therefore greatly resembling males in first-adult plumage.

Further confusion is added by a dark morph, particularly striking in males, which has an entirely brownish-black body plumage and broad black wing-markings.

Note that there may be considerable differences between different populations as to when they moult and how quickly plumages are moulted from one to the next. The biggest differences can be expected between the local, resident populations of W and SW Europe, compared to the migratory birds from further N and E, which migrate as far as sub-Saharan Africa for the winter. The following account refers to the migrant populations.

Juvenile

Uniformly dark, almost blackish-brown upon fledging, but later fading to dark brown. The head varies individually, but most show a yellow-ochre crown and throat, separated by a dark eye-mask, while others only have a yellowish cap, or just a yellowish nape-patch, with a small minority being completely dark-headed with no yellow at all. Unlike older females most juveniles show no yellow markings on the upper forewing or breast, but some do. In fresh plumage told from older females by clearly rusty tips to the greater upperwing-coverts, forming a narrow line, by dark brown uppertail-coverts matching the colour of the back, and by uniformly dark remiges below, save for a small pale crescent at the base of the outer primaries. Also, the tail feathers are uniformly dark brown, not rusty-brown as in older females.

By winter the blackish-brown plumage has faded to a dark brown and the bright yellow-ochre crown has faded to become almost whitish. When juveniles gradually develop a paler crown, from wear and fading, adult females at the same time become increasingly colourful, as their worn crown feathers are replaced by fresh, bright orange feathers during autumn – a good ageing character at this time of year. During winter the body plumage is partially moulted, varying in extent individually.

Some birds return in spring in an almost completely juvenile plumage, while others have replaced considerable parts of their body plumage. Birds with extensively moulted body plumage, rusty-brown uppertail-coverts and light mottling on the forewings and breast can be difficult to tell from older females, but the worn and faded flight feathers with spiky finger-tips and the faded dark brown tail differ from the darker and more recently replaced feathers of adult females. Note also the narrower, juvenile-shaped wings, particularly the narrower hand of the yearlings.

Adult male

First-adult male During the 2nd cy summer, males gradually acquire characters of the adult male. The first signs are found in the newly moulted inner primaries, which are pale below with a distinct black tip. As moult progresses the underwing pattern of an adult male starts to appear, while the upperwing often remains surprisingly dull and brown. By the time of autumn migration (Aug–Sep) birds have replaced about half of their remiges, but the outer primaries are still all-brown juvenile feathers. Also the tail looks greyish by now, with a darker subterminal band, sometimes with finer additional 'ghost-bars'. The body and head remain rather female-like all autumn, and the major change happens during winter: the forebody becomes streaked, contrasting with a uniformly brown lower breast and belly, while the head becomes rusty yellow with darker streaks and darker ear-coverts. Even in fresh plumage in spring the grey areas of the upperwing are duller compared to older males, the black wingtip and trailing edge are wider, and the tail shows a darker, diffuse subterminal band. The iris is yellow.

Definitive adult male Overall lighter and more brightly coloured than younger adults: the grey of the upperwing is more extensive and more silvery in colour and the brown areas are correspondingly smaller. The tail is pale silvery-grey, the uppertail-coverts are rusty-brown to begin with, but may become almost white in old birds. On the underwing the black in the wingtip has shrunk and the trailing edge gets narrower and more diffuse with increasing age, while the underwing-coverts become lighter, from rusty-brown in younger males to nearly white in the oldest males. The head and body plumage becomes lighter and more finely streaked with age, the dark ear-coverts mask dissolves, while the rusty-brown belly-patch recedes gradually towards the vent. The iris remains yellow.

Sometimes a second-adult plumage can be recognised, particularly in autumn when birds are still actively moulting their remiges: the old feathers show dull grey uppersides and a broadly black tip, and the old tail feathers still show the barring typical of the first-adult plumage. However, after completion of the second moult, the huge plumage variation makes a correct ageing practically impossible and one can at best talk about different plumage-types.

Adult female

Very similar to juvenile plumage, with the first-adult plumage in particular being nearly identical. Adults are always told from fresh autumn juveniles by rusty-brown, not dark brown uppertail-coverts and by showing signs of wing-moult.

First-adult female Almost identical to the juvenile plumage, except for the rusty-brown uppertail-coverts, the signs of moult in the wings, and the adult-shaped, broader wings. The underwing still appears very dark, with a pale crescent around the carpal, as in juveniles. Although the plumage may look identical, the silhouette of the bird is different, with a broader and fuller hand compared to the juveniles, giving a broader-winged, more buzzard-like silhouette. Compared to older birds the wing-moult is completed earlier, as early as late Oct, adding further to the confusion with juveniles in late autumn. The iris is still dark.

Definitive adult female Plumage highly variable, partly depending on age, but also owing to individual variation. Time of year also plays an important role, as birds are worn and faded during and just after breeding, but darker and more colourful in winter and early spring. This is important to remember when separating juveniles and older females, as juveniles start to fade from the time they leave the nest, while adults start to fade only after they have completed their moult in late autumn or early winter.

Definitive adult females are generally browner and more variegated compared to the two preceding darker plumages. The tail is often paler than the rest of the upperparts, with the typical rusty colour best seen towards the edges when the tail is fanned, and the underwing is much paler, often with a pale brown hand contrasting with darker fingers and secondaries. Some are similar to younger females, with a pale crown and throat divided by a darker eye-mask, but with increasing age the head pattern dissolves and the upper breast becomes more streaked. Old females look very heavy in flight, with broad wings and a comparatively short tail.

Iris colour varies from medium brown through amber to pale yellow, gradually becoming lighter with increasing age.

SEXING Autumn juveniles are possible to sex by their different size and silhouette, with females being much heavier-bodied, broader-winged and more compact compared to the slimmer, narrower-winged and more agile males, although this can be difficult to judge without previous experience, or unless they are seen together. By the following spring some males develop a lighter iris colour, sometimes even yellow, and any replaced tail feathers will look greyish with a darker subterminal band. Females retain a dark iris and any replaced tail feathers are brown, not grey.

Adults should always be checked for the different proportions, since plumages may in some cases be nearly identical: females appear truly heavy, broad-winged and rather short-tailed, almost buzzard-like at times, compared to the slimmer and lighter males which have narrower wings and a longer, narrow tail.

CONFUSION RISKS Females and juveniles on migration are mostly mistaken for Black Kite or dark morph Booted Eagle, sometimes also dark morph Montagu's Harrier (which see). Locally hunting birds usually identified by plumage and typical harrier behaviour, quartering low over open areas, but dark morph Montagu's may pose a problem, although the latter is much lighter and slimmer with narrower and more pointed wings. Light-coloured adult males may sometimes be mistaken for males of other harrier species, but rufous belly remains diagnostic.

NOTE Birds showing features normally connected with Eastern Marsh Harrier *Circus (aeruginosus) spilonotus*, such as barred remiges and tail, are seen every now and then among normal-looking birds. The female-type birds have shown rather strongly barred remiges and tail feathers, while the adult males appear all white below, save for darker (but brown, not black) head and neck but with normally coloured black wingtips. These birds could either originate from more eastern populations of Western Marsh Harrier, where an exchange of genes with Eastern Marsh Harrier is likely to occur, or it may be a case of old genes from a past common history surfacing every now and then but, for whichever reason, odd birds like this deserve further study. Hybrids and intergrades between Western and Eastern Marsh Harriers may turn up in the Middle East, with chances increasing from west to east.

In a well-studied population in W France up to 30% of the breeding males have shown a brown, female-like plumage, while 'normally coloured' adult males have been in the minority (Blanc *et al.* 2013). The possible occurrence of similar female-like males elsewhere in the region deserves further attention.

References

Blanc, J.-F., Sternalski, A. & Bretagnolle, V. 2013: Plumage variability in Marsh Harriers. *British Birds* 106: 145–158.

Fefclov, I. V. 2001. Comparative breeding ecology and hybridisation of Eastern and Western Marsh Harriers *Circus spilonotus* and *C. aeruginosus* in the Baikal region of Eastern Siberia. *Ibis* 143 (3): 587–592.

375. Western Marsh Harrier *Circus aeruginosus*, juvenile, presumed female by bulky build. Autumn juveniles are very dark brown and the wings show no sign of remex moult. Israel, 4.10.2009 (DF)

376. Western Marsh Harrier *Circus aeruginosus*, juvenile. Israel, 5.10.2009 (DF)

377. Western Marsh Harrier *Circus aeruginosus*, juvenile. Oman, 11.2.2013 (DF)

378. Western Marsh Harrier *Circus aeruginosus*, juvenile. The head pattern varies individually, this juvenile showing only a light nape-patch, but some can be all dark. Israel, 27.9.2008 (DF)

379. Western Marsh Harrier *Circus aeruginosus*, juvenile, presumed male by slim proportions. The amount of light markings on the underwing-coverts varies. Oman, 8.11.2013 (DF)

380. Western Marsh Harrier *Circus aeruginosus*, juvenile, presumed female by heavy build. For a juvenile this bird shows a lot of light markings on the underwings; note also the extensively pale head. Spain, 9.9.2012 (DF)

381. Western Marsh Harrier *Circus aeruginosus*, juvenile (same as **375**). Juveniles are uniformly dark above, but note light tips to larger feathers. Israel, 4.10.2009 (DF)

382. Western Marsh Harrier *Circus aeruginosus*, 2nd cy male, moulting from juvenile to first-adult plumage. Note mix of moulted adult-type flight feathers and retained all-brown outer primaries and inner/median secondaries, and largely brown, female-like body plumage and dark iris. Finland, 30.8.2009 (DF)

383. Western Marsh Harrier *Circus aeruginosus*, 3rd cy male in first-adult plumage. Note variably marked flight feathers, all-brown upperwing-coverts, greyish tail with dark subterminal band and rusty uppertail-coverts and juvenile-like head. Oman, 11.2.2014 (DF)

384. Western Marsh Harrier *Circus aeruginosus*, adult male. The silvery areas are brighter and cleaner compared to first-adult plumage, and the tail lacks the dark subterminal band. Oman, 11.2.2014 (DF)

385. Western Marsh Harrier *Circus aeruginosus*, adult male. Largely brown underbody and light underwings with extensively black tips make adult males unmistakable. Israel, 27.9.2008 (DF)

386. Western Marsh Harrier *Circus aeruginosus*, adult male, breeding. With age, the underwings become lighter. Finland, 22.5.2012 (DF)

387. Western Marsh Harrier *Circus aeruginosus*, adult male. In older males the black area of the wingtip and the rufous of the belly both recede and the bird becomes lighter overall. Oman, 5.11.2004 (DF)

388. Western Marsh Harrier *Circus aeruginosus*, adult male. With age, the dark trailing edge of the wings dissolves and the head and breast become lighter. Israel, 2.10.2008 (DF)

389. Western Marsh Harrier *Circus aeruginosus*, adult male, breeding. Old males are typically bright silvery-grey above, including part of the outer wing-coverts and uppertail-coverts. Spain, 22.4.2010 (DF)

390. Western Marsh Harrier *Circus aeruginosus*, first-adult female. Females in their first adult plumage are still extremely similar to juveniles, and if signs of moult are no longer to be found mistakes can easily happen (this one is still growing its outermost primary p10). Ethiopia, 18.11.2011 (DF)

391. Western Marsh Harrier *Circus aeruginosus*, first-adult female (same as **390**). The upperwing clearly shows feathers of different generations, thus excluding the possibility of a juvenile. Ethiopia, 18.11.2011 (DF)

392. Western Marsh Harrier *Circus aeruginosus*, first-adult female. Very similar to juvenile, but note two generations of secondaries and growing primary p9 and missing p10. Israel, 4.10.2009 (DF)

393. Western Marsh Harrier *Circus aeruginosus*, adult female. Some adult females are largely brown, not dissimilar to juveniles. Israel, 4.10.2009 (DF)

394. Western Marsh Harrier *Circus aeruginosus*, adult female. Older females are much more brightly coloured, with light underwings and clearly darker, but not black fingers. Note the yellow eye of this bird, indicating a greater age. Israel, 5.10.2009 (DF)

395. Western Marsh Harrier *Circus aeruginosus*, adult female, spring migrant. Spring adults are in a reasonably good plumage, while juveniles are often very worn. Note the different shades of the flight feathers, whereas juveniles would look uniform. Israel, 26.3.2008 (DF)

396. Western Marsh Harrier *Circus aeruginosus*, adult male of the dark form, spring migrant. Dark morph males have a dark body plumage and the flight feathers are dark with variable lighter markings. Israel, 26.3.2012 (DF)

397. Western Marsh Harrier *Circus aeruginosus*, adult male of the dark form, spring migrant. The uppertail remains silvery-grey on these otherwise dark birds. Israel, 26.3.2008 (DF)

398. Western Marsh Harrier *Circus aeruginosus*, adult male of the dark form, spring migrant. A slightly browner bird of the dark form. Israel, 26.3.2008 (DF)

How to separate Black Kite, Marsh Harrier and dark morph Booted Eagle

Black Kite, Marsh Harrier and dark morph Booted Eagles can be difficult to tell apart, especially when confronted with distant birds, for example on migration. However, they do all have slightly different proportions and also plumages, so checking for the right features should solve most of the problems.

Looking at the proportions first, Booted Eagle stands out as being the most compact of the trio. The wings are rather broad and short with six clearly-splayed fingers in a rather square wingtip and the tail is rectangular with sharp corners. When gliding and soaring it keeps its wings distinctly arched, a most reliable character once learnt, with the long fingers bent strongly upwards.

Black Kite, which can be rather similar in proportions, is first of all a slimmer bird, with longer and proportionately narrower wings while the tail is narrow, often narrowest at the base but slightly widening towards the forked tip. The tail-fork can be missing due to moult (in autumn) and it can even disappear when the tail is fully fanned, particularly in young birds, which have a shallower fork. The wingtip may appear similar to Booted, but usually it appears narrower, with 5–6 fingered primaries. At this point it is worth mentioning that eastern populations of Black Kites, migrating through and wintering in the Middle East, have broader wings with six fingered primaries, and may therefore appear more similar in silhouette compared to the Black Kites of W Europe. In head-on views gliding Black Kites hold their wings rather flat, often with a slight angle between a short and raised inner arm and a lowered, long outer wing, thus quite different from the Booted Eagle's strongly rounded arch.

The Marsh Harrier differs from the other two by its narrower and more rounded wingtip, with only five fingered primaries. Usually the outer wing (hand) is narrower than the slightly bulging arm, giving it a wing-shape that differs from both Booted Eagle and Black Kite, in which the wings are more parallel-edged.

The tail is narrow, and unlike in the former two, it is clearly rounded at the tip.

In powered flight, between the series of flaps, Marsh Harriers tend to glide on slightly lifted wings, forming a shallow V, but when the bird is descending in a long glide the wings may be held clearly arched, with the wingtips pointing downwards.

Looking at plumage features, Booted Eagle can always be identified by its diagnostic upperparts, with bright diagonal patches across the wing-coverts and scapulars, and its white uppertail-coverts. From below the plumage features can be more difficult to see, but the inner primaries form a pale, translucent window, while the rest of the underwing usually looks very dark. The head is always diagnostic, with a dark face and throat contrasting with a faded brown crown and nape, reminiscent of Golden Eagle.

Black Kite can be very uniform in colour, but from below the inner hand usually contrasts as being paler than the dark fingers and the darker secondaries, the light area thus being wider compared to Booted Eagle. Also, the upperparts are very uniform save for a faintly lighter diagonal area across the upperwing-coverts, more or less in the same place as in Booted Eagle, but much less contrasting. Older adults have light grey or almost white heads, which is diagnostic in this context.

Female-type Marsh Harriers can be rather variable, depending foremost on the bird's age. Young birds can be entirely uniformly dark brown, but these can always be identified by their diagnostic white underwing crescent at the base of the outer primaries. As a rule Marsh Harriers never show fine and regular barring on their wings and tail, like Booted Eagle or Black Kite, but some birds may show sparse and irregular bands. Apart from rare all-dark juveniles most female-type Marsh Harriers show distinct white or yellowish markings on the head, forewings and breast, which are diagnostic compared with both Booted Eagle and Black Kite.

Identification of 'ringtail' harriers

The flight identification of ringtail harriers (harriers with a white 'rump' in juvenile and adult female plumages, i.e. Hen, Pallid and Montagu's Harriers) still remains one of the most challenging identification problems in our region. The adult females, and most autumn juveniles, have their diagnostic plumage characters and can therefore be identified safely, although this requires good views, and sometimes high quality photographs for later investigation. It is the 2nd cy spring juveniles in particular which pose the greatest challenge, owing to their extensive individual plumage variation. During the first winter some of them undergo a partial body moult, which varies greatly in extent, both between species but also between individuals of the same species. The 2nd cy spring Montagu's are the most variable and also the most often misidentified within the entire ringtail harrier complex. For instance, 2nd cy spring Montagu's, regardless of sex, often show an almost identical head pattern to similar-aged Pallid Harriers, a fact that is largely unknown. Another less well-known plumage is the adult female Pallid Harrier, which is also rather variable and may easily be mistaken for a female Hen or Montagu's Harrier. Until recently European birders have had relatively limited experience with adult female Pallid Harriers.

Another season of confusion is late summer–early autumn, when all but juveniles are moulting their flight feathers. This also happens to be the time of autumn migration for all three species, a period when vagrants are expected to turn up. In August and often well into September birds are moulting their longest primaries of the wingtip and this moult creates severe identification problems for the unaware. For instance, moulting male Hen and Montagu's Harriers with missing black primaries may look like Pallid Harriers, showing a pointed wingtip with just a narrow black wedge. Also, adult female Hen Harriers appear similarly pointed-winged and are often misidentified as either Montagu's or Pallid Harriers. In autumn it is therefore important to first check the state of moult and then as many characters as possible before attempting to identify a harrier.

Important features to focus on are, firstly, the underwing pattern, particularly of the primaries, but also not forgetting the secondaries and underwing-coverts, and only second comes the head pattern. Given a good photograph, showing both the underwings as well as the head, every individual should be possible to identify with certainty. It is also worth remembering that the flight behaviour differs between Montagu's and Pallid, especially when comparing hunting birds, although birds on migration can appear very similar. Another point worth noting is that juveniles have a different flight compared to the adults, most clearly so in Montagu's, explained by the shorter wings in juveniles, which gives them a heavier wing-loading and a less graceful flight compared to the feather-light adults.

The recently discovered hybridisation between harrier species takes the identification challenge one step further. As a rule, good quality photographs are needed in order to enable closer study of minor yet important identification features such as the wing-formula, including the emarginations of the fingered primaries.

HEN HARRIER
Circus cyaneus

VARIATION Monotypic. The North American form *hudsonius*, formerly considered conspecific, is now widely regarded a full species, Northern Harrier (Marsh Hawk in the USA) *Circus hudsonius*, and is here treated separately (p. 242).

DISTRIBUTION Breeds in the boreal and temperate zones of Europe and Asia, from the Atlantic coast to the Pacific in the east. Northern populations move south for the winter, the birds from our region reaching S Europe, N Africa and the Middle East.

BEHAVIOUR Mostly seen on the wing as it quarters low over fields and marshes in typical harrier fashion, looking for prey, small birds and rodents. Mostly perches on the ground, where it also roosts for the night, often communally in tall vegetation, but it may also roost in trees.

SPECIES IDENTIFICATION Adult males are quite diagnostic, although moulting birds in Aug–Sep with missing long primaries are regularly mistaken for male Pallid Harriers, but Hen always shows almost *Accipiter*-like broad wings, a more extensive black area to the wingtip and a broader dark trailing edge. Juveniles and adult females are superficially similar to other ringtail harriers, being dark above with whitish uppertail-coverts and straw-coloured to whitish below with darker streaking, but the broader and comparatively shorter wing, with a more rounded wingtip, gives it a heavier look, which is further emphasised by the heavier flight. Separating the ringtails of the three species requires detailed views of wing-formula, underwing pattern and head pattern before they can be safely identified. The behaviour of quartering low over open ground immediately separates harriers from, for example, juvenile Northern Goshawk, which at a quick glance can appear rather similar in plumage. *Of all the harriers covered in this book Hen Harriers proportionately have the shortest and broadest wings*, the compact shape being more obvious in adults than in juveniles, and the flight appearing more heavy in females than in males.

MOULT The entire plumage is replaced in a complete moult between Apr/May and Sep/Oct. Unlike Montagu's, and to a lesser extent Pallid Harrier, juveniles do not undergo a partial body moult during their first winter, but retain their juvenile plumage until the beginning of the first complete moult, in late 2nd cy spring. In late summer or early autumn non-juveniles are in active primary moult, at least until the end of Aug, sometimes into Oct, while juveniles show no signs of moult. Birds moulting the long primaries show an atypical and pointed wingtip and may be mistaken for Pallid or Montagu's Harriers.

PLUMAGES Normally juveniles, adult females and adult males can be separated by plumage, while juveniles can be sexed based on iris colour, and many also by plumage. One-year-old immatures can be aged by their transitional plumage (males more easily than females) as long as they retain juvenile feathers, normally until Aug–Sep of the 2nd cy.

Juvenile

Fairly easy to age by uniform, dark brown appearance above and ochre-coloured underbody and underwing-coverts with dark streaking below. The median upperwing-coverts show a variable lighter area formed by rusty-buffish feather margins and the uppertail-coverts are white. The facial disc is variable, often rather indistinct and streaked but can also be darker than the rest of the head and neck, while a pale collar (always with dark spots) can be discerned from a fair distance on most. The primaries are uniformly barred below with distinct dark tips, but the barring is often incomplete or even missing in juvenile males. The secondaries often appear darker than the primaries, being darkest in young females, while in males the barring of the secondaries is finer on a lighter background. In juveniles the secondary bands are not as black, nor as distinctly defined as in adult females, which show a much more strongly contrasting underwing pattern on both primaries and secondaries.

Juvenile plumage is retained until the late 2nd cy spring, but birds can be notably faded in colour by then, while the secondary bands of the underwing often appear more distinct than in fresh autumn birds. Immatures can be recognised until at least the 2nd cy late summer/autumn by a mixture of juvenile and adult-type feathers in the plumage, e.g. in the secondaries, resulting in a particularly striking appearance in males.

Juveniles can already be sexed at fledging by iris colour: in males pale grey or greenish-yellow to begin with, turning yellow by Aug–Sep, and in females dark brown, making the whole eye look dark in the field. Males also tend to show a greyish wash to the central tail feathers, which are dark brown in females. Also the underwing barring differs on average between the sexes: males tend to have reduced markings in the primaries while the secondaries are paler with distinct dark grey bands, whereas females have completely and boldly barred primaries and the secondaries appear dark with lighter coloured bands.

First-adult male

Until moult is completed males in their first adult plumage can be identified by unmoulted brown and barred juvenile flight feathers, which are occasionally retained until the 3rd cy summer. After moult some birds show dusky tips to the scapulars, and brown areas to the crown and facial disc, while the breast, belly and underwings may show some faint grey markings, making the entire plumage look less clean compared to full adults.

Adult male

Usually easy to identify by pale bluish-grey upperparts, except for white uppertail-coverts and sharply defined black wingtips, the black reaching the primary coverts. Tail appears pale grey when folded, but when fanned becomes gradually whiter towards the edges, with fine greyish barring on the outer rectrices. From below the grey head and upper breast contrast sharply against the white underbody and underwings, the latter with broadly black wingtips and broad dusky trailing edges.

Adult female

Always safely identified from similar-looking juvenile until Aug–Oct by active primary moult. After moult has been completed usually differs from juvenile by more greyish-brown upperwings, with more distinct dark bands across the remiges and greater coverts, making the upperwing as a whole more variegated, while the underbody is whiter with more distinct and broader, spotted or drop-shaped rufous-brown markings. The head pattern is less contrasting, with facial disc and neck paler and clearly streaked,

compared with juveniles. Females moulting into first-adult plumage can be aged until moult is completed by almost uniformly brown juvenile secondaries below. Even after the completed first moult, the secondaries tend to be darker below than in older females, the upperparts are dark brown rather than greyish, and the iris is light to medium brown, not yellow as in old females. Distinguishing adult females from juveniles in autumn is fairly easy, as juveniles are darker and more uniform above compared to the greyer and more variegated adult females. However, by spring the plumages become more similar, as juveniles are dull brown above due to wear while adults are still in fairly good plumage. However, the secondary barring of the underwing is distinct and black in adults, appearing less contrasting in young birds, and the body streaking also differs.

SEXING Adults are easily sexed by highly dimorphic plumage. See under Juvenile for sexing of immatures by iris colour: yellowish in males, brown in females.

CONFUSION RISKS Some brightly coloured and finely streaked juveniles could from a distance be mistaken for juvenile Pallid Harrier, but difference in wing-shape and underwing markings should be obvious. Separation from Northern Harrier is very difficult, particularly in juveniles and females, and requires prolonged close views and preferably good photographs.

NOTE Presumed hybrids between Hen Harrier and Pallid Harrier have been recorded in numbers throughout W Europe, but particularly in the Nordic countries in recent years (see under Pallid Harrier). One record of interbreeding between such an adult hybrid male and a female Hen Harrier has been documented from Finland in 2005 (Forsman & Peltomäki 2007).

It is important to check the state of moult of all autumn harriers, as primary moult may change the appearance of both wing-formula and plumage features in all species.

References

Forsman, D. & Peltomäki, J., 2007: Hybrids between Pallid and Hen Harrier – A New Headache for Birders? *Alula* 13: 178–182.

399. Hen Harrier *Circus cyaneus*, juvenile female, migrant. Juveniles typically have rather dark secondaries contrasting with a lighter hand, the underbody is buffish-yellow and the head markings are distinct. Juvenile females are sexed by their dark eyes (brown iris). Finland, 14.9.2014 (DF)

400. Hen Harrier *Circus cyaneus*, juvenile, migrant. Note broad wings with rounded tip, formed by five clearly fingered primaries. Finland, 13.9.2013 (DF)

401. Hen Harrier *Circus cyaneus*, juvenile male, migrant. Juvenile males have yellow eyes, lighter and more clearly banded secondaries, and the primaries are less regularly marked. Finland, 30.9.2011 (DF)

402. Hen Harrier *Circus cyaneus*, juvenile male, migrant. Finland, 24.9.2011 (DF)

403. Hen Harrier *Circus cyaneus*, juvenile female, migrant. Finland, 27.9.2005 (DF)

404. Hen Harrier *Circus cyaneus*, juvenile male, migrant. Juveniles appear rather dark brown above except for white uppertail-coverts and lighter ochre area of median upperwing-coverts. Juvenile males tend to show a grey hue to the inner tail, while juvenile females are darker and browner. Finland, 13.9.2014 (DF)

405. Hen Harrier *Circus cyaneus*, 2nd cy male starting to moult to adult plumage. This bird has dropped its inner five primaries in rapid succession, but most of the plumage is still juvenile. Note that the left side of the tail has already been moulted into adult-type, probably following an accidental loss of these feathers. Finland, 5.6.2010 (DF)

406. Hen Harrier *Circus cyaneus*, 2nd cy male, migrant, close to completing its first moult to adult plumage. Easily aged by retained brown juvenile flight feathers. Finland, 15.9.2014 (DF)

407. Hen Harrier *Circus cyaneus*, 2nd cy male moulting to adult plumage. During the primary moult a black wedge is formed in the wingtip, which in itself may look narrow and pointed owing to missing feathers. Care is needed not to confuse these moulting young males with the much slimmer and more elegant Pallid Harrier. Finland, 25.8.2008 (DF)

408. Hen Harrier *Circus cyaneus*, 2nd cy female, migrant, moulting to adult plumage. 2nd cy females are much more difficult to age than males, but retained largely dark brown juvenile secondaries and a dark iris indicate a bird moulting into first-adult plumage. Finland, 13.8.2012 (DF)

409. Hen Harrier *Circus cyaneus*, 2nd cy female moulting to adult plumage (same as **408**). The unmoulted secondaries are retained juvenile feathers, uniformly brown above. Finland, 13.8.2012 (DF)

410. Hen Harrier *Circus cyaneus*, adult male. Easily identified by five-fingered and widely black wingtip and distinct dusky trailing edge of the wing. The fine underwing barring and some retained dark greater underwing-coverts indicate that this may be a 3rd cy male. Finland, 5.6.2010 (DF)

411. Hen Harrier *Circus cyaneus*, adult male, moulting. In late summer–early autumn adults are moulting, which can create both strange plumage patterns as well as odd wing-formulae, like a three-fingered wingtip on this adult male. Finland, 25.8.2008 (DF)

412. Hen Harrier *Circus cyaneus*, adult male, migrant, moulting. Finland, 21.8.2013 (DF)

413. Hen Harrier *Circus cyaneus*, adult male. Easily identified by silvery-grey upperparts with a broadly black, five-fingered wingtip and plain white uppertail-coverts. Note dusky trailing edge of wing also visible from above. Finland, 5.5.2012 (DF)

414. Hen Harrier *Circus cyaneus*, adult female. Adult females are whiter on the body than juveniles, their underwing-coverts are rufous brown and the secondaries are distinctly banded. Finland, 4.5.2012 (DF)

415. Hen Harrier *Circus cyaneus*, adult female. Note distinctly banded secondaries and complete black trailing edge of the hand, and compare with other ring-tailed harrier species. Finland, 4.5.2012 (DF)

416. Hen Harrier *Circus cyaneus*, adult female. Finland, 5.4.2012 (DF)

417. Hen Harrier *Circus cyaneus*, adult female. Moulting adult females may show odd, pointed wingtips, but note broad wings typical of Hen Harrier. Spain, 17.9.2012 (DF)

418. Hen Harrier *Circus cyaneus*, 2nd cy female. Adult females have more distinctly barred flight feathers above than juveniles, including the greater coverts. Note the difference between the moulted and the brown retained juvenile secondaries of this 2nd cy female towards the end of its first moult. Finland, 14.9.2014 (DF)

NORTHERN HARRIER
Circus hudsonius

Other names: Marsh Hawk (USA)

VARIATION Monotypic. Formerly considered a subspecies of Hen Harrier, but now widely regarded as a full species.

DISTRIBUTION Across all of N America, with northern populations migratory and dispersive, reaching Central America and northernmost parts of S America in winter. Rare vagrant to westernmost Europe, particularly to UK, Ireland and the Atlantic islands, with an increasing number of records, but may have been overlooked in the past.

BEHAVIOUR Mostly seen hunting low over fields and marshes, much like Hen Harrier. Possibly has a greater preference for perching on low bushes and fence posts.

SPECIES IDENTIFICATION Very similar to Hen Harrier, possibly with narrower wings, but detailed study of plumage is required for reliable identification. Adult females are similar to adult female Hen Harriers, but juveniles and adult males are distinctly different. In typical juveniles the contrasting dark neck and head give a distinctly hooded appearance, the streaking of the richly coloured underbody is restricted to the flanks, and the underwing-coverts are often unmarked along the leading edge of the wing, but other individuals are less typical. Adult males show less black on the wingtips, the upperparts are a duller grey and more mottled (less clean-looking) compared to Hen Harrier, with darker markings on the greater coverts, remiges and tail, while from below the underbody is distinctly spotted rufous and the remiges are barred.

MOULT Complete moult from Apr/May to Oct/Nov.

PLUMAGES Normally juveniles, adult females and adult males can be separated by plumage, while juveniles can be sexed based on iris colour. One-year-old immatures, males in particular, can be aged as long as they retain juvenile feathers, normally until Aug–Sep of the 2nd cy.

Juvenile

Largely similar to juvenile Hen Harrier, with the main differences found in the plumage of the head and underparts. Facial disc and neck dark brown, often forming a conspicuous boa around the neck, making the whole head contrast sharply against the light underbody. Breast and belly mostly uniform orange-buff, deeper in colour compared to most juvenile Hen Harriers, limited streaking confined to the sides of the lower breast, flanks and axillaries in particular. Undertail-coverts often deeper in colour than the belly, appearing somewhat darker. Underwing-coverts often unmarked and pale along the leading edge, contrasting sharply against darker greater coverts, but this character is variable and also occurs in some juvenile Hen Harriers. As a rule the primaries are more densely barred compared to juvenile Hen Harrier, but this is best judged from photographs: the longest primaries show six bars (rarely 5–7) excluding the dark tip in Northern, compared to 4–5 (occasionally six) in Hen Harrier, with variation in both species.

Juvenile plumage is retained until the late 2nd cy spring, when the complete moult starts. Birds in late spring are sometimes extremely faded and can appear almost whitish below. Immatures can be recognised until at least late summer of the 2nd cy by a mixture of juvenile and adult-type feathers in the plumage; this is particularly striking in males.

Juveniles can be sexed by iris colour as early as the first autumn: in males grey or yellowish, later turning yellow, in females dark brown.

First-adult male

Until moult is completed in the 2nd cy autumn, reliably aged by retained brown juvenile feathers. Males with extensively and rather distinctly barred uppertail and remiges, and with a considerable amount of rufous-brown on the head and upper breast are not necessarily first-adults, as this type of plumage has also been recorded in older males (J. Liguori *in litt.*).

Adult male

Compared to adult male Hen Harrier looks a duller grey, with darker mantle and darker diffuse bands across the upperwing-coverts, often with some retained bars across the remiges and tail; similar barring is lacking in adult male Hen Harrier. The primaries are black only distally; normally just the fingers are dark, which is a good character compared to male Hen Harrier. The underbody and underwing-coverts show distinct rufous markings which are lacking in adult male Hen Harrier.

Adult female

Very similar to adult female Hen Harrier, but with a tendency for more strongly and more distinctly marked flanks, axillaries and greater and median underwing-coverts, and with the longest primaries showing more bands compared to Hen Harrier, mirroring the differences of juveniles. Distinguished from juveniles by more distinct markings and the more spotted appearance of the underbody and underwing-coverts, while the head and neck are more clearly streaked, resembling female Hen Harrier. Upperwings are distinctly banded across the remiges and greater upperwing-coverts, with broad dark bands on a greyish ground colour, more striking compared

to the uniformly dark brown appearance of juveniles, but also more distinctly marked in comparison with the average adult female Hen Harrier.

SEXING Adults are easily sexed by highly dimorphic plumage. See under Juvenile for sexing of immatures by iris colour: yellowish in males, brown in females.

CONFUSION RISKS Most likely to be mistaken for a Hen Harrier, but juveniles could also be mistaken for juvenile Pallid Harrier. Juveniles are even more similar to hybrid Hen x Pallid juveniles, which is a possibility that needs to be taken into account in a European context.

References

Mullarney, K. & Forsman, D. 2011. Identification of Northern Harriers and vagrants in Ireland, Norfolk and Durham. *Birding World* 23: 509–523.

419. Northern Harrier *Circus hudsonius*, juvenile male. Structurally similar to Hen Harrier, but note darker head and neck, streaking confined to upper breast, and finer and denser barring of the fingered primaries (see text). USA (Jerry Liguori)

420. Northern Harrier *Circus hudsonius*, juvenile male. Dark head and neck typically contrasts against lighter and plainer lower breast and belly. Note up to seven crossbars to longest primaries and compare with juvenile Hen Harrier. USA (Jerry Liguori)

421. Northern Harrier *Circus hudsonius*, juvenile female. This female has fewer bars across its primaries, but shows typical head and underbody. The secondaries are on average lighter coloured in juvenile female Northern compared to juvenile female Hen. USA (Jerry Liguori)

422. Northern Harrier *Circus hudsonius*, adult male. Adult males are variably marked with rusty-orange spots and bars on breast and flanks. USA (Jerry Liguori)

423. Northern Harrier *Circus hudsonius*, adult male. Male Northern Harriers show less black in wingtips (fingers only) and the dark trailing edge is clearly darkest on the secondaries (cf. adult male Hen Harrier). USA (Jerry Liguori)

424. Northern Harrier *Circus hudsonius*, adult male. Adult males are more mottled above, compared to the clean appearance of adult male Hen Harriers, and the wingtip shows less black. USA (Jerry Liguori)

PALLID HARRIER
Circus macrourus

VARIATION Monotypic.

DISTRIBUTION Has recently expanded its range rapidly towards NW Europe, and it is now a rare but regular breeding species in Finland, and possibly also in N Sweden and Norway.

BEHAVIOUR Hunts low over open areas like other harriers, but flight often faster and more direct, even falcon-like, compared to Hen or Montagu's. Covers much larger areas when hunting than Montagu's, which often patrols the same field over and over again. Often seeks shade in trees for the hottest hours of the day.

SPECIES IDENTIFICATION Pallid Harrier shows the most pronounced sexual size dimorphism of all harriers, with males considerably smaller than females. Males are the smallest of all harriers, while females are the size of female Hen Harrier. Adult males are easy to tell from other male harriers by diagnostic wingtip pattern, but beware of a similar pattern in male Hen and Montagu's Harriers during primary moult in late summer/early autumn. Adult females may greatly resemble both adult female Montagu's and adult female Hen Harrier, and details of wing-formula, underwing markings and head pattern are essential: plumage in many respects is similar to adult female Hen Harrier, but Pallid can be recognised by slimmer body and wings, with narrower and more pointed hand in particular. More similar to female Montagu's in silhouette, but differences in the banding of the axillaries and underwing, particularly of the hand, is diagnostic (see below). Juveniles may be very similar to juvenile Montagu's, and care is needed to separate them. They are best told by differences in the pattern of the primaries and head, both of which need to be checked.

Distant birds can often be identified by flight alone when hunting: Pallid has a fast, strong and direct flight (especially when hunting), covering ground with the speed of a falcon. They appear in the distance, fly straight across the hunting area and disappear at the other end, to return perhaps only hours later. Hunting Montagu's, however, appear slow and lazy, and the flight is swaying and hesitant. They often patrol the same field, ditch or embankment back and forth, making slow progress.

The wing-formula is rather similar to Montagu's, juveniles in particular, with only four outermost primaries fingered, and with no emargination on the outer vane of p6, while Hen Harrier has five fingered primaries and p6 has a distinct emargination on the outer web, about halfway out on the feather.

MOULT The complete moult in adults takes place between May and Aug–Sep, and as a rule is completed earlier than in Montagu's or Hen Harriers. Many adults will have already completed their primary moult before the autumn migration. During their first winter juveniles moult considerably less than young Montagu's and most return in spring in a more or less complete, albeit worn, juvenile plumage. Some first-summer immatures (2nd cy), both males and females, may show some streaks across the upper breast, and only very rarely will some males have replaced the body plumage more extensively, showing some greyish hues above and whitish feathers on the belly and flanks/axillaries. The first complete moult appears to start earlier than in adults, from late Apr onwards.

PLUMAGES Three plumage-types can be easily separated: adult male, adult female and juvenile. Under favourable conditions juveniles can be sexed by iris colour, and adults can be separated into first-adults and older birds.

Juvenile

Very contrasting in fresh plumage, with dark brown upperparts and rich orange-buff underparts. Pale patch in upperwing-coverts usually brighter and more uniform compared to the more broken-up pattern in Hen and Montagu's Harriers. Primary markings below are diagnostic, with banding mostly confined to the median section of each feather, with base and particularly distal parts paler, inner bands broader than the outer bands, and with each individual crossbar irregular in shape, often hooked and bent following the edge of the feather, or dissolved to form irregularly-shaped spots (cf. Montagu's Harrier). *Normally the fingered primaries are only narrowly tipped dark, while most juvenile Montagu's show uniformly dark fingers.* This gives the impression of an open and light-coloured hand in Pallid, while most (but not all) Montagu's show a prominent black triangle on the wingtip. The inner primaries have diffusely darker tips in juvenile Pallid, while in Montagu's the tips are blacker and more contrasting. Note, however, that there is quite some variation in the primary barring in both species, and some individuals can be very similar indeed.

The secondaries appear dark below from a distance, but closer views reveal 1–2 paler bands, usually at least one, running just inside the dark trailing edge and gradually fading out towards the body.

In addition to the underwing pattern the head pattern is also important for identification. Typically, juvenile Pallid shows a much darker face than juvenile Montagu's, with a solid dark facial disc, *dark lores, a smaller white spot below the eye (typically joining the gape-line), and a complete and uniform, distinct pale facial collar, encompassing the head completely from the white nape-patch to the throat.* However, head-markings also vary and some juveniles are lighter with a less striking collar.

Juveniles can safely be sexed by iris colour: dark brown in females, and in males grey at fledging, turning bright yellow during the first autumn. Further, *females tend to show stronger and more complete barring on the underwing primaries, while the secondaries are more uniformly dark and less distinctly marked compared to juvenile males,* much reflecting similar differences in juvenile Hen and Montagu's Harriers.

In the 2nd cy spring, birds still retain their juvenile plumage, although by now it is heavily bleached and worn, differing in this respect from the average juvenile Montagu's Harrier. *Most birds, males as well as females, will not have moulted any feathers during the first winter,* while some show a few new streaked feathers on the collar, boa, upper breast and breast-sides. However, some males undergo a more extensive body moult and show a whitish belly and some grey feathers on the head and upperparts, while the pale collar still remains very distinct, narrow and sharply defined (but *beware of immature Montagu's males with similarly striking collars*).

First-adult male

This plumage can be recognised even after the completed first moult by dusky, rather brownish-grey upperparts, with a poorly defined dark wedge to the wingtip above (although this varies), a diffuse and narrow dark trailing edge to the underwing, and a rather well demarcated greyish breast, often with some fine streaking. The head shows a brownish facial disc bodered by a lighter collar, a browner crown and

prominent white markings above and below the eye. The central tail feathers are diffusely barred, with a slightly broader and darker subterminal band, while the outer tail feathers are whitish with about five grey bands. In this plumage the dark wedge at the wingtip tends to be larger compared to older adults, with p6–9 largely black from below, although this varies individually. In fresh autumn plumage many birds show a rusty wash to the breast, with individually variable rufous streaking, but by next spring only some fine grey streaks may remain and the bird looks 'older'. Many males of this age-class also show a greyish carpal crescent below and diffuse, fine barring on the base of the inner primaries and secondaries, all features which are retained into 3rd cy spring/summer.

Definitive adult plumage is more uniform and paler grey above compared to first-adults and the dark wedge at the wingtip is often smaller, with just pp6–8 largely black, while p9 is dark only at the tip in most and p10 shows no dark at all. The central tail feathers are uniformly grey above, while the outers are whitish with finer grey bars compared to younger males. The underparts are also whiter, with a whitish throat and whitish-grey upper breast, the latter hardly contrasting with the white lower breast and belly.

Note that even adult males can look surprisingly mottled and brownish above, particularly just before and during the summer moult, and because of this they are often misidentified as first-adults. This appearance is explained by moult and plumage wear, where fresh silvery feathers lie next to worn feathers which are browner as the silvery bloom has worn off exposing the brownish lower layer of the feather.

Adult male

Easy to identify by pale pearl-grey upperparts, largely whitish underparts and striking black wedge on the wingtip (note that moulting first-summer Hen and Montagu's males may show a similar wedge in late summer, in Jul–Sep). The uppertail-coverts are white, with fine grey barring. With increasing age adult males become paler and cleaner-looking, particularly on the breast, head and upperparts, and the barring on the outer tail feathers becomes finer and less distinct. Iris is yellow.

First-adult female

(after completed moult in 2nd cy autumn). Coloration still rather dark and similar to juvenile, with darkish brown upperparts and mostly dark, contrasting head pattern, but with underbody showing darker streaking particularly across the upper breast. Secondaries appear dark from a distance, contrasting sharply against the paler hand, but closer views reveal a light band inside the trailing edge, which gradually fades towards the body.

Older females are overall lighter coloured and are therefore easily dismissed as either Hen or Montagu's females, especially distant birds. The head markings are far less contrasting compared to younger birds, the secondaries are lighter below with distinct banding, thus being more similar to adult female Montagu's Harrier, and the underbody is pale with fine brown streaks, but usually still *retaining the division between the more heavily marked upper breast and more finely streaked lower parts.* The upperparts are faded greyish-brown, with a lighter area in the coverts, and the uppertail-coverts are white with greyish barring. Birds in this plumage may appear very similar to adult female Hen Harriers, particularly during their late summer primary moult, when the diagnostic differences in the wing-formula are lost. However, *adult female Pallid Harriers can be recognised by their primary pattern below, with often quite reduced barring, leaving the trailing edge of the hand largely unmarked and pale grey.* The secondaries, although often clearly darker than the primaries below, can be rather pale and clearly banded, thus quite similar to adult female Montagu's or Hen Harriers. In close views the pale, yet dark-spotted collar can be seen to be well-defined, complete and narrow, although it does not stand out as clearly as in younger birds, being in fact very similar to the collar of female Hen Harrier. The dark cheek-patch, so pronounced in juveniles and younger adult females, may by now be rather washed-out in colour, adding to the similarity with Hen Harrier females, but it is more uniform in colour and not clearly streaked as in Hen Harrier.

Adult female Differs from juvenile by streaked underbody, while the trailing edge of the hand is notably pale and translucent; the latter feature is diagnostic compared to any ringtail Hen or Montagu's Harriers. Secondaries are dark with 1–2 narrow pale bands, but from a distance they look dark, contrasting strongly with the lighter primaries, a feature typical of adult female Pallid, but not found in adult female Montagu's. Underwing-coverts are clearly two-toned, with pale and almost unmarked lesser coverts contrasting with darker and more strongly marked median and greater coverts, the latter with pale spotting (cf. female Montagu's). Breast streaking is heavier on the upper breast, contrasting with the finely streaked lower breast and belly, while in adult female Montagu's the streaking appears uniform over the entire underbody. Axillaries are often rather dark chestnut with pale spotting, differing from the boldly barred (chess-board pattern) axillaries of female Montagu's.

SEXING Adults are easily sexed by highly dimorphic plumage. Juveniles can be sexed by iris colour as soon as they fledge, with the difference becoming more marked as autumn progresses: iris is grey in males at fledging, gradually turning yellow, while females show a dark brown iris from the beginning. A simple rule of thumb is that if the black pupil can be made out, it's a male, while in females the eye looks all dark. Juvenile females also tend to have darker secondaries, more completely and more regularly barred primaries, a darker neck-boa and darker head-markings compared to juvenile males, but there is some overlap in these characters.

CONFUSION RISKS Adult males are rather straight-forward to identify when seen well, but ringtails, particularly adult females can be tricky to tell from other ringtail harriers, which see.

NOTE Hybridises successfully with Hen Harrier in N Europe, where ranges nowadays overlap widely. Has also successfully hybridised with Montagu's Harrier, raising young. Hybrid Hen x Pallid Harriers can be very similar to juvenile Northern Harriers, with wing-formula and underwing markings being the most important features to focus on. Many hybrids/intergrades are probably not possible to recognise as such in the field, without the aid of good quality photographs. Since the successful hybridisation between these harriers has been verified only fairly recently, very little is known about the magnitude of this phenomenon, but it appears to happen on a rather large scale in N Europe. Although first-hand information of actual mixed pairs is scanty, it appears that first generation hybrids (F1) are relatively easy to identify, when seen well, since they feature a mosaic of Pallid and Hen Harrier characters. As hybrids appear to be fertile, difficult-to-identify back-crosses should also be expected. The offspring of such a pair, where one parent is a hybrid and the other is either of the two 'pure' species, could apparently be almost identical to the 'pure' parental species involved, and would most probably escape detection in normal field conditions.

Interestingly, most documented hybrids have been males, which could mean that females have a reduced viability, or that they could be more easily overlooked.

References

Forsman, D. 1993. Hybridising Harriers. *Birding World* 6: 313.

Forsman, D. 1995. Male Pallid and female Montagu's Harrier raising hybrid young in Finland in 1993. *Dutch Birding* 17: 102–106.

Forsman, D. & Peltomäki, J. 2007. Hybrids between Pallid and Hen Harrier – A New Headache for Birders? *Alula* 13: 178–182.

Forsman, D. & Erterius, D. 2012. Pallid Harriers in northwest Europe and the identification of presumed Pallid Harrier x Hen Harrier hybrids. *Birding World* 25: 68–75.

425. Pallid Harrier *Circus macrourus*, juvenile female, migrant. A typical juvenile in all respects. Dark iris in juvenile indicates female. Finland, 20.9.2014 (DF)

426. Pallid Harrier *Circus macrourus*, juvenile female. Oman, 9.12.2012 (DF)

427. Pallid Harrier *Circus macrourus*, juvenile male, spring migrant. Note reduced primary-barring. Israel, 20.3.2007 (DF)

428. Pallid Harrier *Circus macrourus*, juvenile female, migrant (same as **425**). Compare wing-shape, wing-formula and head markings with **404**. The ochre upperwing-coverts patch is usually more solid in juvenile Pallid Harrier compared to a more broken-up, scaly pattern in juvenile Montagu's. Finland, 20.9.2014 (DF)

429. Pallid Harrier *Circus macrourus*, 2nd cy male, migrant. Note dusky flight feathers and more extensively black wingtips compared to adult males, but the variably streaked breast is the best ageing feature of 2nd cy autumn males. Spain, 8.12.2012 (Javier Elorriaga)

430. Pallid Harrier *Circus macrourus*, second plumage male (3rd cy), spring migrant. Note dusky trailing edge of wing and retained streaking on breast and underwing-coverts, while head often shows retained markings of light collar, darker cheeks and white areas around the eye. Israel, 26.3.2008 (DF)

431. Pallid Harrier *Circus macrourus*, adult male, spring migrant. Slim and white with a black wedge to the wingtip makes an adult male Pallid Harrier unmistakable. Israel, 26.3.2010 (DF)

432. Pallid Harrier *Circus macrourus*, adult male, spring migrant. When hunting, Pallid Harriers are faster than any other harrier, their behaviour being almost falcon-like at times. Israel, 20.3.2010 (DF)

433. Pallid Harrier *Circus macrourus*, adult male, autumn migrant. Oman, 24.11.2014 (DF)

434. Pallid Harrier *Circus macrourus*, younger adult female. Note typical underwing barring and black tail-bands (chestnut-red in adult female Montagu's). Aged as a younger adult by dark iris. Cameroon, 4.4.2006 (Ralph Buij)

435. Pallid Harrier *Circus macrourus*, adult female, migrant. Note typical head-markings, dark secondaries and translucent trailing edge of hand, and compare with adult females of Hen and Montagu's Harriers. Finland, 11.9.2013 (DF)

436. Pallid Harrier *Circus macrourus*, younger adult female. Typically darker secondaries contrast with lighter hand. Note uniformly dark ear-coverts patch framed by spotted collar. Cameroon, 14.1.2007 (Ralph Buij)

437. Pallid Harrier *Circus macrourus*, adult female, spring migrant. Older females have lighter underwings, but secondaries remain darker and trailing edge of hand is grey and indistinct (cf. adult female Montagu's). Israel, 29.3.2012 (DF)

438. Pallid Harrier *Circus macrourus*, older adult female. Note diagnostic light trailing edge to hand and dark primaries with one lighter band. Aged as older female by yellow iris. Cameroon, 27.2.2008 (Ralph Buij)

439. Pallid Harrier *Circus macrourus*, adult female, migrant. Another, presumably older, female showing typical contrast between darker arm and lighter hand. Finland, 14.9.2014 (DF)

440. Pallid Harrier *Circus macrourus*, older adult female. Some adult females show rather light underwings and may therefore recall adult female Montagu's, but note light trailing edge and typical barring of hand, dark tail-bands and typically spotted underwing-coverts. Aged as older female by yellow iris. Cameroon, 22.2.2008 (Ralph Buij)

441. Pallid Harrier *Circus macrourus*, adult female, migrant (same as **435**). Adult females have rather dark secondaries above and, if anything, the trailing edge is darkest (cf. adult female Montagu's). Finland, 11.9.2013 (DF)

442. Pallid Harrier *Circus macrourus*, adult female, migrant (same as **439**). Older females tend to get a clearly greyish uppertail with distinct dark banding. Finland, 14.9.2014 (DF)

253

MONTAGU'S HARRIER
Circus pygargus

VARIATION Monotypic. A conspicuous dark morph occurs in Spain and Portugal, but is rare elsewhere (see below).

DISTRIBUTION Widespread but patchy distribution across Europe reaching W Siberia and C Asia in the east. Winters in sub-Saharan Africa and S Asia.

BEHAVIOUR Mostly seen quartering low over fields and meadows in typical harrier fashion. Hunting flight often lazier and more swaying than Pallid Harrier's, the difference being more obvious in adults than in juveniles, but when migrating both species can appear very similar. *When hunting appears to make slow progress, patrolling the same field, ditch or embankment back and forth, in contrast to the dashing appearance of Pallid Harrier.*

SPECIES IDENTIFICATION A long-winged and long-tailed raptor with a slim body. Typically, in side views the body is widest around the chest, gradually tapering towards belly and vent, and appears to merge into the tail level with the trailing edge of the wings. The head appears small compared to Pallid and Hen Harriers. Wings are long and narrow, with a long and pointed hand, but note that *autumn juveniles are shorter-winged and shorter-tailed than adults*, greatly resembling Pallid Harrier in proportions.

Seen well, adult males are unmistakable, but adult females and juveniles are easily confused with other ringtail harriers and multiple plumage characters are needed to confirm identification. The rare dark morph could easily be mistaken for a Marsh Harrier by the unwary, although the slim silhouette and the graceful flight should soon reveal the mistake.

MOULT A long-distance migrant with moult typically starting during breeding, but arrested for the autumn migration, to be resumed and completed in the winter quarters. Juveniles undergo an individually variable body moult during their first winter, which can be extensive. Accidentally lost feathers are replaced by adult-type feathers, even in fresh autumn juveniles. Unlike juvenile Pallid Harrier, *most immatures returning in spring show a mixture of moulted adult-type and retained juvenile body feathers*, resulting in an individually highly variable transitional plumage, which is often very different from the autumn juvenile. Despite the sometimes extensive body moult the juvenile remiges are retained, the primaries often providing a reliable clue to distinguish between spring immatures of Pallid and Montagu's Harriers.

PLUMAGES Two age-classes, juveniles and adults, are easily separated by plumage alone, while retained juvenile remiges separate 2nd cy birds from adults until late summer and autumn, with males in particular appearing in a striking plumage. Under favourable conditions adult-type birds can be further divided into young adults and older adults.

Juvenile

The juvenile plumage is a striking combination of dark upperparts with white uppertail-coverts, and rich rusty-yellow or coppery underparts. The diagonal ochre patch across the upperwing-coverts appears scaly compared to the more solid and brighter patch of juvenile Pallid Harrier. The head pattern is individually variable and in particular the pale facial collar varies from practically non-existent to an *almost Pallid-like full collar*, albeit mostly broken by darker feathering at the top, which is crescent-shaped (being widest mid-way and tapering towards the tips) and often also dark-spotted (cf. juvenile Pallid Harrier). The dark ear-coverts often appear as an isolated dark patch, but in some darker birds the patch is more extensive and reaches the gape. *Most juveniles show much more white around the eye and also a lighter loral area compared to juvenile Pallid Harriers*, giving them a more open-faced and friendlier look, but darker birds may show just small whitish flecks above and below the eye (fortunately these dark birds seem to lack a prominent pale collar).

The underparts tend to be more coppery in colour compared to more yellowish juvenile Pallids, but this tanned colour fades during autumn making the difference less useful as the season progresses. Juvenile Montagu's often show faint, diffuse darker streaks on the upper breast and flanks, sometimes only visible in the best of views, while Pallids appear uniform, although the odd diffuse and narrow shaft-streak may be visible in some, usually on the flanks next to the axillaries.

The underwing barring is probably the single most reliable character to differentiate between juveniles of Pallid and Montagu's, although this character should preferably be used alongside the head pattern. Most juvenile Montagu's show all-dark fingers below, something that practically never occurs in juvenile Pallid, and in many birds the barring of the hand is confined to the inner primaries. However, some birds have entirely barred primaries (possibly more common in females), in which case other features need to be consulted. The dark trailing edge of the hand is darker and more distinct in Montagu's than in young Pallid, while the barring tends to be finer and more regular in Montagu's, or it may be almost missing.

The secondaries also vary in pattern on the underside, from practically all-dark (mostly females) to more slate-grey with darker bands (typically males), and some are quite similar to juvenile Pallid.

Juveniles can easily be told at distance from older female Montagu's throughout the autumn by their very dark upperparts and the rich colour of their underparts. Adult and 2nd cy females are much lighter, faded greyish-brown above and lighter below.

Unlike Pallid and Hen Harriers the juvenile plumage is only retained for the first autumn, and the partial body moult in winter changes the plumage considerably.

Juveniles and first-winter birds can be reliably sexed by iris colour: in males greyish at fledging, turning yellow before autumn, and dark brown in females.

Transitional plumage (first-winter to 2nd cy autumn)

An immensely variable plumage, depending on the extent of the body moult during the first winter. In extreme cases birds return in a fully juvenile plumage (rare), while others can look almost adult but, importantly, *they always retain their juvenile primaries and secondaries*. In this plumage many birds sport a distinct pale collar, which can be misleadingly similar to that of juvenile Pallid Harriers.

First-summer male

In 2nd cy spring recognised by combination of yellow iris and juvenile remiges. Most birds have acquired some lead-grey on the head and breast and the underparts may be partly white with distinct chestnut streaks. Axillaries are often moulted, being white with a diagnostic rufous chess-board pattern. Many birds have also moulted the central tail feathers, which are clearly longer than the juvenile feathers and grey in colour.

By late 2nd cy summer/autumn most birds appear largely grey and adult-like in the field, but are always identified by retained brown secondaries and outer primaries, and the breast streaking tends to be broader and more distinct compared to older males.

First-summer female

In 2nd cy spring most females show a dark iris and, importantly, juvenile remiges. Most also show some streaking below, at least on the upper breast, while others are more adult-like with completely streaked underparts. Any moulted tail feathers stick out from the tail-tip, as they are noticeably longer than the retained juvenile feathers. Head pattern varies, *but many show a distinct pale, Pallid-like collar.*

By late summer/autumn ageing becomes more difficult with the progressing moult, but immature females can always be aged by their retained juvenile secondaries, which appear very faded brown above and rather uniformly dark below. The difference between dark moulted and retained faded feathers in the upperwing is more striking than in older females, and the wings are clearly shorter. Note that the new secondaries will be very dark above, lacking the adult female's greyish colour with dark bands. The iris colour varies from brown to amber, but can in some birds already be nearly yellow.

Adult male

Males have tricoloured upperparts with lead-grey mantle and wing-coverts, while the remiges are silvery-grey and the wingtip is extensively black. A diagnostic black line runs across the secondaries and onto the inner primaries. The head and upper breast are grey and the belly is white with fine rusty streaks, broader in younger adults. The underwing-coverts are white with prominent chestnut markings, gradually becoming finer and less obvious with age. The central tail feathers are grey, while the outers are white with broad chestnut bands.

Adult female

Compared to adult female Pallid Harrier the remiges of Montagu's appear rather pale below when seen from a distance, while in Pallid the secondaries stand out as darker, contrasting with a clearly lighter hand. The primaries are evenly barred and the secondaries show a dark trailing edge and two black bands further in. Unlike adult female Pallid the underbody appears uniformly streaked and the underwing-coverts evenly marked, lacking the underwing contrasts of the former. The upperparts are greyish-brown, with a clearly visible dark band at the base of the secondaries and the hand is greyer with visible dark barring.

Females in first-adult plumage differ from older females in having darker, browner, less grey upperparts, with rather dark secondaries, usually lacking a clear secondary band. The uppertail is darker brown, not greyish, and the primaries lack the greyish cast and distinct barring of older females. Iris colour varies from speckled amber to bright yellow in older birds.

Dark morph

A rare dark morph occurs in the Iberian peninsula, where dark birds may locally comprise up to 10% of the population. All plumages lack the white uppertail-coverts of normally coloured birds, but the markings of the underwing remiges are the same, although less distinct.

Juvenile The plumage is overall blackish-brown when fresh, as in juvenile Marsh Harrier, except for a pale nape-patch. The remiges show the normal juvenile pattern below, although less contrasting. Juveniles can be sexed by iris colour as described above. During winter and spring the plumage fades to become dark brown, as in juvenile Marsh Harrier.

Transitional plumage (first-winter to 2nd cy autumn) Birds can be aged using the same criteria as for normally coloured birds, with retained juvenile remiges being the key to ageing and the iris colour the key to sexing.

Adult male Males are overall dark slate-grey, with black primaries and a black band visible across the underwing secondaries. Iris is bright yellow, legs and cere almost orange-yellow.

Adult female Females are dark brown, darker in fresh plumage, but can be recognised by the diagnostic wing-barring, both above and below, similar to normal adult females. Iris is pale brown at first, gradually turning yellow with increasing age.

SEXING Size difference is minimal between the sexes, unlike Hen and Pallid Harriers, but the plumages differ in adults, while iris colour differs in male and female in juveniles as well as adults.

CONFUSION RISKS Adult males are distinctive, while juveniles and females could easily be mistaken for other ringtail harriers. A combination of slow and lazy hunting behaviour, sleek proportions, particularly noticeable in the long-tailed and long-winged adults, and diagnostic plumage features, such as the type of barring of the underwing primaries and secondaries and the head markings allow for a safe identification of all plumages.

NOTE Known to have hybridised with Pallid Harrier in Finland in 1993, where an adult male Pallid and an adult female Montagu's Harrier fledged three hybrid chicks (Forsman 1993, 1995).

References
Forsman, D. 1993. Hybridising Harriers. *Birding World* 6: 313.
Forsman, D. 1995. Male Pallid and female Montagu's Harrier raising hybrid young in Finland in 1993. *Dutch Birding* 17: 102–106.

443. Montagu's Harrier *Circus pygargus*, juvenile, migrant male. Typically juveniles show black secondaries, all-dark fingers and a rather distinct dark trailing edge to the hand, while the barring of the primaries is confined to the proximal parts of the inner primaries. Note also typical head markings, with poorly defined collar and extensive white areas around the eye. Light iris denotes a male. Spain, 5.9.2013 (DF)

444. Montagu's Harrier *Circus pygargus*, juvenile, migrant male. Note typical partly unmarked primaries with extensively dark tips as well as typical head pattern. Spain, 5.9.2013 (DF)

445. Montagu's Harrier *Circus pygargus*, juvenile, migrant male. Extensively dark wingtips and trailing edge of hand are typical of juvenile Montagu's. This bird also shows a typical head. Israel, 27.9.2008 (DF)

446. Montagu's Harrier *Circus pygargus*, juvenile, migrant female. This bird shows the other type of primary barring in juvenile Montagu's, with regular barring throughout. Note uniformly dark secondaries. Dark iris denotes a female. Israel, 2.10.2008 (DF)

447. Montagu's Harrier *Circus pygargus*, juvenile male, migrant. Despite its young age this bird has replaced some accidentally lost feathers with adult-type feathers (one primary, one tail feather and some breast feathers). Spain, 3.9.2013 (DF)

448. Montagu's Harrier *Circus pygargus*, juvenile male, migrant (same as **443**). Upperparts of a typically marked juvenile. Spain, 5.9.2013 (DF)

449. Montagu's Harrier *Circus pygargus*, juvenile female, migrant. The light patch of the upperwing-coverts is typically scalloped in juvenile Montagu's, as here, compared to the more uniformly coloured area of most juvenile Pallid. Spain, 3.9.2013 (DF)

450. Montagu's Harrier *Circus pygargus*, 2nd cy male in transitional plumage, breeding. First-summer males turn up in a highly variable plumage. This one has replaced most of the head and upper breast feathers, some axillaries and underwing-coverts, and has just shed its two innermost primaries. Spain, 5.5.2008 (DF)

451. Montagu's Harrier *Circus pygargus*, 2nd cy male in transitional plumage, spring migrant (with Blackcap). This male has replaced most of its body plumage, partly with adult-type feathers, but also with feathers intermediate in colour between juvenile and adult. Israel, 17.4.2009 (DF)

452. Montagu's Harrier *Circus pygargus*, 2nd cy male, spring migrant (same as **451**). Except for retained juvenile primaries, secondaries, primary coverts and alula, this bird has also replaced most of its upperparts, including its tail. Israel, 17.4.2009 (DF)

453. Montagu's Harrier *Circus pygargus*, 2nd cy male. Another rather extensively moulted juvenile in transitional plumage, with retained juvenile flight feathers and tail only. Spain, 25.4.2006 (DF)

454. Montagu's Harrier *Circus pygargus*, 2nd cy male, autumn migrant in transitional plumage. Easily aged by retained juvenile primaries and secondaries. Spain, 3.9.2013 (DF)

455. Montagu's Harrier *Circus pygargus*, 2nd cy male, autumn migrant in transitional plumage (same as **454**). Spain, 3.9.2013 (DF)

456. Montagu's Harrier *Circus pygargus*, 2nd cy female, autumn migrant in transitional plumage. Transitional females are best aged by their mixture of retained brown and fresh barred secondaries, while differentiating between juvenile and adult-type primaries can be difficult. Spain, 4.9.2013 (DF)

457. Montagu's Harrier *Circus pygargus*, 2nd cy female in transitional plumage, autumn migrant (same as **456**). The upperparts are more worn compared to adults and the faded juvenile secondaries stand out clearly. Spain, 4.9.2013 (DF)

458. Montagu's Harrier *Circus pygargus*, adult female and male soaring together. Adults are much longer-winged and longer-tailed compared to juveniles. Spain, 17.4.2010 (DF)

261

459. Montagu's Harrier *Circus pygargus*, adult male, autumn migrant. Note the suspended moult in the primaries of this fully adult male. Spain, 7.9.2013 (DF)

460. Montagu's Harrier *Circus pygargus*, adult male, autumn migrant. In older adults, such as this bird, the grey of the breast stretches further down and the streaking is finer compared to first-adults. Spain, 7.9.2013 (DF)

461. Montagu's Harrier *Circus pygargus*, adult male, autumn migrant (same as **460**). Spain, 7.9.2013 (DF)

462. Montagu's Harrier *Circus pygargus*, breeding adult male in spring. In fresh plumage males are silvery-grey above, brightest on the inner hand, with slightly darker grey arm and mantle, and largely black primaries. Spain, 20.4.2010 (DF)

463. Montagu's Harrier *Circus pygargus*, adult female in spring, breeding. Note distinctly barred flight feathers throughout, while underwing-coverts are boldly marked with chestnut, and compare with female Pallid. Spain, 25.4.2006 (DF)

464. Montagu's Harrier *Circus pygargus*, adult female, autumn migrant. Told from younger females in transitional plumage by clearly barred secondaries throughout. Spain, 5.9.2013 (DF)

465. Montagu's Harrier *Circus pygargus*, adult female, autumn migrant. With increasing age, as in this older female, the banding of the flight feathers and the markings of the underwing-coverts become finer and the iris turns yellow. Spain, 15.9.2012 (DF)

466. Montagu's Harrier *Circus pygargus*, adult female in spring, breeding. Adult female Montagu's is more faded and greyer above compared to adult female Pallid, and shows a distinct dark band across the base of the secondaries. Spain, 25.4.2006 (DF)

467. Montagu's Harrier *Circus pygargus*, dark morph, juvenile male, autumn migrant. Note all-dark plumage, apart from normally patterned yet darker remiges. Told from Marsh Harrier by four-fingered wing-formula, more blackish colour and faintly barred primaries and tail. Spain, 9.9.2004 (DF)

468. Montagu's Harrier *Circus pygargus*, dark morph, juvenile female, autumn migrant. Similar to juvenile male in **467**, but remiges more clearly barred. Spain, 8.9.2004 (DF)

469. Montagu's Harrier *Circus pygargus*, dark morph, juvenile female, autumn migrant. Spain, 17.9.2011 (DF)

470. Montagu's Harrier *Circus pygargus*, dark morph, juvenile male (same as **467**), autumn migrant. All dark, but faint barring visible on underwing and tail. Spain, 9.9.2004 (DF)

471. Montagu's Harrier *Circus pygargus*, dark morph, 2nd cy transitional male. Grey uppertail-coverts are typical of dark morph adult males, as are all-black primaries. Grey greater upperwing-coverts are only found in males. Spain, September 2009 (Markku Saarinen)

472. Montagu's Harrier *Circus pygargus*, dark morph, 2nd cy transitional female, autumn migrant. Note moulted inner primaries with female-type barring; secondaries are retained juvenile feathers. Dark eye, as in juvenile female. Spain, 8.9.2013 (DF)

473. Montagu's Harrier *Circus pygargus*, dark morph, adult male. Identified by all-grey plumage with uniformly dark flight feathers. Spain, September 2009 (Markku Saarinen)

474. Montagu's Harrier *Circus pygargus*, dark morph, adult female, autumn migrant. More clearly barred secondaries, above and below, compared to juvenile. Note also yellow iris, in female indicating an old adult. Spain, 14.9.2011 (DF)

Identification of hybrid harriers

The identification of hybrid harriers has become topical since the rapid breeding range extension of Pallid Harrier into Northern Europe, followed by peaking numbers of presumed hybrids over much of Europe, particularly in autumn 2011. So far very few successful interbreedings have been documented (Forsman 1993, 1995, Forsman & Peltomäki 2007), but the number of presumed hybrids, especially between Hen and Pallid Harriers and particularly autumn juveniles, indicates that hybridisation occurs frequently.

The most common interbreeding appears to involve Pallid and Hen Harriers, with tens of presumed hybrids documented from the Nordic countries through Western Europe south to Spain. Most of the records have been discovered only thanks to good quality photographs and very few have been identified as hybrids/intergrades in the field. It further appears that hybrid offspring are fertile and capable of reproducing, leading to an array of intergrades more or less approaching the phenotype of either parent species (Forsman & Peltomäki 2007). These highly variable backcrosses (intergrades) naturally add to the identification difficulties.

In order to be able to identify a hybrid or intergrade harrier one needs to know the involved parent species extremely well, and even then good quality photographs are required to enable a thorough scrutiny of plumage and structural details. Important plumage details to note include head pattern, occurrence of breast-streaking and the underwing pattern of the primaries and secondaries, while structural features include general proportions, such as shape of head, and in particular details of the wing-formula. Many hybrids/intergrades look like good juvenile Pallid Harriers at first glance, but they have a more or less streaked breast, the head markings show more white around the eye and importantly, the wing-formula shows a longer, slightly protruding p6 (fifth finger counting in from the outermost). The underwing barring of the primaries as well as secondaries is often intermediate between Hen and Pallid Harrier.

Possible hybrids between Pallid and Montagu's Harriers would be extremely difficult to prove because of the similarity of the juveniles. In the documented interbreeding in Finland between a female Montagu's and a male Pallid (Forsman 1993, 1995) the two female chicks looked very much like Pallid Harriers, while the male chick had a head pattern more similar to a juvenile Montagu's. The underwing pattern, the little that was seen of it, appeared intermediate between the two species. It is possible that hybrids between Montagu's and Pallid are more common than so far thought, but the hybrid juveniles go largely undetected owing to their great similarity to either parent species. However, since no older hybrid individuals, which probably would be more easily identified, have been claimed so far, hybridisation may in fact be exceptional.

In this context it may be appropriate also to mention the American Northern Harrier. Juvenile Northern Harriers are very similar in plumage to many hybrid/intergrade Pallid x Hen Harriers, except for their Hen Harrier-like wing-formula with five distinct fingers (Mullarney & Forsman 2011). Whenever confronted with a potential hybrid/intergrade Hen x Pallid Harrier, particularly if seen along the western seaboard of Europe, the possibility of a Northern Harrier should also be kept in mind.

References

Forsman, D. 1993. Hybridising Harriers. *Birding World* 6: 313.

Forsman, D. 1995. Male Pallid and female Montagu's Harrier raising hybrid young in Finland in 1993. *Dutch Birding* 17: 102–106.

Forsman, D. & Peltomäki, J. 2007. Hybrids between Pallid and Hen Harrier – A New Headache for Birders? *Alula* 13: 178–182.

Forsman, D. & Erterius, D. 2012. Pallid Harriers in northwest Europe and the identification of presumed Pallid Harrier x Hen Harrier hybrids. *Birding World* 25: 68–75.

Mullarney, K. & Forsman, D. 2011. Identification of Northern Harriers and vagrants in Ireland, Norfolk and Durham. *Birding World* 23: 509–523.

475. Presumed hybrid/intergrade Hen x Pallid Harrier *Circus cyaneus x C. macrourus*, fresh juvenile female. Note striking pale collar and plain underbody suggesting Pallid, while underwing markings and extensive white areas around the eye are more similar to Hen. Note diagnostic intermediate wing-formula, with p6 (fifth finger) longer than Pallid but shorter than Hen. Finland, 30.8.2009 (DF)

476. Presumed hybrid/intergrade Hen x Pallid Harrier *Circus cyaneus x C. macrourus*, fresh juvenile female (same as **475**). General appearance is slimmer than Hen, approaching Pallid in proportions, but wing-formula diagnostic with clearly emarginated p6 (see text). Finland, 30.8.2009 (DF)

477. Possible hybrid/intergrade Montagu's x Pallid Harrier *Circus pygargus x C. macrourus*, fresh juvenile male. Identified as Pallid in the field, but head-markings, finely streaked breast/flanks and details of primary emarginations (not visible in this image) and overall very slim proportions may indicate introgression of Montagu's genes. Finland, 2.9.2012 (DF)

478. Possible hybrid/intergrade Montagu's x Pallid Harrier *Circus pygargus x C. macrourus*, fresh juvenile male (same as **477**). Finland, 2.9.2012 (DF)

479. Presumed hybrid/intergrade Hen x Pallid Harrier *Circus cyaneus x C. macrourus,* moulting 2nd cy male. Best identified by details of wing-formula and primaries, with clearly emarginated p6 and aberrant distribution of black in wingtips compared to both Hen and Pallid Harriers (note that p10 is still in pin and not visible). Finland, 16.9.2012 (Janne Riihimäki)

480. Presumed hybrid/intergrade Hen x Pallid Harrier *Circus cyaneus x C. macrourus,* 2nd cy male (same as **479**). Note largely dark p6, which is also clearly emarginated well inside of the tip, and compare with Pallid and Hen. In general proportions appears broader-winged than Pallid. Finland, 16.9.2012 (Janne Riihimäki)

Identification of *Accipiter* hawks

Identifying *Accipiter* hawks can be a real challenge, depending on where in the region one is birding. In the west you only need to distinguish Eurasian Sparrowhawk from Northern Goshawk, which is pretty straightforward, given some experience. Problems usually arise from lack of experience with Northern Goshawk, but if you are at all familiar with that species, its bigger size, different proportions and stronger flight soon become apparent. A simple rule is that if you struggle with the identification, it's a Eurasian Sparrowhawk. Also the plumages are different enough to guarantee a safe identification given decent views. However, if you only get a fleeting glimpse of a stooping bird, or a bird shooting past in a fast glide, the situation can be challenging to anyone.

Moving further east in the region the problems become more of a reality, with Levant Sparrowhawk joining the scene. Again, plumage is always different when comparing Eurasian and Levant Sparrowhawks, and the quicker wingbeats and narrower wing with a more pointed wingtip are further pointers of the latter

(but be aware of moulting Eurasian Sparrowhawks in early autumn). In Levant Sparrowhawk the wingtip is always darker than the rest of the underwing, whereas it appears uniform in Eurasian.

Moving further east still a third small *Accipiter*, the Shikra, adds further to the problem. Shikras are intermediate between Eurasian and Levant Sparrowhawks when it comes to structure, but also in plumage. Shikras are rare in the region, and are confined to the far eastern and southeastern corners, where they may occur side by side with the other two smaller *Accipiter* species. The adults are generally very pale and faintly marked below, except for darker fingers, while the juveniles are rather similar to juvenile Levants, but the markings below are lighter brown and the iris is pale yellow. In fact, the juvenile plumage is more similar to juvenile Northern Goshawk than to either of the smaller *Accipiter* species. The wing-formula is diagnostic in all three species and remains a reliable character whenever it can be confirmed.

NORTHERN GOSHAWK
Accipiter gentilis

Other names: Goshawk

VARIATION The nominate subspecies is found over most of the region, with birds becoming larger and paler from south to north. Larger and paler ssp. *buteoides* of N Russia occasionally turns up southwest of its normal range (juveniles light and strikingly mottled above, adults paler grey above and finely barred below). In Corsica and Sardinia ssp. *arrigonii* is smaller and darker than the nominate. North American ssp. *atricapillus* is a potential vagrant to W Europe, but juveniles are probably inseparable from European birds. Adult *atricapillus* are darker above, the head pattern is strong and the underparts show fine grey vermiculations with pronounced black shaft-streaks.

Note that birds from S and C Europe are on average smaller and darker, and juveniles are more deeply coloured below compared to larger and paler Scandinavian birds.

DISTRIBUTION Breeds over much of Europe. Adults sedentary; juveniles and 2nd cy birds of northern populations are partly migratory.

BEHAVIOUR A bird of forest and edges. Normally not aerial, with exceptions for migration and territorial display, in autumn and spring. Hunts mostly by ambushing from cover, but from time to time also indulges in impressive aerial pursuits.

SPECIES IDENTIFICATION A large and heavy *Accipiter* with a distinct shape and flight compared to Eurasian Sparrowhawk. The body is stronger, the neck and head are more protruding and the tail is broader at the base compared to Sparrowhawk. The head looks longer with a deep bill and a flatter forehead, compared to the small rounded head and tiny bill of Sparrowhawk. In active flight wingbeats are powerful, clearly slower and stiffer than Sparrowhawk's, giving the feeling of a heavy and strong bird. Soaring birds have longer wings, with a clearly broader arm than hand, and a shorter, more rounded tail compared to Sparrowhawk. In adults the underwing barring is rather diffuse, distinct only in the primaries, while juveniles are more mottled above with a brighter tail-pattern compared to Sparrowhawks in corresponding plumage.

MOULT Northern Goshawks moult the entire plumage in an annual complete moult during the breeding season, from Apr to Sep/Oct. Occasionally some wing-coverts, but also some remiges, are left unmoulted, mostly s4 and/or s7–9, varying considerably between individuals and years, and possibly related to food stress. Retained juvenile feathers are essential markers when ageing birds in first-adult plumage.

PLUMAGES Normally three plumages can be separated: juvenile, first-adult and definitive adult plumages. While separating juveniles from older birds is rather straightforward, recognising the two age-classes of adults in the field requires good views and even then females can be difficult.

Juvenile

The ground colour of the underparts varies from white to tawny-buff, and markings vary from narrow streaks to broader spear-heads or tear-drops. Upperparts vary from uniformly brown to more light-mottled, with buffish markings on the bases of some greater and median upperwing-coverts, tertials and uppertail-coverts; as a rule the upperparts tend to be darker and more uniform in southern populations, while the frequency of birds with pale markings increases northwards. Also the underparts are deeper tawny in juveniles from S Europe. Juveniles show a markedly S-curved trailing edge to the wing, particularly pronounced in females, which have a broad arm with strongly bulging secondaries.

First-adult

Rather similar to definitive adult, but best told by irregular breast-barring and poorly defined head markings. The upperparts are more brownish-grey, often with some retained worn and brownish juvenile coverts, or even secondaries. The breast is coarsely barred with an irregular-looking pattern of rather prominent arrowheads, instead of fine arches, as in older adults. The irregular pattern is clearer in males, while many first-adult females may look rather similar to older females. The head pattern is rather diffuse, often with some streaking still visible on the crown and neck-sides, and the darker auricular patch is equally streaked and poorly defined.

Definitive adult

Upperparts uniformly grey, more bluish in males, more brownish-grey in females, the intensity of blueness increasing with age. Underparts are whitish with fine, even and dense darker barring, whiter on the throat and vent to undertail-coverts. The barring becomes finer with age and males are more finely barred than females of similar age. *The barring of the underwing remiges is typically diffuse, save for more clearly barred primaries, a major difference compared to Eurasian Sparrowhawk.*

Adult males tend to have a more contrasting head, with a striking white supercilium between the dark crown and the dark ear-coverts compared to females, which are lighter and lack the strong contrasts of males, but brightly marked females do occur.

SEXING The notable size difference between the sexes, with females much heavier but also clearly bigger than males, is easy to see in direct comparison. However, seen singly in normal field conditions sexing is difficult, and assessing the small differences in proportions between males and females requires previous experience with the species. However, females are broader-winged, with a particularly broad arm compared to males of the corresponding age-class.

Old adult males show on average more bluish-grey upperparts than adult females, and their heads are much more distinctly patterned, but in first-adult plumage the difference is less clear.

CONFUSION RISKS Eurasian Sparrowhawks are frequently mistaken for Northern Goshawks, especially by people with limited experience of the latter. True Northern Goshawks are more similar to Gyrfalcon or female Hen Harrier and, depending on the situation, these three can appear very similar.

481. Northern Goshawk *Accipiter gentilis*, juvenile female, in territorial display. Overall rather light underparts, with finely streaked body and wing-linings. Note boldly banded tail compared to finely barred underwing remiges, and narrow hand compared to bulging arm. Finland, 6.3.2013 (DF)

482. Northern Goshawk *Accipiter gentilis*, juvenile male, migrant. Uniformly streaked below, with markings changing to arrowheads and chevrons towards flanks and axillaries. Note rather bland face markings and narrow hand compared to arm. Finland, 20.9.2014 (DF)

483. Northern Goshawk *Accipiter gentilis*, juvenile female. Females are bulkier with a comparatively smaller head, and the wings are broader. This bird belongs to the more spotted rather than streaked type. Finland, 14.10.2011 (DF)

484. Northern Goshawk *Accipiter gentilis*, juvenile female (same as 483). Young Goshawks are more variegated above compared to Eurasian Sparrowhawk; note boldly banded uppertail and visible barring in remiges. Finland, 14.10.2011 (DF)

485. Northern Goshawk *Accipiter gentilis*, juvenile male (same as **482**). The chequered upperwing-coverts are a common feature in northern populations, while birds from further south are darker and more uniform. The head lacks strong markings, except for the light supercilium above the eye. Finland, 20.9.2014 (DF)

486. Northern Goshawk *Accipiter gentilis*, first-adult plumage (2nd cy), migrant. Differs from full adult by sparser and more irregular barring, with individual markings more like arrowheads compared to finer chevrons or crescents in older adults. Note strongly graduated (rounded) tail-tip. Finland, 16.10.2010 (DF)

487. Northern Goshawk *Accipiter gentilis* adult male, territorial. Light grey below with finely barred body and underwing-coverts. Remiges diffusely barred, apart from more boldly marked outer primaries. Finland, 31.3.2011 (DF)

488. Northern Goshawk *Accipiter gentilis* adult male, breeding. Males show on average more distinct head markings than females, with darker crown and ear-mask. Finland, 22.3.2014 (DF)

489. Northern Goshawk *Accipiter gentilis* adult female, breeding. Appearance rather uniformly light grey except for more boldly marked outer wings and tail. Finland, 2.4.2013 (DF)

490. Northern Goshawk *Accipiter gentilis*, adult female, breeding. Compare the muted underwing barring of adult Northern Goshawk with the distinct barring of Eurasian Sparrowhawk. Finland, 19.3.2014 (DF)

491. Northern Goshawk *Accipiter gentilis* adult female, breeding (same as **490**). During territorial display flies with fanned-out undertail-coverts. Finland, 19.3.2014 (DF)

EURASIAN SPARROWHAWK
Accipiter nisus

Other names: Sparrowhawk

VARIATION The nominate subspecies is the dominant breeding form throughout the region, with two isolated island populations, both smaller and darker: in Corsica and Sardinia (ssp. *wolterstorffi*) and in Madeira and the Canary Islands (ssp. *granti*). Birds of NW Africa have sometimes been separated as ssp. *punicus*. Larger and paler ssp. *nisosimilis* of C and E Siberia is a potential vagrant to the Middle East, but possibly not safely identified owing to intergradation with nominate *nisus* resulting in overlapping features.

DISTRIBUTION Widely distributed across the region and also one of the most common raptors.

BEHAVIOUR Mostly hunts from cover ambushing prey in a short dashing spurt. Also hunts from high up in the air, particularly when migrating, seizing any arising opportunity. Inexperienced juveniles sometimes attack prey the size of themselves or even bigger.

SPECIES IDENTIFICATION Easily recognised as an *Accipiter* by its characteristic silhouette, with rather short and broad wings with rounded tips, and a long tail with narrow base and rather square tip. Typical active flight consists of a series of floppy wing-beats followed by a short descending glide. Soars with wings fully open and pressed forward and the long tail fanned. In all plumages the wingtip shows six well-fingered primaries and is of same colour as the rest of the underwing, lacking Levant Sparrowhawk's dark fingertips. *The barring of the underwing is always distinct throughout*, which is a diagnostic feature compared to adults of Levant Sparrowhawk, Shikra and Northern Goshawk, while juveniles of all these species share a rather similar barring on the underwing.

MOULT The moult is complete and annual and takes place during the breeding season, between Apr and Sep/Oct. Migrating adults in autumn have often not completed their primary moult, and show half-grown outermost primaries which could lead to misidentifications as other *Accipiter* species. Occasionally the odd secondary is left unmoulted and retained until the next moult.

PLUMAGES Normally two age-classes can be recognised in the field: adults and juveniles. Only rarely can the first-adult plumage be separated from older adults. Adult male and female differ by plumage.

Juvenile

Fresh juveniles are dark brown above with rufous fringes and tips, from a distance appearing rather uniformly brown with a rufous cast. The head is rather strongly streaked, with a *prominent whitish supercilium, a darker ear-coverts patch* and streaky sides to the neck and lower cheeks. The ground colour below varies from white to a deep rusty-buff and the markings on the breast vary from fine barring to broader spots or even streaks (rarer). Usually the upper breast is more irregularly patterned with spots or arrowheads, while the lower breast, belly and flanks are finely and more regularly barred. By spring, juveniles have lost the rufous fringes and the upperparts appear more greyish-brown, not that different in colour from adult females.

Sexes differ on average, but cannot be reliably sexed by plumage alone. Males tend to be more extensively rufous and the underside is more spotted compared to females, which show less rufous and are on average more neatly barred.

Adult

Grey or brownish-grey above, lacking the extensive rufous feather margins of the juveniles. Adults also show more uniformly coloured heads, with finely streaked cheeks and throat, and lack the eye-catching dark ear-coverts patch and the boldly streaked cheeks and neck-sides of the juveniles. In adults the sexes can be told apart by plumage.

Adult male is strikingly bluish-grey above, often with a poorly defined supercilium, or even lacking a pale supercilium. The underparts are finely barred, usually with rufous, rarely with grey, and often with extensively rufous lower cheeks, upper breast and flanks. On average, males appear rufous across the upper breast, while females are grey.

Adult female is duller, more brownish-grey above, with a more pronounced pale supercilium. Underparts can be rather similar to adult male, but as a rule they are never as intensely rufous as most males, nor is the rufous colour as uniform across the upper breast. Most females are barred grey below, with the rufous colour restricted to the lower cheeks and sides of throat.

SEXING When seen together in direct comparison females are considerably bigger than males, and the flight silhouette can appear up to 30% larger in females, but single birds can be surprisingly difficult to sex by size alone.

CONFUSION RISKS Most often mistaken for Northern Goshawk, which is a much bigger and more heavily built bird, with a stronger and steadier flight and more powerful wing action. The flight silhouette of Northern Goshawk shows longer wings, with a narrower hand than arm when soaring, and a proportionately shorter and more rounded tail that appears thickset at the base. It is also longer-necked with a more protruding head and shows a longer and deeper bill.

More similar to Levant Sparrowhawk, which has longer and slimmer wings with more pointed, dark wingtips and a comparatively shorter tail, and its flight appears hurried; the wingbeats are rapid, almost like a Kestrel in display flight, and when flapping the glides are shorter. On migration Levant Sparrowhawks often glide for long distances when leaving a thermal before starting to flap. Eurasian Sparrowhawk is very similar to Shikra in general shape, and great care is needed to separate the two. Juvenile Shikra is lighter, both above and below, with drop-marks or spots below and a uniformly streaked, less distinctly marked head. Adults look generally pale and bleached, with darker fingers but otherwise faintly marked underwings.

NOTE Adults, possibly of eastern origin and approaching *nisosimilis* in appearance (paler and less rufous below compared to birds of western and northern Europe), have been seen on migration in the Middle East (pers. obs.).

492. Eurasian Sparrowhawk *Accipiter nisus*, juvenile, migrant. Typical silhouette of a gliding bird, with rather short and rounded wings, long tail and small head. Finland, 2.9.2009 (DF)

493. Eurasian Sparrowhawk *Accipiter nisus*, juvenile, migrant. Many juveniles are finely barred below, but told from adult females by the coarser rufous markings across the upper breast and by more distinct dark ear-mask. Finland, 26.8.2013 (DF)

494. Eurasian Sparrowhawk *Accipiter nisus,* juvenile male, migrant. A rather spotted individual. Finland, 3.10.2012 (DF)

495. Eurasian Sparrowhawk *Accipiter nisus,* juvenile male, migrant. Some juveniles, more frequently males than females, are intensely rusty-yellow below, some even more colourful than this bird. Finland, 2.9.2009 (DF)

496. Eurasian Sparrowhawk *Accipiter nisus,* juvenile, migrant. Juveniles are dark brown above with rather uniform upperwing-coverts. The tail is distinctly banded, but remiges less clearly so. Finland, 25.8.2005 (DF)

497. Eurasian Sparrowhawk *Accipiter nisus,* juvenile female, migrant. The rufous feather fringes of fresh juveniles are visible in close views only. The distinct dark eye-mask and streaked neck are typical of a juvenile. Finland, 21.8.2014 (DF)

498. Eurasian Sparrowhawk *Accipiter nisus*, adult male, migrant. The amount of rufous on the throat and breast of adult males is highly variable and some, as this male, are not that different from the most colourful adult females. Finland, 3.10.2012 (DF)

499. Eurasian Sparrowhawk *Accipiter nisus*, adult male, migrant. A more colourful and hence more easily identified adult male. Finland, 6.10.2010 (DF)

500. Eurasian Sparrowhawk *Accipiter nisus*, adult male, migrant. Note the uniformly orange areas of the cheek and neck typical of adult males and compare with adult females. Finland, 28.9.2014 (DF)

501. Eurasian Sparrowhawk *Accipiter nisus*, adult male, migrant. Some males are almost uniformly orange on the upper breast. Finland, 6.10.2010 (DF)

502. Eurasian Sparrowhawk *Accipiter nisus*, adult male, migrant. Adult males are beautifully blue-grey above, while females always appear browner-grey. From certain angles the entire hand may appear darker than the arm, but adult male Eurasian Sparrowhawk never shows contrasting dark fingers on the primaries like adults of Shikra or Levant Sparrowhawk. Finland, 28.9.2014 (DF)

503. Eurasian Sparrowhawk *Accipiter nisus*, adult female, spring migrant. Typical silhouette, with broad wings showing six fingered primaries (cf. Shikra and Levant Sparrowhawk). Note contrasting underwing barring and compare with adult Northern Goshawk. Finland, 5.4.2011 (DF)

504. Eurasian Sparrowhawk *Accipiter nisus*, adult female, autumn migrant. In adult females the rufous of the underparts is mostly confined to the lower cheeks, while the barring may be either rufous or grey. The white supercilium stands out more clearly than in adult males. The outermost primary p10 is not yet fully grown. Finland, 27.9.2005 (DF)

505. Eurasian Sparrowhawk *Accipiter nisus*, adult female, migrant. A more average bird, with limited rufous restricted to the cheeks, while the fine barring of the breast is dark grey. Finland, 12.10.2005 (DF)

506. Eurasian Sparrowhawk *Accipiter nisus*, presumed first-adult (2nd cy) female, migrant. The first-adult plumage can rarely be identified in the field, usually only by retained juvenile feathers. However, note the rather brownish, not greyish upperparts of this fresh female after completed moult, with rufous rather than whitish feather-edges on the upperparts. Finland, 25.9.2010 (DF)

507. Eurasian Sparrowhawk *Accipiter nisus*, adult female, migrant. Before completion of the primary moult the wingtip may be atypical, and similar to Levant Sparrowhawk or Shikra. Finland, 12.9.2013 (DF)

LEVANT SPARROWHAWK
Accipiter brevipes

VARIATION Monotypic.

DISTRIBUTION SE Europe, from the Balkans to W Black Sea coasts, Moldova, Ukraine and S Russia north of the Black and Caspian Seas into NW Kazakhstan.

BEHAVIOUR A bird of mixed deciduous woodland, with largely skulking habits. Rather secretive during the breeding season and seen in flight only when moving between the nest and hunting grounds. Migratory, leaving the breeding areas in Aug–Sep, returning in late Apr–early May. Gregarious on migration, with major concentrations along the eastern flyway in late Sep and late Apr. Winter quarters probably in C and E Africa, but still largely unknown.

SPECIES IDENTIFICATION Similar to other small *Accipiter* hawks and juvenile in particular is similar to juvenile Shikra. *Wing-formula, with only 4 visible fingers, is diagnostic among sympatric Accipiter species.* Dark wingtips, most obvious in adults, is a good feature compared with Eurasian Sparrowhawk. Flight silhouette shows narrower and more pointed wings and comparatively shorter tail compared to Eurasian Sparrowhawk. Bill deeper and heavier than Eurasian Sparrowhawk's, with swollen yellow cere; iris dark in adults, turning from greyish to dark in juveniles. Flapping flight with more rapid series of wingbeats, almost winnowing (like Kestrel in display flight); often glides for long distances, especially when leaving a thermal on migration. Unlike other *Accipiters* of the region occurs on migration in big and dense flocks, often several hundred strong.

MOULT Moult begins in the breeding range but is suspended for the autumn migration, although signs of moult or differences between new and old feathers are difficult to see in normal field conditions. Birds seen on migration mostly show 2–4 retained unmoulted outer primaries (pers. obs.). Moult is resumed in the winter quarters and completed there.

PLUMAGES Normally two age-classes can be separated: adults and juveniles. Occasionally 2nd cy birds can still be separated in autumn. Adult male and female differ by plumage.

Juvenile

Underparts pattern diagnostic, with variable dark spotting on the breast turning into broad cross-bars on the flanks, the colour of the spotting darker brown and more contrasting than in juvenile Shikra, which can appear otherwise rather similar. Remiges of underwing evenly and boldly barred, the underwing-coverts appearing spotted. In fresh plumage fingers are darker than rest of underwing, unlike in Eurasian Sparrowhawk, creating a narrow dark leading edge to the wingtip. Fanned tail shows 4 dark bands outside of the undertail-coverts, but the outermost tail feathers are paler, with 6–7 fine bars.

During their first winter juveniles undergo a partial body moult, which varies individually in extent. *By spring migration most juveniles have moulted the majority of their body plumage above and below,* but some retained juvenile streaked/spotted breast feathers can usually still be seen. However, the juvenile wing feathers (including coverts) and tail are retained, serving as reliable ageing characters: the wings are brown above and the barring below is complete and regular.

As a result of wear and bleaching juveniles in spring show hardly any dark at all in their wingtips, and may therefore more easily be mistaken for a Eurasian Sparrowhawk.

Transitional plumage (2nd cy autumn)

Some birds can still be aged in their second autumn by retained juvenile remiges, showing the regular and complete barring of the juvenile, but these feathers are usually very difficult to spot in the field, although can be seen easily in good photographs.

Adult

Uniform above and finely barred below, with a distinctly dark wingtip visible both above and below. The iris is red, but appears dark in the field, and the bill is black with a strikingly yellow cere. Males and females differ in plumage.

Adult males are beautifully bluish-grey above, with contrasting black wingtips. The head is pale grey, with a big dark (red) eye, while the underparts are whitish with a faint peachy blush to the upperbreast. The pinkish barring of the breast and underwing-coverts is fine and rather soft and the individual bars are mostly impossible to make out. The underwing looks very pale with a contrasting black leading edge to the hand, and the barring of the remiges is faint and usually confined to the bases of the outer long primaries.

Adult females are darker grey above and the darker wingtip is far less obvious than in males. The head is darkish grey with a dark eye and the underparts are whitish with rather contrasting and fine dark barring across the breast. The remiges are finely barred throughout, unlike in adult males, but the barring is softer and less contrasting compared to the juveniles. The leading edge of the hand is dark, as in adult males.

SEXING Adult males and females clearly differ by plumage, whereas juveniles cannot be sexed in the field.

CONFUSION RISKS Similar to Eurasian Sparrowhawk and Shikra, but the combination of a four-fingered wing-formula and dark wingtips is diagnostic, as is the dark iris.

508. Levant Sparrowhawk *Accipiter brevipes*, juvenile, autumn migrant. Boldly streaked breast, changing to chevrons and crossbars towards the flanks and axillaries. Wing-linings densely spotted and remiges distinctly barred, with darkish, fingered primaries. Egypt, 12.10.2010 (DF)

509. Levant Sparrowhawk *Accipiter brevipes*, juvenile, autumn migrant. Compared to Eurasian Sparrowhawk, note the slimmer proportions, with more uniformly broad wings and narrower wingtips, with four fingered primaries only and outer hand clearly darker than the rest of the underwing. Egypt, 12.10.2010 (DF)

510. Levant Sparrowhawk *Accipiter brevipes*, juvenile, autumn migrant. Note contrasting pattern of underparts and details of wing-formula, and compare with Shikra and Eurasian Sparrowhawk. Egypt, 11.10.2010 (DF)

511. Levant Sparrowhawk *Accipiter brevipes*, juvenile, autumn migrant (same as **509**). Upperparts rather similar to juvenile Eurasian Sparrowhawk, but note poorly marked head and deeper black bill with prominent yellow cere. Egypt, 12.10.2010 (DF)

512. Levant Sparrowhawk *Accipiter brevipes*, transitional 2nd cy male, spring migrant. Most birds easy to age by mix of adult-type and retained juvenile feathers on underbody and entirely juvenile wings. Also note dark iris and black bill with swollen yellow cere. Israel, 22.4.2009 (DF)

513. Levant Sparrowhawk *Accipiter brevipes*, transitional 2nd cy male, spring migrant (same as **512**). Upperparts look dull and faded while head is already moulted. Note diagnostic, pointed wingtips formed by longest three primaries. Israel, 22.4.2009 (DF)

514. Levant Sparrowhawk *Accipiter brevipes*, adult male, migrant. Adult males are distinct, with a pinkish blush to the underbody and wing-linings, faintly barred remiges and distinctly black wingtips. Israel, 6.11.2009 (DF)

515. Levant Sparrowhawk *Accipiter brevipes*, adult male, migrant (same as **514**). Note typical silhouette with narrower and more pointed wings, and compare with otherwise rather similar adult Shikra. Israel, 6.11.2009 (DF)

516. Levant Sparrowhawk *Accipiter brevipes*, adult male, migrant. Distant males look very pale with a faint peachy blush to the breast and dark wingtips. Egypt, 12.10.2010 (DF)

517. Levant Sparrowhawk *Accipiter brevipes*, adult male, migrant. Adult males are rather pale bluish-grey above with clearly darker wingtips, and thus rather similar to adult male Shikra, but shape of wings and wingtips differ. Egypt, 10.10.2010 (DF)

518. Levant Sparrowhawk *Accipiter brevipes*, adult female, migrant. Adult females are somewhat darker and more distinctly barred below compared to adult males (including remiges and tail), and the head contrasts as darker still. Also note dark iris and compact black bill with conspicuous yellow cere. Egypt, 11.10.2010 (DF)

519. Levant Sparrowhawk *Accipiter brevipes*, adult female, migrant. Typical silhouette, with rather long and narrow wings and comparatively short tail for an *Accipiter*. Although adult females don't show the distinct dark wingtips of the males, the bolder barring of the longest primaries creates the impression of darker wingtips in most field situations. Egypt, 12.10.2010 (DF)

SHIKRA
Accipiter badius

VARIATION Complex taxonomy, with several subspecies across its vast range, which spans most of sub-Saharan Africa and southern Asia. Possibly up to three subspecies may occur within the Western Palearctic region: ssp. *cenchroides* of C Asia is partly migratory and may occur as a vagrant and winter visitor to the Middle East; Indian ssp. *dussumieri* is a possible vagrant to the Arabian Gulf and Arabia; and African ssp. *sphenurus* is a breeding resident in parts of S Arabia. Further studies are required to clarify the occurrence of the different subspecies in the Middle East.

DISTRIBUTION Migratory *cenchroides* breeds in forest steppe and open woodland of C Asia, reaching Armenia in the west, and migrates mostly to the Indian subcontinent for the winter. The birds which have recently colonised the UAE, however, look different and probably belong to *dussumieri* (males clearly bluish above), but further studies are needed to confirm their taxonomic affinity.

BEHAVIOUR Like other *Accipiter* species, hunts largely from cover, ambushing small birds, reptiles and large insects. Behaviour in flight is very similar to Eurasian Sparrowhawk and it is mostly seen on the wing when hunting, or when moving between hunting grounds. Adults of ssp. *sphenurus* typically soar with tail closed.

SPECIES IDENTIFICATION Similar in size and behaviour to the other small *Accipiter* species of the region and identification is always a challenge. Wing-formula intermediate between Eurasian Sparrowhawk and Levant Sparrowhawk, with five clearly separated fingers, giving a more rounded wingtip compared to Levant, being thus more similar to Eurasian Sparrowhawk in silhouette. Adults are pale grey above, males more bluish than females (but see below for subspecific differences), and the underbody is faintly barred pinkish, the barring being often difficult to discern in males, but slightly better defined in females, though not as distinct as in adult female Levant. The underwing is pale with clearly darker fingers, giving a darker leading edge to the wingtip. *Juveniles resemble young Levant Sparrowhawks, but the markings below are a lighter, more rufous-brown on a buffish ground colour, compared to the much darker, liver-brown markings on a* whiter ground colour in Levant, and the upperparts are a lighter brown with paler fringes and markings; thus the plumage of a juvenile Shikra is rather similar to a juvenile Northern Goshawk. *The iris is pale grey in juveniles, turning yellow by the first autumn, as in Eurasian Sparrowhawk, while Levant Sparrowhawk has dark eyes.*

Subspecies identification

African/Arabian *sphenurus* show slightly longer central tail feathers, which protrude from the tip of the fanned tail. Males are pale bluish-grey above with a faint peachy wash to the upper breast, while females are less blue above and the breast is more clearly barred pink.

Adults of *cenchroides* are very pale grey above with a distinct sandy tone; some males are as pale as adult male Pallid Harriers above with very faint barring below, hardly discernible in many adult males, but slightly more distinct in females. The closed tail is uniformly grey above in males, but adult females show a dark subterminal band on the uppertail.

Adult Indian *dussumieri* are deeper bluish-grey above and the breast is more distinctly barred rufous-brown, again with more contrasting markings in females than males; the females of this subspecies are the most distinctly barred of all adults.

Juveniles of the various subspecies are more similar and thus difficult to separate, but ssp. *cenchroides* is overall paler with less contrasting markings below.

MOULT Moulting dates vary depending on area and subspecies, but takes place during the breeding season. In migratory *cenchroides* moult is suspended for the autumn migration, when about half of the primaries have been replaced. The moult is then resumed and finished on the wintering grounds (pers. obs).

PLUMAGES Moults directly from juvenile into adult plumage at the age of one year, thus only two plumage types can be described, adults and juveniles. However, a transitional plumage can be recognised by retained juvenile feathers until completion of the first moult. Adult females and males differ in plumage.

Juvenile

Brown above with buffish fringes and tips, the ground colour being a lighter brown compared to juvenile Levant Sparrowhawk. Upperparts are also more variegated than in Eurasian Sparrowhawk, more similar to juvenile Northern Goshawk. Head is rather pale, evenly streaked, lacking prominent features; iris is greyish at first (may appear darkish in the field) turning yellow during the first autumn. The underbody is buffish with rufous-brown streaking on the breast, changing to broader and sparser spots and bands towards the flanks; the overall appearance is not dissimilar to a miniature juvenile Northern Goshawk. Flight feathers are finely and evenly barred with a broader diffuse subterminal band, and the underwing-coverts are diffusely spotted. Tail is greyish-brown above with three darker bands, pale below with dense but fine barring on the outermost tail feathers, the others showing 3–4 dark bands including the subterminal band.

Transitional plumage

In the 2nd cy summer/autumn this plumage can be recognised for as long as the birds are in moult. The plumage is a mix of newly moulted adult-type feathers and retained juvenile feathers. In migratory *cenchroides* the last remiges are not moulted until on the wintering grounds, but the body plumage already looks adult on departure from the breeding grounds. The other two resident subspecies probably commence and also finish their moult earlier in the season.

Adult

Adults are grey above and whitish below, with faint barring across the breast. Flight feathers are finely and sparsely barred and the underwing-coverts are only faintly barred, barely visible.

Adult male

Head and upperparts are a delicate bluish-grey in ssp. *sphenurus* and *dussumieri*, pale ashy-grey in *cenchroides*, often with a sandy wash. The underwing shows distinctly darker fingers, creating a dark leading edge to the hand. Iris is deep orange to bright red, but the colour is brilliant and vivid, not appearing

dark as in Levant Sparrowhawk. The underbody is whitish, with peachy or light orange barring across the upper breast; the pattern is often inconspicuous and appears more like a wash, particularly faint in adult male *cenchroides*. The uppertail may appear uniformly grey when closed, but faint barring is exposed when fanned.

Adult female

Grey above, but less colourful than the male, lacking the blue tones, while the barring below is slightly more distinct, orange or rufous, and the iris is yellow or orange. In ssp. *cenchroides* adult females show a broad dark subterminal tail-band, which is lacking in males.

SEXING Males are smaller than females, but this can be difficult to judge in the field without direct comparison. In adults, males are paler grey above (also more bluish in *dussumieri* and *sphenurus*), their breast-barring is more diffuse, often appearing as a wash, and the iris is deeper orange or red compared to adult females, which mostly show a yellow iris and slightly more distinct barring to breast, which is particularly contrasting in female *dussumieri*.

CONFUSION RISKS Distant birds are truly difficult to identify and a suite of plumage details are normally required. Eurasian Sparrowhawk has six fingers, which are never dark below, and juveniles are finely barred on the flanks and lower breast. Levant Sparrowhawks only have four fingers, giving them an almost falcon-like wingtip, but remember that moulting Shikras can look the same in autumn. Levant Sparrowhawks are also darker in all plumages with more contrasting plumage.

NOTE Iris colour is not enough to separate Shikra and Levant Sparrowhawk, as miscoloured irises do occur. Identification should always be based on a suite of characters, such as wing-formula, plumage details of breast and upperparts, and the iris colour.

Hybrids between Shikra and Levant Sparrowhawk have been reported, but further evidence is needed.

520. Shikra *Accipiter badius*, fresh juvenile of the smaller African ssp. *sphenurus*. Note the rather sparse markings of streaked upper breast, barred flanks and spotted lower breast, the five-fingered wingtips and the wedge-shaped tail-tip, with elongated central rectrices; the latter is diagnostic of this subspecies. Senegal, 31.1.2006 (DF)

521. Shikra *Accipiter badius*, fresh juvenile of African ssp. *sphenurus* (same as **520**). Note narrower hand compared to Eurasian Sparrowhawk (only five fingers) and more densely barred tail feathers, in particular the outermost. Senegal, 31.1.2006 (DF)

522. Shikra *Accipiter badius*, juvenile of African ssp. *sphenurus*. From certain angles the wingtips are clearly dark unlike in Eurasian Sparrowhawk. This bird is moulting into adult plumage, showing fresh barred feathers on the breast. The Gambia, 5.2.2008 (DF)

523. Shikra *Accipiter badius*, fresh juvenile of C Asian ssp. *cenchroides*. Note five-fingered wingtips, streaked underbody and pale iris. Kazakhstan, 3.9.2009 (René Pop)

524. Shikra *Accipiter badius*, fresh juvenile of C Asian ssp. *cenchroides*. Note light head and iris, and general impression similar to juvenile Northern Goshawk. Kazakhstan, 3.9.2009 (René Pop)

525. Shikra *Accipiter badius*, adult of African ssp. *sphenurus*. Breast and wing-linings are finely barred in pink while remiges are faintly barred save for dark fingers. The Gambia, 18.4.2007 (DF)

526. Shikra *Accipiter badius*, adult male of African ssp. *sphenurus*. Breast and wing-linings are finely barred with pink while remiges are faintly barred apart from dark fingers. Outer tail feathers are pale and practically uniform, with inner feathers showing four dark bands. Note elongated central tail feathers in *sphenurus*. Ethiopia, 18.2.2012 (DF)

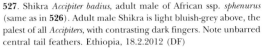

527. Shikra *Accipiter badius*, adult male of African ssp. *sphenurus* (same as in **526**). Adult male Shikra is light bluish-grey above, the palest of all *Accipiters*, with contrasting dark fingers. Note unbarred central tail feathers. Ethiopia, 18.2.2012 (DF)

528. Shikra *Accipiter badius*, adult male of C Asian ssp. *cenchroides*, breeding. Note very pale and faintly marked plumage with five darker fingered primaries. Kazakhstan, 29.5.2008 (Tom Lindroos)

529. Shikra *Accipiter badius*, adult male of C Asian ssp. *cenchroides*. Males are paler and more faintly marked than adult females. Kazakhstan, 29.5.2008 (Tom Lindroos)

530. Shikra *Accipiter badius*, adult female of C Asian ssp. *cenchroides*. One of the larger and heavier subspecies. Underbody and wing-linings finely barred greyish-pink. Remiges with fine and rather light barring, except fingers, which are boldly barred and hence make the wingtips look darker. Note yellow iris. Kazakhstan, 7.5.2010 (DF)

531. Shikra *Accipiter badius*, adult female of C Asian ssp. *cenchroides* (same as **530**). Rather uniformly sandy-grey above except for darker fingers and subterminal tail-band, typical of females of this subspecies. Kazakhstan, 7.5.2010 (DF)

DARK CHANTING GOSHAWK
Melierax metabates

VARIATION An African species, with several subspecies throughout sub-Saharan Africa, reaching S Arabia in the northeast of its range. Formerly occurred as ssp. *theresae* in the Souss valley of SW Morocco, but this is now believed to be extinct. Nominate *metabates* occurs in semi-dry areas north of Equator from the Atlantic to the Red Sea coasts and ssp. *mechowi* is in similar dry bush country south of the Equator in E and S Africa.

DISTRIBUTION Rare vagrant to the Middle East from S Arabia and NE Africa.

BEHAVIOUR Perches prominently on tops of trees, poles and other vantage points. When hunting moves from perch to perch in low flapping flight interspersed by short glides. During display may circle to considerable heights.

SPECIES IDENTIFICATION Medium-sized raptor, with a quite diagnostic silhouette characterised by parallel-edged, broad and rather short wings and a longish tail. Six fingered primaries produce a broad and rounded wingtip. Adults are easy to identify by diagnostic behaviour, silhouette and plumage, while juveniles may be mistaken for a Northern Goshawk in brief views (although most unlikely to ever occur together), or a juvenile Gabar Goshawk which is only half the size. Tarsus is very long, with feet reaching beyond the tips of the undertail-coverts in flight.

MOULT Commences at the age of one year and comprises the entire plumage. The last traces of juvenile plumage should be looked for among the flight feathers and wing-coverts.

Plumage

Two plumages can be identified, juvenile and adult, which are clearly different.

Juvenile

Uniformly brown above, gradually turning greyish-brown with increasing wear. Uppertail-coverts appear white from a distance, but show fine dark barring on close views. Brown tail with 4–5 broad, dark bands, best visible when the tail is fanned, the tip broadly white when fresh. Head shows a prominent pale supercilium, darker ear-coverts and a pale staring eye. Underbody shows darker upper breast contrasting with paler, finely barred lower breast, belly and undertail-coverts. Underwings appear pale from a distance, but show fine and rather dense barring on closer views. Legs are extremely long with the feet reaching beyond the undertail-coverts in flight. Cere and feet are pale yellowish.

Transitional plumage

Recognised during the first moult (approximately one year old) by a mixture of brown juvenile and grey adult-type feathers.

Adult

Appears rather uniformly grey from above, with darker mantle, paler innerwing and broadly blackish wingtips. The closed tail is black above with a broad white tip, but shows striking black and white bands when fanned. Head and upper breast are grey, lower breast is whitish, and belly and undertail-coverts are finely and densely barred black. Base of bill, cere and legs deep red, but tending towards orange in many birds. Iris is dark, deep red-brown.

SEXING Males are smaller than females, but the difference is slight and not of use in the field unless both sexes are seen together for direct comparison.

CONFUSION RISKS Juveniles may be mistaken for juvenile Northern Goshawk, but show more parallel-edged wings with fuller hand compared to that species, a colour contrast across the mid-breast, and longer legs. Adults vaguely resemble adult male harriers in plumage, but silhouette and behaviour in flight are different, more recalling a large *Accipiter*.

532. Dark Chanting Goshawk *Melierax metabates*, juvenile. Note typical silhouette with broad and rounded wings and a fairly long tail. Underwings are finely banded, wing-linings and lower breast/belly finely barred, upper breast darker with bold streaks down the middle. Note yellow iris and long exposed legs. The Gambia, 9.2.2008 (DF)

533. Dark Chanting Goshawk *Melierax metabates*, juvenile, about one year old. Still in juvenile plumage, but has started its first moult and is growing an inner primary. Ethiopia, 8.2.2012 (DF)

534. Dark Chanting Goshawk *Melierax metabates*, juvenile. Rather uniformly greyish-brown above, but note darker secondaries, pale primary flash and white, strongly barred uppertail-coverts. The Gambia, 5.1.2012 (DF)

535. Dark Chanting Goshawk *Melierax metabates*, juvenile (same as **534**). Note typical heavily banded tail with pale edges. The Gambia, 5.1.2012 (DF)

536. Dark Chanting Goshawk *Melierax metabates*, adult. Diagnostic shape with broad black-tipped wings and a longish tail. The Gambia, 5.12.2013 (DF)

537. Dark Chanting Goshawk *Melierax metabates*, adult. Note grey body contrasting with lighter underwing with broadly dark tips. The fine barring of the underbody and underwing-coverts is difficult to see in the field. Ethiopia, 6.2.2012 (DF)

538. Dark Chanting Goshawk *Melierax metabates*, adult. Close views reveal the contrast between uniformly grey head and upper breast and finely barred wing-linings and rest of underbody, as well as the red cere and legs and the dark eye. The Gambia, 1.12.2013 (DF)

539. Dark Chanting Goshawk *Melierax metabates*, adult. The upperwing is grey with a slightly lighter inner hand and dark fingers. Note long legs, red cere and dark iris. The Gambia, 10.2.2008 (DF)

540. Dark Chanting Goshawk *Melierax metabates*, adult. The fine barring of the uppertail-coverts can be hard to discern. Ethiopia, 18.2.2012 (DF)

GABAR GOSHAWK
Micronisus gabar

VARIATION Monotypic, but has two colour morphs, a common grey morph and a rarer black morph.

DISTRIBUTION Widespread throughout Africa in semi-dry thornbush savanna, reaching the SW Arabian Peninsula as a breeding species. Extremely rare vagrant elsewhere in the Western Palearctic.

BEHAVIOUR Spends most of the time perched inside cover. Flies mostly low, keeping close to vegetation when moving between perches, but from time to time also circles higher in the sky.

SPECIES IDENTIFICATION A small and dashing *Accipiter*-like hawk, with short rounded wings (five fingered primaries) and a very long, strongly barred tail. Diagnostic face, with orange-red base of bill and naked loral area, rather similar to the much larger Dark Chanting Goshawk. Grey morph shows a narrow white trailing edge to the wing in juvenile and adult plumages. In our region most similar to Eurasian Sparrowhawk and Shikra, but wings are shorter and more parallel-edged, the wingtip more evenly rounded, and the tail is comparatively longer. The grey morph shows white uppertail-coverts in both juvenile and adult plumages, but this can be difficult to confirm in side views (note that the white undertail-coverts of *Accipiter* species may bend around the tail-base to give an impression of white uppertail-coverts). The grey morph also shows contrast between the more uniform upper breast and barred lower breast, both in adult and juvenile plumages. Black morph has uniformly dark body plumage, but the flight feathers and tail appear normally barred from below, with a prominent pale and translucent window on the inner hand. Legs and cere are orange in young birds, deep red in adults. Iris is yellow in young birds, and deep red in adults, looking dark from a distance.

MOULT Juvenile plumage is replaced by adult plumage during the first moult, at an age of approximately one year.

PLUMAGES Only two separable age-classes: juveniles and adults can be separated in both grey and black morphs.

Juvenile

Grey morph Brownish above with white uppertail-coverts and trailing edge to the wings; tail darker with 3–4 blackish bands. Head shows rather prominent pale supercilium and streaky ear-coverts and neck. Upper breast is densely streaked, giving a rather uniformly brown impression in flight, contrasting strongly with the paler, finely barred lower breast and belly. Undertail looks pale when closed, but shows 4–5 striking black bands when fanned. Underwings are pale, including the wingtips, with fine and rather dense barring.

Black morph Uniformly blackish-brown, except for the distinctly barred tail and underwings, as in the grey morph, but lacks the white uppertail-coverts of the grey morph. Upperwing shows a translucent pale window on the inner hand.

In juveniles of both colour morphs the legs and cere are yellowish to begin with, later turning orange, and the iris is pale yellow.

Transitional plumage

Immatures during their first moult can be recognised by a mixture of juvenile and adult-type features which, however, can be difficult to make out in the dark morph.

Adult

Grey morph Uniformly dove-grey above with white uppertail-coverts, tail-tip and trailing edge of the wings. The tail is darker grey than the back, whiter towards the edges, with 4–5 contrasting broad dark bands. Head and upper breast are uniformly grey, lacking markings, and lower breast and belly are whitish with fine and dense dark barring. Underwings appear uniform and pale, with fine even barring extending to the tips of the feathers, and the undertail is whitish with 3–4 distinct black bands.

Black morph Similar to juvenile of black morph, but cere and legs red and iris dark.

In both colour morphs legs and cere are red and the iris is deep red, often appearing dark in the field.

SEXING Although females are almost twice as heavy as males, the visible size difference is negligible and not of use for field identification purposes, unless the birds of a pair are seen together for direct comparison.

CONFUSION RISKS In the current context only likely to be mistaken for a small *Accipiter*, but the white uppertail-coverts, the pale upperwing window and the long, distinctly barred tail, as well as the more brightly coloured bare parts, should avoid confusion. Much smaller, with more dashing appearance, compared to the similar-looking Dark Chanting Goshawk.

541. Gabar Goshawk *Micronisus gabar*, juvenile. Rather sparrowhawk-like silhouette, but wings are more evenly broad. Note densely barred underbody contrasting with heavily streaked upper breast, white trailing edge to arm and boldly banded tail. Ethiopia, 8.3.2006 (DF)

542. Gabar Goshawk *Micronisus gabar*, juvenile. Note heavily streaked (full) crop contrasting sharply with barred underparts, white tail-base and red legs. Underwings are uniformly and densely barred. Ethiopia, 8.1.2010 (DF)

543. Gabar Goshawk *Micronisus gabar*, juvenile. Streaked crop contrasts with finely barred lower body. Note distinctly barred tail and orange cere and iris. Cameroon, 22.4.2007 (Ralph Buij)

544. Gabar Goshawk *Micronisus gabar*, adult. Broader-winged and longer-tailed than juvenile. Underwings are finely barred throughout and tail is boldly banded, while details of grey underbody are difficult to make out. Ethiopia, 8.3.2006 (DF)

545. Gabar Goshawk *Micronisus gabar*, adult. Plumage may recall adult Dark Chanting Goshawk, but note sparrowhawk-like proportions and finely barred underwings. Cameroon, 25.3.2007 (Ralph Buij)

546. Gabar Goshawk *Micronisus gabar*, adult. Note finely barred underwings, bold black and white bands of tail, and sharp contrast between grey head and breast and finely barred underbody. Cere and legs are bright red, iris is dark. The diagnostic white tips of the secondaries are not visible against the light sky. The Gambia, 5.12.2013 (DF)

547. Gabar Goshawk *Micronisus gabar*, adult, robbing a weaver's nest. Ethiopia, 13.11.2011 (DF)

548. Gabar Goshawk *Micronisus gabar*, adult (same as **545**). Cameroon, 25.3.2007 (Ralph Buij)

549. Gabar Goshawk *Micronisus gabar*, adult. The white trailing edge to the arm and the white uppertail-coverts are a diagnostic combination. Note also the colourful cere and feet. Cameroon, 22.4.2007 (Ralph Buij)

550. Gabar Goshawk *Micronisus gabar*, adult, dark morph. All-black body plumage, finely barred remiges, and red feet and cere make dark morph birds unmistakable. Cameroon, 25.3.2007 (Ralph Buij)

551. Gabar Goshawk *Micronisus gabar*, adult, dark morph. Dark morph birds lack the white uppertail-coverts. Cameroon, 25.4.2007 (Ralph Buij)

Separating Common Buzzard and European Honey-buzzard

Despite being two of the region's commonest birds of prey, the Common Buzzard and the European Honey-buzzard are still often confused and mis-identified. It is the juvenile honey-buzzard, in particular, which remains unknown to many birders and causes a lot of confusion. In fact, this is probably the most often misidentified raptor in the Western Palearctic. The explanation is probably the short period that juvenile European Honey-buzzards are available to European birders. They are visible only for a few weeks each autumn before they leave Europe, not to return again until they are in full adult plumage nearly two years later.

Buzzards in general are extremely variable, with a lot of different plumage types. While the adults and the juveniles are pretty similar in Common Buzzard, with only small differences in proportions and plumage, the situation in European Honey-buzzard is quite the opposite. Adults and juveniles are so different, in shape, proportions and plumage, that they could easily be seen as belonging to different species by the unaware. While the adult European Honey-buzzard is a most distinctive bird, with diagnostic shape and plumage, and easily told from Common Buzzard, autumn juvenile honey-buzzards appear much more similar to Common Buzzards, especially from a distance when plumage details cannot be seen.

To an experienced observer the flat wings and the relaxed and somewhat slower wing-beats of European Honey-buzzard catches the eye, but also the habit of flying with a protruding chest and slightly lifted head in active flapping flight are good pointers compared to Common Buzzard, which always shows a kink in the wing in frontal views and the wing-action appears a bit more hurried in comparable situations. Even the silhouette of a circling juvenile European Honey-buzzard differs from Common Buzzard by its clearly pinched-in wing-base, more bulging arm and clearly narrower hand, fuller tail and with wings pressed more forward in full soar. *The diagnostic difference is found in the underwing pattern, where the barring of the remiges, as well as the markings of the coverts, are clearly different between the species.* For closer encounters, it is worth remembering the juvenile honey-buzzard's bright yellow bill with just a dark tip, and its dark brown eye, while young Common Buzzards show less yellow on the bill-base and the iris is often a lighter brown with a clearly visible black pupil.

Identification of *Buteo* buzzards

The *Buteo* buzzards are a challenging group to identify. The huge individual plumage variation within each species is further complicated by notable geographical variation between populations of the same species, added to plumage variation depending on age and sex.

Although the nominate Common Buzzard *B. b. buteo* is a highly variable bird, the major identification problem is distinguishing Steppe Buzzard *B. b. vulpinus* and Long-legged Buzzard *B. rufinus*, which in some plumages can be practically identical. In this particular case the size and structure of the bird are even more important to note than details of plumage. Some first-year birds can be almost identical down to the smallest plumage detail, while structural differences can be easy to see.

Important details to register are:

- the age of the bird

- general colour of upperparts and underparts, with special attention to the actual *colour* of the light and dark areas

- colour and pattern of lesser underwing-coverts (patagium)

- colour of uppertail and type of barring.

- iris colour

- structure of feet and bill

- type of barring of underwing remiges

Identification of *Buteo* hybrids

Identifying any interspecific hybrid between two *Buteo* buzzards would be a difficult, if not an impossible, task owing to the immense plumage variation within each species. Perhaps because of this variation and the similarity of different species, possible hybrids may have gone undetected, as very few documented cases are known. To date, hybrids between Rough-legged Buzzard and Common Buzzard (*sensu lato*) are known, with one documented case each for Norway and Finland. In both cases the mixed pair successfully managed to fledge young. In Hungary Long-legged and Common Buzzards are known to have hybridised, and in Italy and Spain mixed pairs between Common and (Atlas) Long-legged Buzzards have bred successfully raising chicks. From further afield, outside our region, mixed pairs between Long-legged Buzzard and Upland Buzzard *B. hemilasius* occur quite frequently where the two species meet.

A presumed hybrid *Buteo* buzzard should look somehow odd, not matching any described taxon of buzzard, and should preferably combine features of two parental species or taxa. It goes without saying that such a hybrid would escape the notice of most birders as 'just another buzzard'. In fact, most proven cases of *Buteo* hybridisation are based on nest-finds, where the identity of the breeding adults has been confirmed. The plumages of the known hybrids between Common and Rough-legged Buzzards have been very similar to certain pale juveniles of Common Buzzard, but the tarsi have been half-feathered. In the field these

hybrids are likely to be overlooked as pale Common Buzzards, as would probably any hybrid between Common and Long-legged Buzzards.

For images of presumed hybrids between Common Buzzard and Atlas Long-legged Buzzard, see under Atlas Long-legged Buzzard.

References

Corso, A. 2009. Successful breeding of Atlas Long-legged Buzzard and Common Buzzard on Pantelleria, Italy, in 2008. *Dutch Birding* 31: 226–228.

Elorriaga, J. & Muñoz, A.-R. 2013. Hybridisation between the Common Buzzard *Buteo buteo buteo* and the North African race of Long-legged Buzzard *Buteo rufinus cirtensis* in the Strait of Gibraltar: prelude or preclude to colonization? *Ostrich* 84: 41–45.

Forsman, D. & Lämsä, E. 2007. Successful interbreeding between Common Buzzard and Rough-legged Buzzard in Finland. *Linnut* 42: 36–37 [in Finnish].

Gjershaug, J. O., Forset, O. A., Woldvik, K. & Espmark, Y. 2006. Hybridisation between Common Buzzard *Buteo buteo* and Rough-legged Buzzard *B. lagopus* in Norway. *Bull. Brit. Orn. Club* 126: 73–80.

Pfander, P. & Schmigalew, S. 2001. Umfangreiche Hybridisierung der Adler, *Buteo rufinus* Cretz. und Hochlandbussarde *B. hemilasius* Temm. et Schlegel. *Ornithol. Mitteil.* 53: 344–349.

Rodriguez, G., Elorriaga, J. & Ramirez, J. 2013. Identification of Atlas Long-legged Buzzard and its status in Europe. *Birding World* 26: 147–173.

COMMON BUZZARD
Buteo buteo buteo

Other names: Fomerly known simply as Buzzard in the British Isles

VARIATION Highly variable, with several subspecies described for the Western Palearctic, exhibiting extensive but poorly understood geographical variation in plumage, size and structure. Nominate *buteo* from the British Isles and across most of mainland Europe, gradually intergrades with eastern *vulpinus* (for full treatment of *vulpinus*, see under Steppe Buzzard) roughly at longitude 20–30 degrees east. Marked geographical variation also found within the nominate subspecies with, for instance, Spanish birds being darker and smaller than birds from continental W Europe. Other subspecies, some of disputed validity, include island forms, such as *arrigonii* (Corsica, Sardinia), *rothschildi* (Azores), *insularum* (Canary Islands), and *bannermani* (Cape Verde Islands), which all are rather poorly differentiated from nominate *buteo*. Resident (?) south-eastern *menetriesi*, from the Caucasus and Asia Minor, is rather similar to *vulpinus*, but fractionally larger and on average darker rufous-brown, thus easily confused with the dark rufous morph of the sympatric Long-legged Buzzard *B. rufinus*. The Socotra Buzzard *B. socotraensis*, endemic to the island of Socotra, is another poorly defined island form, which has recently been given species rank.

Nominate birds from S Scandinavia and W and C Europe can be divided into three main colour morphs, dark, intermediate and light, with a wide range of intermediates. Largely white birds are most frequent in the northern parts of mainland Europe, particularly in southern Scandinavia, Germany and the Low Countries. These birds look quite exotic, compared to ordinary buzzards, and are notorious for causing identification problems to the unaware.

Because of extensive plumage variation in both *buteo* and *vulpinus*, only the most typical individuals can be separated in the field, and adults more easily than juveniles. The two forms and their intergrades coexist in a wide intergradation zone extending from N Sweden through Finland, east of the Baltic States and down to E Greece, but because of identification difficulties the area of overlap is poorly known and further studies are required (see under Steppe Buzzard). Recent information (pers. obs.) from breeding areas in Finland suggests that the Steppe

Buzzard population has been declining for some time, while nominate-type birds are expanding north and east into the former range of *vulpinus*.

DISTRIBUTION A common breeding species over most of Europe. Mostly resident or dispersive, but northern populations from Scandinavia, Finland and the Baltic states are migratory, wintering in C and S Europe and Asia Minor. Some enter Africa, possibly birds with partial *vulpinus* ancestry.

BEHAVIOUR When hunting perches openly on posts by roadsides, or partly hidden in trees, overlooking open fields and meadows. Also hunts inside the forest and on the wing, often stopping to hover, but the frequency of hovering seems to vary geographically.

SPECIES IDENTIFICATION A rather stocky raptor, with broad wings, and a broad and short tail. Best identified by clearly darker patagium, contrasting against pale median underwing-coverts, and by its pale breast-band, which together are usually enough to separate it from its nearest allies, Long-legged and Rough-legged Buzzards and the honey-buzzards. *The barring of the secondaries is broader on a duskier background, compared to the finer and more distinct barring on a whiter background in typical vulpinus. In the primaries the barring continues further out in many buteo, while in vulpinus the outer 4–5 primaries tend to be unbarred and white with dark fingers only.* The body plumage is highly variable, depending on colour morph, but most birds show a darker, heavily streaked upper breast, a pale breast-band and darker flanks and trousers, while the belly is usually lighter with variable darker mottling or barring. The darkest birds only show a hint of a pale breast-band, while even the whitest individuals usually retain some dark markings on the head, breast-sides and flanks. Contrary to most other raptors with a largely whitish plumage, pale Common Buzzards also tend to retain a darker malar stripe and darker crown, while the face is white with a conspicuous dark eye (lighter grey in juveniles).

MOULT Moults during the breeding season, from Apr to Sep–Oct, with first-timers starting on average before the breeding adults. In the first moult usually 2–4 outer primaries are regularly left unmoulted, and serve as a reliable ageing character from 2nd cy autumn until the next moult in 3rd cy spring (see below). In older birds the odd primary or secondary may be left unmoulted, and frequently also some upperwing-coverts. Although the plumage of adults always comprises feathers of different age and wear, it is never as faded and worn as in the case of typical adult Steppe Buzzards.

There are probably differences in moulting strategies between birds from different parts of Europe, relating to their corresponding migration strategies, but further studies are required.

PLUMAGES Common Buzzards can be divided into three main colour morphs, dark, intermediate and light, with a continuous series of intergrades, particularly between dark and intermediate forms.

Normally three age-classes can be separated: juveniles, first-adults and full adults. At times a second-adult plumage can be defined. Important features to note are signs of moult, type of dark trailing edge to the wings, type of tail-barring, iris colour and general shape of the bird. Juveniles tend to be more streaked on the breast and underwing-coverts; adults are more barred, with first-adults often appearing intermediate (mottled), but this tends to vary geographically, with more clearly streaked juveniles and more clearly barred adults in the northern populations. Juveniles have narrower wings and longer tails than adults, but this difference is more marked in the migratory populations of the east and north.

Juvenile

Ageing can be difficult, as some juveniles can appear rather similar to adults (particularly the darkest birds), but most birds are easily aged. In juveniles the upperparts show neat pale fringes to most wing-coverts and scapulars, forming a regular pattern. The tail is densely and finely barred, often becoming paler and less marked towards the base, but lacking the adult's broad black subterminal band. The head, neck and underparts are streaked in most, more irregularly spotted in others, and the underwing-coverts in particular appear streaked, compared to the more barred pattern in adults. The underwing remiges are barred with broadly dark tips, but the subterminal band is less sharply defined (with a 'bleeding' inner margin), less contrasting and narrower, compared to the often very distinctly marked adults, but some dark juveniles can be surprisingly adult-like in this respect; they may even show a broad dark subterminal tail-band similar to the adults. Compared to juvenile *vulpinus*

the barring of the underwing remiges is broader, more diffuse and often with a duskier background colour making the difference often quite striking compared to the more brightly marked Steppe Buzzards.

Dark and intermediate juveniles can often be aged from quite some distance by their diagnostic upperparts. In fact, ageing is often easier when looking at the better lit upperparts, compared to the shadowed and darker underparts: *the upperwing-coverts stand out as being lighter brown contrasting sharply against the uniformly dark greater coverts and secondaries.* In adults the upperparts look much more uniform, with just a darker trailing edge to the wing.

First-adult (2nd cy summer–3rd cy spring)

The juvenile plumage is largely replaced during the 2nd cy summer. After the completed moult in autumn, superficially similar to adult but usually easily aged by the retained outer 2–4 juvenile primaries, by now worn and faded brown in colour, clearly appearing too short to neatly match the other primaries forming the wingtip. The band along the trailing edge of the wing is not as broad, nor as sharply defined along its inner border as in full adults, and the eye is a medium brown with a discernible black pupil, not all dark as in full adults. The barring of the underbody is not yet as fine and regular as in full adults, appearing more like irregular mottling, but this character is individually variable in all age-classes.

Second-adult (3rd cy summer–4th cy spring)

This age-class can tentatively be identified from good photographs, but only rarely in the field. Rather similar in most respects to a first-adult, but differs most clearly in details of primary moult. Birds showing the following characters can tentatively be aged as second-adult plumage: an adult-type bird with a paler brown iris (black pupil still discernible), and often with some remaining irregularities to the inner margin of the dark trailing edge of the wing, but *lacking the faded brown outer primaries of the previous age-class. The outer and inner primaries are the freshest with dark tips, while the median primaries (usually pp4–6) are more worn.* Due to extensive individual variation in moult progress and iris coloration the ageing of this age-class may not be reliable, and some birds may no longer be told from full adults.

Definitive adult

Usually darker and more solidly coloured compared

to juveniles. Underparts and underwing-coverts often rather finely barred rather than streaked, but some adults appear blotchy and probably the type of marking is geographically determined, with barring increasing in Europe from W to NE (increasing *vulpinus* influence?). Upperparts appear uniform, save for some patchiness caused by feathers of different wear. Adults can often easily be told from juveniles from afar by the upperwing alone: *the entire upperwing looks uniformly brown (except for the normal slightly paler panel in the primaries) with a distinctly darker trailing edge*; it thus lacks the marked contrast between lighter coverts and darker secondaries typical of all juveniles. From below adults show a broad black and sharply defined black trailing edge to the wing and the tail has a broad dark subterminal band, best visible from above. Full adults have a very dark brown iris, and the pupil is not discernible as in juveniles and younger adults.

Extreme pale morph

The nominate Common Buzzard is known for its striking pale morph, which is most frequent in the areas around the southern Baltic Sea, from where birds move S and SW in severe winters. Individuals are highly variable, and one could say that no two pale morphs birds look the same. The head and underbody are mostly creamy white, with some dark markings retained on the crown, malar area, breast-sides and flanks in most. The upperparts vary from near-normal to striking combinations of variably white upperwing-coverts and white inner tail, the latter often showing white longitudinal wedges along the feathers. From below the *dusky secondaries* are apparent, a good character to distinguish it from the otherwise similar Rough-legged and Long-legged Buzzards. At first sight birds like this can be overwhelming, but the partly white upperparts are usually enough to exclude other species. Only adult Egyptian Vultures are partly white above, but their unique shape and structure, as well as plumage pattern leave no room for confusion.

Ageing pale morph birds is a lot harder, since the plumage characters normally used for ageing do not apply, but some general features can be helpful: birds in active moult cannot be juveniles, but apart from this, only iris colour is a safe ageing feature: dark in adults, greyish-brown to pale with a black pupil in juveniles. Also the different structure can be helpful, with adults appearing heavy and broad-winged compared to the more sleek juveniles.

SEXING It is usually not possible to sex Common Buzzards in the field, although males are fractionally smaller and lighter on the wing compared to females. In adult birds males tend to be more finely marked below than females, but this is not a constant trait, as it is also linked to age. The silhouette is also somewhat different, with males showing a proportionately larger head, narrower wings and shorter tail compared to females.

CONFUSION RISKS Because of their endless plumage variation, Common Buzzards are often confused with other medium-sized, largely dark and broad-winged raptors. Seen well, the underwing- and tail-barring, as well as the body plumage are diagnostic compared with most other species, but certain light-coloured individuals can be almost identical to similar-looking Long-legged Buzzards. This similarity is even more pronounced between Steppe and Long-legged Buzzards, which see.

552. Common Buzzard *Buteo buteo buteo*, 2nd cy immature (left) and breeding adult. Note difference in proportions, trailing edge of wings, tail-tip and general markings below in these two rather average representatives for their respective age-classes. Finland, 27.5.2012 (DF)

553. Common Buzzard *Buteo buteo buteo*, juvenile, resident. Trailing edge of wing is typically poorly marked in this rather pale juvenile. Note also pale iris colour. S Sweden, 5.9.2010 (DF)

554. Common Buzzard *Buteo buteo buteo*, juvenile, autumn migrant. Note streaked body and underwing-coverts, typical of juvenile. Finland, 11.10.2013 (DF)

555. Common Buzzard *Buteo buteo buteo*, juvenile, autumn migrant (same as **554**). Juveniles show marked contrast between dark secondaries and lighter coverts above. Finland, 11.10.2013 (DF)

556. Common Buzzard *Buteo buteo buteo*, juvenile, wintering. Dark juveniles may show a rather contrasting trailing edge, and dusky and broadly banded underwings. Finland, 5.3.2012 (DF)

557. Common Buzzard *Buteo buteo buteo*, juvenile. Another rather average juvenile. Finland, 7.4.2013 (DF)

558. Common Buzzard *Buteo buteo buteo*, juvenile (same as **557**). Typical upperside of a juvenile (tail damaged). Finland, 7.4.2013 (DF)

559. Common Buzzard *Buteo buteo buteo*, juvenile, recently fledged. Hungary, 21.7.2011 (DF)

560. Common Buzzard *Buteo buteo buteo*, juvenile, autumn migrant. Finland, 11.9.2014 (DF)

561. Common Buzzard *Buteo buteo buteo*, juvenile, autumn migrant. Juveniles usually lack contrast on their trailing edge, unlike adults. S Sweden, 6.10.2014 (DF)

562. Common Buzzard *Buteo buteo buteo*, 2nd cy, autumn migrant. Typically cross-barred below like adults, but aged by retained short and faded juvenile feathers in the wings. S Sweden, 10.9.2014 (DF)

563. Common Buzzard *Buteo buteo buteo*, 2nd cy after suspended first moult, autumn migrant. Barred like adult on body and underwing-coverts, but aged as a 2nd cy by short and faded juvenile remiges lacking the black tips. S Sweden, 6.10.2014 (DF)

564. Common Buzzard *Buteo buteo buteo*, 2nd cy after suspended first moult, autumn migrant. This rather pale bird has retained two of its outer juvenile primaries but no secondaries. Note the thin subterminal band in the tail, indicating a younger bird. S Sweden, 6.10.2014 (DF)

565. Common Buzzard *Buteo buteo buteo*, 2ⁿᵈ cy after suspended first moult, autumn migrant. Like an adult, uniformly brown with darker trailing edge of wing, but told by a few retained, short and faded juvenile secondaries and outer primaries. S Sweden, 6.10.2014 (DF)

566. Common Buzzard *Buteo buteo buteo*, 2ⁿᵈ cy after suspended first moult, autumn migrant. Like a dark adult, but note juvenile s4 and s9 and pp7–10, which are shorter and faded brown juvenile feathers. S Sweden, 6.10.2014 (DF)

567. Common Buzzard *Buteo buteo buteo*, adult, breeding. Rather brightly marked remiges on this darkish breeding adult from the interbreeding zone of *buteo* and *vulpinus*. S Finland, 21.4.2013 (DF)

568. Common Buzzard *Buteo buteo buteo*, adult female, breeding. Typical dusky markings of nominate subspecies on this bird, from the neighbouring territory of **567**. S Finland, 8.4.2012 (DF)

569. Common Buzzard *Buteo buteo buteo*, adult. S Sweden, 8.9.2014 (DF)

570. Common Buzzard *Buteo buteo buteo*, adult (same as **569**). Note uniformly brownish upperparts with slightly darker tips to remiges and broad dark subterminal band on tail. S Sweden, 8.9.2014 (DF)

311

571. Common Buzzard *Buteo buteo buteo*, adult. A dark adult with dusky and broadly barred remiges typical of nominate *buteo*. S Sweden, 5.9.2010 (DF)

572. Common Buzzard *Buteo buteo buteo*, adult. A lighter and smaller migrant adult. S Sweden, 3.9.2010 (DF)

573. Common Buzzard *Buteo buteo buteo*, light morph juvenile. Birds like this could be mistaken for Rough-legged or Long-legged Buzzard, but upper breast and head are too heavily marked. S Finland, 11.9.2009 (DF)

574. Common Buzzard *Buteo buteo buteo*, light morph juvenile. Barring of wings and tail are typical of Common Buzzard, as is the darker patagium on the forewing. S Sweden, 5.9.2010 (DF)

575. Common Buzzard *Buteo buteo buteo*, light morph juvenile. Another Rough-legged type bird, but barring of underwing and tail differs. Pale Common Buzzards have a tendency to retain the dark markings around the neck and upper breast, even when the belly-patch and flanks are dissolving, which is diagnostic compared with Rough-legged and Long-legged Buzzards. S Sweden, 3.9.2010 (DF)

576. Common Buzzard *Buteo buteo buteo*, light morph juvenile. Birds in this plumage tend to have very pale eyes and extensively white-fringed upperwing-coverts, although the latter varies a lot individually. S Sweden, 4.10.2013 (DF)

577. Common Buzzard *Buteo buteo buteo*, light morph juvenile (same as **576**). Although a very pale bird, note the retained dark markings around the head and neck. S Sweden, 4.10.2013 (DF)

578. Common Buzzard *Buteo buteo buteo*, light morph juvenile. S Finland, 7.10.2008 (DF)

579. Common Buzzard *Buteo buteo buteo*, light morph juvenile. This bird is as pale as they can get. Note wide dusky trailing edge of the wings and tail, and black carpal crescents. S Sweden, 3.9.2010 (DF)

580. Common Buzzard *Buteo buteo buteo*, light morph adult male, breeding. Pale underbody except for dark neck-sides and carpal commas. S Finland, 12.4.2008 (DF)

581. Common Buzzard *Buteo buteo buteo*, light morph juvenile. Although extremely variable, light morph birds tend to show partly white upperwing- and uppertail-coverts, and partially white tail. S Finland, 26.8.2008 (DF)

STEPPE BUZZARD
Buteo buteo vulpinus

VARIATION The Steppe Buzzard, a subspecies of Common Buzzard, replaces the nominate form roughly east of longitude (20°–) 30° east. However, where the two forms meet there is thought to be a 400–500 km wide zone of overlap, which runs roughly from N Sweden (long. 20° E) and Finland to the Black Sea (long. 30° E). Because of the extensive plumage variation in Steppe Buzzard, and its general similarity to Common Buzzard, particularly in the western part of its range, only the most typical individuals can be identified, adults more easily than juveniles. Although the two subspecies are known to occur side by side in many areas, e.g. in parts of Finland, the area of overlap is poorly known, and further research is needed.

DISTRIBUTION According to ringing recoveries of Finnish breeding birds most migrate south taking a route east of the Black Sea, but some, particularly Swedish but also some Finnish birds, follow a more westerly route down through W and SW Europe. True Steppe Buzzards winter in Africa, from East Africa to the southernmost regions of the continent, but some birds, many showing Common Buzzard features (probably intergrades), are also found in winter further north, in eastern Europe, Asia Minor and the Middle East.

BEHAVIOUR More of a forest dweller compared to the western Common Buzzard and often found in areas dominated by forests interspersed by bogs and small clearings. Because of its smaller size and greater agility it moves with ease through the trees, even in relatively dense forest. Hunting behaviour is similar to Common Buzzard, either from an elevated perch or from the air, hovering frequently, at least regionally.

SUBSPECIES IDENTIFICATION Owing to the immense plumage variation Steppe Buzzards can be extremely difficult to identify to subspecific level, some being very similar to Common Buzzard, others to Long-legged Buzzard. Because of intergrades occurring in the overlap areas between Steppe and Common Buzzards, individual birds away from their normal range cannot be identified with certainty, unless carefully measured and photographed.

The most typical forms are rather diagnostic, particularly as adults, and uniformly fox-red, deep rufous or black birds can safely be identified as Steppe Buzzards, as these forms are lacking in nominate Common Buzzard (provided that the similar forms of Long-legged Buzzard have been excluded first). *Steppe Buzzards tend to have whiter remiges below than Common Buzzards, with finer and more distinct barring and with the outer 4–5 primaries practically unbarred, while the upperparts in adults are a faded greyish shade of brown and the underbody is more finely barred.*

MOULT Steppe Buzzards have a complex moult, as part of the plumage is moulted in the breeding areas and part in the wintering grounds, a strategy shared by many other long-distance migrant raptors. Adults commence their moult during breeding, but arrest the moult for the autumn migration. Moult is resumed and completed in the winter quarters. This moulting strategy explains why *older Steppe Buzzards always show a mixture of fresh and worn feathers in their plumage*, an important difference compared to most nominate Common Buzzards.

Immatures start to moult for the first time in the 2nd cy spring, some already while still on the wintering grounds, and 2nd cy birds with fresh inner primaries are frequently seen on spring migration. Moult is arrested in autumn, when the outermost primaries are still retained juvenile feathers, a good feature of the age-class. Many immatures can still be aged in the 3rd cy spring by their retained juvenile outermost primaries (pp9–10).

PLUMAGES Immensely variable, with several recognised colour morphs and their intergrades. Although the classic colour morphs, the grey, the fox-red, the dark rufous and the black, are easy to tell from each other when typical, the various intergrades between the morphs, particularly between grey and fox-red, between fox-red and dark rufous, and between dark rufous and black, are common and add to the variation. Adults are often easier to assign to a certain colour morph, while juveniles are more similar, for instance between the grey and fox-red morphs.

Most birds are easy to recognise as belonging to the Common Buzzard group, based on typical underwing- and tail-barring, darker lesser underwing-coverts and a lighter breast-band, but some are confusingly similar to Long-legged Buzzards, with the most similar birds perhaps not separable on plumage at all, particularly some juveniles.

Normally three age-classes can be separated: juveniles, first-adults and definitive adults, just as in Common Buzzard. As a result of spending the winter in dry and sunny conditions the plumage of Steppe Buzzards is often notably faded and worn compared to resident Common Buzzards of Europe.

Juvenile

Best aged by streaked, not barred underbody, rather diffuse dark trailing edge to the underwings (but may be dark and adult-like in some) and by rather pale greyish-brown iris. Upperparts diagnostic, with brown upperwing-coverts contrasting with darker secondaries and greater coverts, and with tail uniformly and finely barred, lacking the adult's distinct and broad, dark subterminal band. The upperwing-coverts and mantle feathers are edged rufous or orange when fresh, but these markings will wear off by first spring, by which time juveniles appear very drab and faded greyish-brown above.

Juveniles of the uniformly dark morph are best aged by their diffuse trailing edge to the underwing, lack of distinct dark subterminal tail-band and their pale iris.

First-adult

Birds in their 2nd cy autumn are best identified by their retained juvenile outer primaries, sometimes also by retained median secondaries, which are faded brown in colour, shorter and more pointed compared to the fresh adult-type feathers. The underbody pattern is often intermediate between juvenile and definitive adult, being blotchy or coarsely barred or spotted on the breast, compared to being clearly streaked in the juvenile or finely barred in definitive adult. The iris colour is not yet completely dark as in full adults, but clearly a lighter brown colour.

Adult

Identified by all-dark iris, distinct, broad and black trailing edge to the underwing and finely barred tail with broad and dark subterminal band, although some red-tailed adults may be lacking tail-barring altogether, save for a narrow black subterminal band (cf. first-adult plumage of Long-legged Buzzard). In closer views the adult plumage always shows a mixture of old and fresh feathers above, and the upperparts appear rather uniformly faded greyish-brown, with a distinctly darker trailing edge to the wing from

above, a good ageing character compared to juveniles. The tail is reddish, brownish or greyish, usually with fine barring, which often dissolves towards the base, making the inner half of the tail lighter, while the tip shows a variably broad and blackish subterminal band, sometimes poorly defined from below. The lighter breast-band and the underwing-coverts are finely barred in most, while uniformly rufous or dark birds lack any kind of markings to their body plumage.

SEXING Steppe Buzzards cannot be sexed in the field, although the males have on average narrower wings, a shorter tail, a comparatively larger head and they are more finely marked below than females.

CONFUSION RISKS Some birds can be very similar to either Long-legged or Common Buzzards (which both see). Claiming a Steppe Buzzard away from the subspecies' normal range is usually not possible, as intergrades with Common Buzzard are common and stragglers superficially looking like *vulpinus* may in fact come from the wide interbreeding zone in NE Europe.

References

van Duivendijk, N. 2011. Steppebuizerd in Nederland: herziening, status en determinatie. *Dutch Birding* 33: 283–293.

582. Steppe Buzzard *Buteo buteo vulpinus*, fresh juvenile, autumn migrant. Note yellow and rufous-brown tones of body plumage, as well as slim build and white, distinctly marked remiges. Egypt, 11.10.2010 (DF)

583. Steppe Buzzard *Buteo buteo vulpinus*, fresh juvenile, autumn migrant. The body is more uniformly streaked in many juvenile *vulpinus*, compared to juveniles of nominate *buteo*. Egypt, 10.10.2010 (DF)

584. Steppe Buzzard *Buteo buteo vulpinus*, fresh juvenile, autumn migrant from S Finland, an interbreeding zone between *vulpinus* and *buteo*. Despite showing typical *vulpinus* structure and plumage, an intergrade with *buteo* cannot be excluded. Finland, 14.9.2013 (DF)

585. Steppe Buzzard *Buteo buteo vulpinus*, fresh juvenile, autumn migrant, possibly an intergrade with *buteo*. Note narrow wings and tail, strongly yellowish ground colour and uniformly streaked underparts of this rather typical southern Finnish juvenile. Finland, 14.9.2013 (DF)

586. Steppe Buzzard *Buteo buteo vulpinus*, fresh juvenile, autumn migrant. Apart from *vulpinus*-type plumage this bird also shows a light iris colour, another trait of juvenile *vulpinus*; an intergrade is not possible to exclude. Finland, 18.9.2010 (DF)

587. Steppe Buzzard *Buteo buteo vulpinus*, fresh juvenile, autumn migrant (same as **582**). The upperparts show extensively fringed coverts and scapulars, and the tail is distinctly barred in typical juvenile *vulpinus*. Egypt, 11.10.2010 (DF)

588. Steppe Buzzard *Buteo buteo vulpinus*, juvenile, 2nd cy spring migrant. A typical lightly streaked juvenile. Israel, 19.4.2009 (DF)

589. Steppe Buzzard *Buteo buteo vulpinus*, juvenile, 2nd cy spring migrant. Breast and flanks partly moulted; note also fresh inner primary p1. Israel, 19.4.2009 (DF)

590. Steppe Buzzard *Buteo buteo vulpinus*, juvenile, 2nd cy spring migrant. Steppe Buzzard does not have a pale form, but some juveniles can appear very light. Note typically pale and finely barred underwing remiges compared to juvenile nominate *buteo*. Israel, 2.4.2008 (DF)

591. Steppe Buzzard *Buteo buteo vulpinus*, juvenile, 2nd cy spring migrant. Some juveniles resemble juvenile Long-legged Buzzard, with dark belly and dark carpal patches. Israel, 19.4.2009 (DF)

592. Steppe Buzzard *Buteo buteo vulpinus*, juvenile, 2nd cy spring migrant, dark form. Dark birds vary from deep rufous to nearly black, but will fade during winter to become browner by spring. They have duskier remiges below compared to lighter birds. Israel, 19.4.2009 (DF)

593. Steppe Buzzard *Buteo buteo vulpinus*, juvenile, 2nd cy spring migrant, dark form. A more rufous, uniformly coloured juvenile. Israel, 29.3.2012 (DF)

594. Steppe Buzzard *Buteo buteo vulpinus*, juvenile, 2nd cy spring migrant, dark form. This dark morph juvenile has moulted its inner primary in both wings. Israel, 18.4.2009 (DF)

595. Steppe Buzzard *Buteo buteo vulpinus*, juvenile, 2nd cy spring migrant. Spring juveniles are extremely worn above. Note rather light tail and sharp contrast between coverts and secondaries typical of juvenile. Israel, 19.4.2009 (DF)

596. Steppe Buzzard *Buteo buteo vulpinus*, juvenile, 2nd cy spring migrant. Typical upperparts of a spring juvenile. Israel, 19.4.2009 (DF)

597. Steppe Buzzard *Buteo buteo vulpinus*, first-adult plumage, 3rd cy spring migrant. Reliably identified if, as here, outermost juvenile primaries are still retained, as they seem to be in most. This bird has also retained s9 in its right wing. By and large like adult, but iris not yet dark and belly not as finely barred as in many older birds. Israel, 3.4.2008 (DF)

598. Steppe Buzzard *Buteo buteo vulpinus*, first-adult plumage, 3rd cy spring migrant. This rather *buteo*-like immature can be aged by retained juvenile outer primaries. Seen away from the main *vulpinus* flyway this bird would probably be classed as a nominate *buteo* based on its contrasting plumage. Israel, 24.3.2012 (DF)

599. Steppe Buzzard *Buteo buteo vulpinus*, first-adult plumage, 3rd cy spring migrant. This average-looking *vulpinus* is aged by its outermost juvenile primary. Note medium-brown iris, lighter than in full adults. Israel, 26.3.2010 (DF)

600. Steppe Buzzard *Buteo buteo vulpinus*, first-adult plumage, 3rd cy spring migrant (same as **599**). Adults are greyish-brown above with a darker trailing edge to the wings and a variably coloured tail. Israel, 26.3.2010 (DF)

601. Steppe Buzzard *Buteo buteo vulpinus*, first-adult plumage, 3rd cy spring migrant, dark morph. Similar to adult dark morph, but told by a few retained juvenile flight feathers. Israel, 19.4.2009 (DF)

602. Steppe Buzzard *Buteo buteo vulpinus*, first-adult plumage, 3rd cy spring migrant, dark morph. A uniformly dark rufous-brown bird with retained juvenile outermost primary. Israel, 26.3.2008 (DF)

603. Steppe Buzzard *Buteo buteo vulpinus*, first-adult plumage, 3rd cy spring migrant, dark morph (same as **602**). Note grey and poorly marked tail in this bird. Israel, 26.3.2008 (DF)

604. Steppe Buzzard *Buteo buteo vulpinus*, first-adult plumage, 3rd cy spring migrant. Note typically coloured and barred tail, and retained outer primaries and s9 in this rather average *vulpinus*. Israel, 26.3.2008 (DF)

605. Steppe Buzzard *Buteo buteo vulpinus*, full adult plumage, spring migrant. A full adult of the so-called fox-red form. The tail markings can be difficult to see from below. Israel, 22.3.2012 (DF)

606. Steppe Buzzard *Buteo buteo vulpinus*, full adult plumage, spring migrant. Adult fox-red morph. Note dark iris. Israel, 22.3.2012 (DF)

607. Steppe Buzzard *Buteo buteo vulpinus*, full adult plumage, spring migrant. Another fox-red adult. Israel, 20.3.2013 (DF)

608. Steppe Buzzard *Buteo buteo vulpinus*, full adult plumage, spring migrant. This lighter adult has similarities with adult Long-legged Buzzard, but differs by having a barred belly. Israel, 19.4.2009 (DF)

609. Steppe Buzzard *Buteo buteo vulpinus*, full adult plumage, spring migrant. A deeply coloured adult with darker belly and prominent dark carpal patches, but told from Long-legged Buzzard by broad dark subterminal band of tail. Israel, 24.3.2012 (DF)

610. Steppe Buzzard *Buteo buteo vulpinus*, full adult plumage, spring migrant. A browner bird, but tail typically rufous with fine barring and distinct subterminal band. Israel, 2.4.2008 (DF)

611. Steppe Buzzard *Buteo buteo vulpinus*, full adult plumage, spring migrant. A typical adult, greyish-brown with darker trailing edge to wings; tail is a mix of red and grey with fine, distinct barring and a wider subterminal band. Israel, 21.3.2012 (DF)

612. Steppe Buzzard *Buteo buteo vulpinus*, full adult plumage, spring migrant. Adults are typically faded greyish-brown above, much lighter than typical nominate adult ssp. *buteo*. Note diagnostic tail. Israel, 24.3.2012 (DF)

613. Steppe Buzzard *Buteo buteo vulpinus*, adult, spring migrant. A red morph adult from above, with fox-red tail and broad rusty fringes to upperparts. Israel, 3.4.2008 (DF)

614. Steppe Buzzard *Buteo buteo vulpinus*, adult, spring migrant. A light grey, almost whitish tail, similar to Long-legged Buzzard, is also common, but barring is more complete in Steppe Buzzard. Israel, 23.3.2010 (DF)

615. Steppe Buzzard *Buteo buteo vulpinus*, adult, spring migrant. Deep rufous adults can be very similar to corresponding plumage of Long-legged Buzzard. Israel, 28.3.2008 (DF)

616. Steppe Buzzard *Buteo buteo vulpinus*, adult, spring migrant. Dark birds look black from a distance but are often dark brown when seen close. The plumage is almost identical to the corresponding plumage of Long-legged Buzzard. Israel, 26.3.2013 (DF)

617. Steppe Buzzard *Buteo buteo vulpinus*, adult, spring migrant. Most dark birds are actually brown in close views. Israel, 24.3.2012 (DF)

EASTERN LONG-LEGGED BUZZARD
Buteo rufinus rufinus

VARIATION A highly variable species, with two recognised subspecies, the larger nominate *rufinus* (Eastern Long-legged Buzzard) breeding from E Europe and the Balkans to C Asia and wintering from the Middle East and down to the equator in E Africa, and the smaller and mostly resident ssp. *cirtensis* (Atlas Long-legged Buzzard) of N Africa, from Morocco to Libya and W Egypt, with non-breeding immatures undertaking mostly local movements. The birds breeding in Arabia and S Israel require further studies in order to clarify their taxonomic affinities.

This account refers to the nominate Eastern Long-legged Buzzard *B. r. rufinus*, while the Atlas Long-legged Buzzard *B. r. cirtensis* is treated separately on page 335.

DISTRIBUTION Wide breeding range extends from E Europe, including the Balkans and major E Mediterranean islands, through the Levant, the Black Sea and Caspian Sea areas reaching the foothills of the C Asian mountain massifs. Winters to the south, reaching sub-Saharan Africa and the Indian subcontinent.

BEHAVIOUR Hunts mostly from a perch, but with favourable winds often seen hunting on the wing, including longer periods of hovering over open hillsides and other suitable habitat. Also perches freely on the ground, particularly in open flat areas.

SPECIES IDENTIFICATION Easily recognised as a *Buteo* buzzard by largely pale and finely barred underwing remiges, with a dark carpal patch and a distinct dark trailing edge, but plumage variation and similarity with other buzzards, most notably the Steppe Buzzard, adds to the identification challenge. The Eastern Long-legged Buzzard is by far the largest *Buteo* of the region, heavy and long-winged, and often the size is conveyed through the bird's movements in the air: when soaring the circles are slower and wider compared to other buzzards and the active flight is heavier with a slower wing action, at times appearing almost eagle-like. When soaring the wings are typically held lifted in a shallow V, as in Rough-legged Buzzard, and this is often enough to single it out in a flock of migrating buzzards. Although a suite of characters is usually enough to identify the most commonly occurring pale form correctly, the uniformly dark rufous and black morphs can be confusingly similar

to the corresponding forms of Steppe Buzzard, and structural differences need to be assessed alongside plumage features.

In close-up views the bill appears heavier and deeper than in Steppe Buzzard, with a noticeably long and curved tip, and the gape-line is long and fleshy, giving the bird a vicious grin, lacking in other buzzards.

MOULT As in Common and Rough-legged Buzzards, the annual moult is usually complete in adults, which moult their remiges at several simultaneously active foci, while immatures moulting for the first time only show one moult front in the primaries and always retain a few outer juvenile primaries, and sometimes also some median secondaries, after moult is suspended. Juveniles often start to moult their primaries prior to spring migration, as birds with fresh innermost primaries can be seen from Mar onwards. Adults commence moulting during early breeding, females before males. Moult is completed by Sep–Oct.

PLUMAGES Plumages can be grouped into four main types regardless of age: pale, intermediate, dark rufous and black morphs, although the variation seems continuous between pale and intermediate types, and between the dark rufous and black types.

The pale and intermediate morphs are straw-coloured below on the body and underwing-coverts with distinctly darker carpal and belly patches, while the dark forms are uniformly dark on the body and underwing-coverts.

Juveniles and adults are normally easily separated if seen well. Also, a first-adult plumage can often be recognised by retained juvenile feathers, until these are lost early in the second moult (3rd cy late spring/early summer).

Juvenile light and intermediate morphs

Best aged by their uppertail pattern, their dark brown secondaries from above, their diffuse trailing edge to the underwing and their pale iris.

The light and intermediate morphs are rather similar, and only the amount of streaking on body and underwing-coverts differs. The body is pale with a contrasting dark belly, sometimes reduced to dark flanks and trousers only, and the head and upper breast are variably streaked. In the lightest birds, unlike in whitish Common Buzzards, the head and neck are usually the lightest part of the body and lack prominent dark markings, save for a dark nape-patch.

The underwing-coverts are pale straw-coloured and variably streaked, but usually the lesser coverts do not stand out as being darker than the rest, as is the case in otherwise similar-plumaged Steppe Buzzards. The underwing also shows a contrasting round, dark carpal patch while the darker greater coverts usually form a noticeable wing-band. The remiges are whitish with fine and rather sparse barring throughout, but normally the secondaries appear duskier than the primaries, enhancing the white primary window. The primaries have dark fingers, while the trailing edge of the wing is narrower and rather diffuse compared to the more distinctly marked adults, although more distinctly marked adult-like juveniles occur.

From a distance the upperparts normally show a very light head and a sandy-brown area on the coverts, contrasting sharply with dark greater coverts and secondaries, the latter a good ageing feature. The primaries appear lighter brown than the secondaries, often with an individually variable white patch at the bases of the outer primaries, although this distinctive feature may be missing in many birds.

The tail is typically rather pale greyish-brown with a whiter base, overall paler and greyer than in most young Steppe Buzzards, and with fine darker greyish-brown barring. Others have a whiter tail with some orange admixed, and fine browner barring only distally.

Iris colour in juveniles is often noticeably pale, pearl-grey to almost white, with a clearly discernible black pupil. It is obviously paler and brighter than in most juvenile Steppe Buzzards and together with the prominent gape-flange gives the juveniles a fierce look.

Juvenile dark rufous and black morphs

In the darker forms the entire body plumage is uniformly dark, either rufous brown or black, which in juveniles will fade to a dark brown by the 2nd cy spring.

The remiges are markedly dusky below, with rather broadly, irregularly and diffusely barred secondaries, a diagnostic difference compared to the dark form of Steppe Buzzard, where the barring is fine and distinct, while the outer primaries show a lighter flash, often clearest in the outermost primaries, and the whole wing has a blackish tip and broad dark trailing edge. The tail is dark brown or dark grey above with a few broad dark bands and sometimes a broader, similarly coloured not black subterminal band.

The pale iris and the fleshy gape are more outstanding in dark birds, making them look even more vicious than the light morph birds.

First-adult (second) plumage

Safe to age by retained worn and faded juvenile feathers in the wings. Even after moult is suspended, from late 2nd cy autumn, this plumage is recognised by retained outermost juvenile primaries, sometimes also a few median secondaries, which differ in length, colour and pattern from the replaced ones. These last remaining juvenile remiges are kept until the following moult in the 3rd cy spring, when they are gradually replaced.

Light morph birds are still rather juvenile-like in their second plumage with extensive streaking below, and the tail may show a fine black subterminal bar and additional, partly fragmented narrow distal ghost-bars on an otherwise clean-looking white tail with orange outer half. The trailing edge of the wing is still rather diffuse and also narrower compared to full adults, the secondaries are less clearly grey above and the iris is still pale.

Dark birds differ from dark juveniles by their blacker and more distinct trailing edge to the wings and the more distinctly marked uppertail, which by now is clearly grey, with variable black barring and a broad black tip.

Birds of this age-class still have a noticeably pale iris, not that different from juveniles.

Adult light and intermediate morphs

Light morph adults are best told from juveniles of the pale morph by their greyish and clearly barred secondaries above, their unmarked white and orange tail, their distinct black trailing edge to the wing, and their dark iris. In moulting birds, old and new secondaries and primaries are the same shape, colour and pattern.

Adult dark rufous and black morphs

Darker morph adults tend to be blacker compared to the more brownish-looking darker morph juveniles and the bars in the wings and tail are more distinct, but the underwing barring is still broader and more diffuse and the ground colour of the remiges is smokier compared to the distinctly marked underwings of darker adult Steppe Buzzards. From above the remiges are clearly greyish with darker barring, just as in adults of the lighter forms (darker, not grey, in dark morph

Steppe Buzzards). Iris is dark brown, and the black pupil is not discernible.

SEXING As a rule, sexing in the field is not possible, although females are larger and heavier than males when seen side by side.

CONFUSION RISKS Easily confused with other *Buteo* buzzards, some of the plumages being almost identical to the corresponding plumages of Steppe Buzzard. Also difficult to tell from the N African subspecies of Long-legged Buzzard, the Atlas Long-legged Buzzard, *B. r. cirtensis,* and although their ranges are widely separated during the breeding season, they overlap in winter when ssp. *rufinus* moves south. Young Rough-legged Buzzards may also look superficially similar to certain light morph juvenile Long-legged Buzzards, particularly in worn plumage in the late 2nd cy spring, but can always be told by their white tail with a contrasting, broadly dark tip, reduced secondary barring below, their fine bill, and tiny feet with feathered legs.

NOTE Hybrids between Common Buzzard and Eastern Long-legged Buzzard have been reported from Hungary, where mixed pairs have successfully raised hybrid chicks. For the situation in Atlas Long-legged Buzzard, see below.

618. Long-legged Buzzard *Buteo rufinus* (subspecies uncertain), juvenile. Note strong bill, fleshy gape and light iris, all typical of the species. This rather small and pale individual has a prominent dark carpal patch, while the dark flanks are just barely indicated. Oman, 8.12.2012 (DF)

619. Long-legged Buzzard *Buteo rufinus* (subspecies uncertain), juvenile (same as **618**). The bright upperwing-coverts contrast clearly with darker greater coverts and flight feathers. The pale tips to the greater coverts indicate a fresh juvenile. Oman, 8.12.2012 (DF)

620. Long-legged Buzzard *Buteo rufinus rufinus,* juvenile, migrant. A rather typical individual in most respects. Note the long wings held in a shallow V on this soaring bird. Egypt, 22.11.2008 (DF)

621. Long-legged Buzzard *Buteo rufinus rufinus* (same as **620**). A classic juvenile showing a light head and breast, solid dark carpal patches, light lesser underwing-coverts, a dark greater underwing-coverts band, and solid dark thighs and flanks. Also note finely barred light tail. Egypt, 22.11.2008 (DF)

622. Long-legged Buzzard *Buteo rufinus rufinus*, juvenile, wintering. Note uniformly streaked underwing-coverts, dark flanks and pale primary patch. Dark upperwing secondaries and pale iris identify this bird as a juvenile. India, 13.12.2007 (DF)

623. Long-legged Buzzard *Buteo rufinus rufinus*, juvenile, migrant. Israel, 4.4.2008 (DF)

624. Long-legged Buzzard *Buteo rufinus rufinus*, juvenile, migrant. Note light iris and diffuse trailing edge of underwing in this juvenile. Israel, 19.4.2009 (DF)

625. Long-legged Buzzard *Buteo rufinus rufinus*, juvenile, migrant (same as **623**). The sharp contrast between lighter upperwing-coverts and darker secondaries is a common feature of young *Buteo* buzzards, clearly shown by this rather faded spring bird. Also note light and finely barred tail and pale primary patch, but compare with some Steppe Buzzards in corresponding plumage. Israel, 4.4.2008 (DF)

626. Long-legged Buzzard *Buteo rufinus rufinus*, juvenile, migrant (same as **624**). A greyish-brown tail with distinct barring is typical of more strongly pigmented juveniles. Wings are regularly held in a harrier-like fashion when gliding. Israel, 19.4.2009 (DF)

627. Long-legged Buzzard *Buteo rufinus rufinus*, 2nd cy immature, summering. This light morph bird, harassed by a Black Kite, shows an almost unbarred tail and striking white primary patches. Note freshly moulted inner primaries and emerging new scapulars. Kazakhstan, 6.5.2010 (DF)

628. Long-legged Buzzard *Buteo rufinus rufinus*, darker morph juvenile, wintering. Apart from the often dusky underwing barring, the eagle-like head also differs from the otherwise similar dark morph Steppe Buzzard. India, 13.12.2007 (DF)

629. Long-legged Buzzard *Buteo rufinus rufinus*, darker morph juvenile, wintering (same as **628**). Compared with the previous image, this photo shows how the appearance of the bird depends on the light conditions. India, 13.12.2007 (DF)

630. Long-legged Buzzard *Buteo rufinus rufinus*, darker morph juvenile, migrant. In certain angles juveniles can look surprisingly slim, with narrow wings and long tail. When backlit, as here, dark brown birds appear almost black. Israel, 19.4.2009 (DF)

631. Long-legged Buzzard *Buteo rufinus rufinus*, darker morph juvenile, wintering (same as **628** and **629**). Dark juveniles appear rather uniform above, but the tail-barring varies individually. India, 13.12.2007 (DF)

632. Long-legged Buzzard *Buteo rufinus rufinus*, adult, on migration. This probable first- or second-adult plumage bird is rather similar to many Steppe Buzzards, but note typical light and unbarred tail, and broad pale upperwing primary flash. Israel, 19.4.2009 (DF)

633. Long-legged Buzzard *Buteo rufinus rufinus*, adult, on migration (same as **632**). Note diagnostic adult tail and bright rufous upperwing-coverts on this rather faded bird. Israel, 19.4.2009 (DF)

634. Long-legged Buzzard *Buteo rufinus rufinus*, adult, wintering. A typical light morph adult with pale head, neck and breast, contrasting with dark thighs. The uppertail is uniformly pale grey. India, 10.12.2007 (DF)

635. Long-legged Buzzard *Buteo rufinus rufinus*, adult, wintering. A most typical light morph adult showing contrasting dark thighs and a translucent tail. India, 15.12.2007 (DF)

636. Long-legged Buzzard *Buteo rufinus rufinus*, adult, wintering. Another typical light morph (but younger) adult. India, 15.12.2007 (DF)

637. Long-legged Buzzard *Buteo rufinus rufinus*, dark rufous morph adult, migrant. Birds of this colour morph often show fine, black barring on the tail, with the subterminal band only marginally broader than the others (cf. Steppe Buzzard). This individual is best told from similar plumage of Steppe Buzzard by dusky underwing flight feathers, with contrasting lighter outermost primaries. Israel, 24.3.2012 (DF)

638. Long-legged Buzzard *Buteo rufinus rufinus*, dark rufous morph (young) adult, migrant. Almost identical Steppe Buzzards occur and when diagnostic uppertail cannot be seen, best identified by size and proportions. Israel, 24.3.2010 (DF)

639. Long-legged Buzzard *Buteo rufinus rufinus*, dark rufous morph (young) adult, migrant (same as **638**). Differs from similar Steppe Buzzards by diagnostic plain uppertail. Note also strongly curved, long bill, as well as eagle-like head shape and vicious-looking gape flanges. Israel, 24.3.2010 (DF)

640. Long-legged Buzzard *Buteo rufinus rufinus*, black morph adult, migrant. Black morph adult differs from similarly black form of Steppe Buzzard by broader and duskier underwing barring, and they often show dusky-marbled outer primaries, as here. Note rather Steppe Buzzard-like tail, completely barred with a broad subterminal band. Israel, 18.4.2009 (DF)

ATLAS LONG-LEGGED BUZZARD
Buteo rufinus cirtensis

VARIATION Atlas Long-legged Buzzard is regarded as a subspecies of Long-legged Buzzard *Buteo rufinus*, but is here given a full treatment in order to encourage further studies and discussions on the identification and taxonomy of the *Buteo* complex.

DISTRIBUTION The Atlas Long-legged Buzzard *B. r. cirtensis* is the N African subspecies of Long-legged Buzzard. It is a resident of mountains and semi-deserts of northern Africa, with a rather patchy distribution north of the Sahara, from Mauritania and Morocco to Libya and western Egypt. The breeding birds of Arabia and S Israel require further studies in order to clarify their taxonomic affinities in relation to *cirtensis* and *rufinus*.

BEHAVIOUR Perches prominently on rocks, poles and pylons, but in more open country frequently also on the ground or on low bushes. Often hunts from a perch, but also from the air, and is capable of hovering when looking for prey, which mostly comprises various rodents and reptiles.

SPECIES IDENTIFICATION Superficially similar to Eastern Long-legged Buzzard, but appears smaller with slightly different proportions: the wings are shorter and more rounded and the head and legs are comparatively larger in comparison with nominate *rufinus*. Although typical individuals are easily recognised as Long-legged Buzzards on plumage, certain individuals are almost identical to Steppe Buzzards and are not possible to identify with certainty from the latter by plumage alone. This identification problem is further complicated in the area of the Strait of Gibraltar, where birds with mixed characters occur on both sides of the Strait. These buzzards may in fact comprise a hybrid population originating from hybrid offspring between resident Atlas Long-legged Buzzards and wintering Steppe Buzzards and/or vagrant Common Buzzards. The hybrid theory also explains their extreme individual plumage variation, ranging from typical, blond Long-legged Buzzards to birds practically identical to Steppe Buzzard.

The main differences compared to Eastern Long-legged Buzzard, apart from smaller size and more compact proportions, are the often clearly more marked and therefore darker brown head and neck,

the more patterned breast (at times showing a pale U as in Common Buzzard), a poorly defined dark belly-patch, often light-coloured trousers, often a poorly defined dark carpal patch on the underwing (sometimes just a dark comma), often clearly darkish or dark-streaked lesser underwing-coverts, often partially retained tail-barring even in adults, darker brown upperparts in juveniles, including the uppertail, and lack of pure grey on the upperwing secondaries in adults.

MOULT The complete moult takes place during the breeding cycle. Birds studied in N Morocco (pers. obs.) showed a similar moulting strategy to Steppe Buzzard, with multiple simultaneous moult fronts in the primaries in adults, showing flight feathers of different age and wear at any one time, and a relatively late moulting season, with birds still actively moulting in Sep.

PLUMAGES In terms of age-classes two main plumage-types can be recognised, juveniles and adults. Until all the juvenile remiges have been replaced a second plumage can be recognised by retained juvenile feathers in the wings and tail, at least until the 2nd cy autumn/3rd cy spring.

Adults and juveniles both occur in a light and an intermediate colour morph, with a continuous scale of intergrades, while a dark form, as known from Eastern Long-legged and Steppe Buzzard, is lacking.

Juvenile

Similar to juvenile Eastern Long-legged, but shows on average darker and browner upperparts, with narrower and less pronounced orange feather margins to the upperwing-coverts and scapulars, and a darker and browner, often distinctly barred uppertail. Uppertail-coverts are rusty-coloured and contrasting in many, but brown in others.

Head, neck and upper breast variably streaked, often creating a darker breast-shield similar to many juvenile Steppe Buzzards. In fresh juveniles the head often stands out as being clearly orange-brown, warmer in colour than, for example, the mantle. The belly-patch is usually broken in the middle, rarely as distinct and uniform as is typical in juveniles of the nominate form. The underwing can be very similar to the nominate, but many birds show a poorly defined, or strongly reduced dark carpal patch, while others show Steppe Buzzard-like darker lesser underwing-coverts contrasting with paler median coverts. The greater underwing-coverts usually stand out as a darker band.

The iris colour is noticeably pale in juveniles, thus differing from most (but not all) juvenile Steppe Buzzards, being similar to Eastern Long-legged Buzzard.

Second plumage

Birds of this age-class may appear adult-like at first, but they can be identified until the completion of the first moult by their retained juvenile remiges and tail feathers, which stand out as being extremely worn and faded brown in colour and also shorter, narrower and more pointed compared to the replaced ones. Birds of this age still have a noticeably pale iris.

Adult

The light and the rufous morphs are rather similar to the corresponding plumages of the nominate form, but the tail can be more extensively barred, often with very fine bars and a slightly broader, yet often incomplete subterminal band (still narrower than in most adult Steppe Buzzards). Other adults are more like Steppe Buzzards in showing more streaked underparts and heavily marked underwing-coverts. In all forms the tail can be of the typical Long-legged type, with orange outer tail and whitish inner tail, but some birds (hybrids?) show greyer or browner tails with variably distinct and broad barring, thus more similar to Steppe Buzzard.

In difficult cases structural differences may offer some help when separating birds from Steppe Buzzard: Atlas Long-legged Buzzards appear comparatively big-headed for their size, the bill is deep and heavy, but not necessarily long, and the feet are large with powerful toes and a long tarsus; as in Eastern Long-legged Buzzard the wings are held in a marked dihedral when soaring.

SEXING Usually not possible under normal field circumstances.

CONFUSION RISKS Easy to confuse with Eastern Long-legged and Steppe Buzzards, some even with certain light-coloured plumages of Common Buzzard. The yellowish and rufous tones dominating in the plumage should be enough to exclude any Common Buzzard, but certain Atlas Long-leggeds can be

practically identical to the former two in terms of plumage characters, and out-of-range vagrants in particular need to be scrutinised with extreme care, and may still not always be possible to identify with certainty.

NOTE Some of the buzzards breeding in the area around the Strait of Gibraltar can be very difficult to identify, as many of them feature intermediate or mixed characters between *B. r. cirtensis* and Common/Steppe Buzzard *B. buteo buteo/vulpinus*, with considerable individual plumage variation (Javier Elorriaga *in litt.* and Rodriguez *et al.* 2013). These birds are thought to represent a hybrid swarm originating from past hybridisation between Atlas Long-legged Buzzards and Common/Steppe Buzzards. In the past southern Spain was also an important migration bottleneck for Common Buzzards (*sensu lato*) (Bernis 1975), presumably involving Scandinavian (mostly N Swedish) *vulpinus*-type long-distance migrants. Some Swedish ringing records indicate that W Africa may have been an important wintering area for these birds in the past (Fransson & Pettersson 2001) and possibly the origin of hybridisation may date back to that time. Nowadays, when the range of *vulpinus* is receding and the numbers are declining in N Europe, further introgression of *vulpinus* genes into the Atlas Long-legged Buzzard populations seems less likely. Hybridisation between Common Buzzard *B. buteo buteo* and Atlas Long-legged Buzzard *Buteo rufinus cirtensis* has also been reported from the island of Pantelleria between mainland Italy and Tunisia (Corso 2009).

References

Bernis, F. 1975. Migración de falconiformes y *Ciconia* spp. por Gibraltar II. Análisis descriptive del verano-otoño 1972. *Ardeola* 21: 489–580.

Corso, A. 2009. Successful breeding of Atlas Long-legged Buzzard and Common Buzzard on Pantelleria, Italy, in 2008. *Dutch Birding* 31: 226–228.

Fransson, T. & Pettersson, J. 2001. *Swedish Bird Ringing Atlas. Vol 1.* Stockholm.

Rodriguez, G., Elorriaga, J. & Ramirez, J. 2013. Identification of Atlas Long-legged Buzzard and its status in Europe. *Birding World* 26: 147–173.

641. Atlas Long-legged Buzzard *Buteo rufinus cirtensis*, juvenile. A rather typical Moroccan juvenile. Note rather distinct pale breast-band and darker lesser underwing-coverts, both features recalling juvenile Steppe Buzzard, but note very pale iris, faintly barred remiges and solid brown flanks. Morocco, 8.1.2015 (DF)

642. Atlas Long-legged Buzzard *Buteo rufinus cirtensis*, juvenile (same as **641**). Upperparts already rather worn by Jan. Note pale iris and faintly barred tail with lighter inner-tail, but generally very similar to many juvenile Steppe Buzzards. Morocco, 8.1.2015 (DF)

643. Atlas Long-legged Buzzard *Buteo rufinus cirtensis*, juvenile. Another, somewhat darker and even more Steppe Buzzard-like individual, but note rather diffuse wing-barring and solid, dark flanks. Morocco, 6.1.2015 (DF)

644. Atlas Long-legged Buzzard *Buteo rufinus cirtensis*, juvenile (same as **643**). Very similar to many juvenile Steppe Buzzards and probably not possible to identify from this image. Morocco, 6.1.2015 (DF)

645. Atlas Long-legged Buzzard *Buteo rufinus cirtensis*, juvenile. This bird is very similar to some light Steppe Buzzards and would be difficult to identify away from its normal range. An intergrade *B. r. cirtensis x B. buteo* 'Gibraltar Buzzard' cannot be excluded. Morocco, near Tangier, 14.9.2012 (DF)

646. Atlas Long-legged Buzzard *Buteo rufinus cirtensis*, juvenile. Note distinct difference between light upperwing-coverts and darker flight feathers. An intergrade *B. r. cirtensis x B. buteo* 'Gibraltar Buzzard' cannot be excluded. Morocco, near Tangier, 14.9.2012 (DF)

647. Atlas Long-legged Buzzard *Buteo rufinus cirtensis*, second plumage. A rather typical adult bird of the light form, but can be aged by retained juvenile outer two primaries. Note less contrasting carpal patch and more chubby proportions compared to nominate. Morocco, near Agadir, 21.1.2014 (DF)

648. Atlas Long-legged Buzzard *Buteo rufinus cirtensis*, adult. A typical light adult, very similar in plumage to nominate adults of the same colour type, but note more compact body and rounded head, and faint subterminal tail-band. Morocco, High Atlas, 26.3.2009 (DF)

649. Atlas Long-legged Buzzard *Buteo rufinus cirtensis*, adult. A typical light adult from the deserts of SW Morocco. Western Sahara, Morocco, 18.1.2014 (DF)

339

650. Atlas Long-legged Buzzard *Buteo rufinus cirtensis*, adult. A typical light adult from the coast near Agadir. Morocco, 11.1.2015 (DF)

651. Atlas Long-legged Buzzard *Buteo rufinus cirtensis*, adult (same as **649**). Note typical upperwing-coverts, light-coloured head, plain tail and barred remiges above, all rather similar to nominate *rufinus*. Western Sahara, Morocco, 18.1.2014 (DF)

652. Atlas Long-legged Buzzard *Buteo rufinus cirtensis*, juvenile. These two birds from N Morocco, near Tangier, differ from the average juvenile *cirtensis*. Note especially the dark head and neck of the right-hand bird, reminiscent of *B. b. vulpinus*, while the left bird may recall a juvenile Bonelli's Eagle in its uniform light plumage. In neither case can an intergrade *B. r. cirtensis x B. buteo* 'Gibraltar Buzzard' be excluded. Morocco, near Tangier, 14.9.2012 (DF)

340

653. Atlas Long-legged Buzzard *Buteo rufinus cirtensis*, juvenile. The dark head and neck make this bird even more Steppe Buzzard-like. Note also light thighs (= rear flanks), which is unusual in the nominate race. An intergrade *B. r. cirtensis x B. buteo* 'Gibraltar Buzzard' cannot be excluded. Morocco, near Tangier, 14.9.2012 (DF)

654. Presumed 'Gibraltar Buzzard' *B. r. cirtensis x B. buteo*, juvenile. Another Steppe Buzzard-like bird from the same area in northern Morocco. Morocco, near Tangier, 14.9.2012 (DF)

655. 'Gibraltar Buzzard' *B. r. cirtensis x B. buteo*, juvenile. Similar to Common Buzzard, but note distinct rufous fringes above, rusty uppertail-coverts, diffuse tail-barring and light-coloured iris. Spain, near Tarifa, 11.09.2012 (DF)

656. Atlas Long-legged Buzzard *Buteo rufinus cirtensis*, or perhaps more probably a *B. r. cirtensis x B. buteo* intergrade 'Gibraltar Buzzard', adult. This light adult shows a typical Long-legged Buzzard tail, while the underwing and the body markings suggest genetic introgression from Common Buzzard. Ceuta (Spain), NW Africa, 12.9.2012 (DF)

657. 'Gibraltar Buzzard' *B. r. cirtensis x B. buteo*, adult female, breeding. Rather Common Buzzard-like plumage, but note light crop and upper breast. Ceuta (Spain), NW Africa, 12.9.2012 (DF)

658. 'Gibraltar Buzzard' *B. r. cirtensis x B. buteo*, adult female, breeding (same as 657). Note Long-legged-type colour and markings of uppertail. The upperparts are darker than in typical adult Atlas Long-legged Buzzards. Ceuta (Spain), NW Africa, 12.9.2012 (DF)

659. 'Gibraltar Buzzard' *B. r. cirtensis x B. buteo*, adult male, breeding, paired to 657/658. Similar to a darker rufous Atlas Long-legged Buzzard from below. Ceuta (Spain), NW Africa, 12.9.2012 (DF)

660. 'Gibraltar Buzzard' *B. r. cirtensis x B. buteo*, adult male, breeding (same as 659). Note dull greyish-brown Common Buzzard-like upperparts, including tail-colour and tail-barring. Ceuta (Spain), NW Africa, 12.9.2012 (DF) .

ROUGH-LEGGED BUZZARD
Buteo lagopus

Other names: Rough-legged Hawk (North America)

VARIATION Three subspecies recognised, with nominate *lagopus* breeding in the Western Palearctic. North American ssp. *sanctijohannis* is a rare vagrant to the Atlantic islands and the western seaboard of Europe, yet hardly identifiable, and ssp. *menzbieri* of N Siberia is a potential winter visitor to C Asia. All three subspecies are rather similar, but *sanctijohannis* and the overall lighter-coloured *menzbieri* also have a dark morph in all age-classes.

DISTRIBUTION Breeds on the tundra and boreal parts of N Europe, wintering in C Europe from the British Isles to the Black Sea, rarely reaching as far south as the Mediterranean.

BEHAVIOUR Most often seen in flight above open areas, stopping every now and then to hover when hunting. Perches openly on rocks or treetops, but also uses open, flat ground, more often than, for example, Common Buzzard. Prefers to perch on the top of a tree, while Common Buzzards often perch lower down.

SPECIES IDENTIFICATION A typical buzzard, but appears longer-winged and longer-tailed compared to most populations of Common Buzzard, approaching nominate Long-legged Buzzard in proportions. Also, flight much more elegant, with slower and more buoyant wing-beats, compared to the flappier flight of the shorter-winged and more compact Common Buzzard. Always identified by bicoloured tail, with white base and broad dark subterminal band, but details of tail-pattern vary greatly according to age and sex, and considerable individual variation. Underbody and underwings highly variable depending on age and sex, but dark carpal patch mostly clear and complete, and rounded. Feet noticeably small, in flight only reaching to the base of the tail; bill also comparatively small and fine (cf. Common and Long-legged Buzzards).

From a distance paler and greyer above than Common Buzzard, with clearly white inner tail and sharply defined dark tip. Head-on view of approaching bird reveals the diagnostic white leading edge of the wing, in juveniles most pronounced around the carpal area. Owing to the comparatively longer wings,

flies with slower and more relaxed wing-beats than Common Buzzard. Soars on raised wings, like Long-legged Buzzard or Golden Eagle, with a notable kink between lifted arm and more level hand.

MOULT Annual moult occurs during the breeding season, but some flight feathers are often left unmoulted, possibly in response to the food situation, as the number of unmoulted feathers seems to vary from year to year. Old adults show multiple moult fronts in the primaries, enabling them to replace the entire set of primaries in one favourable season. The first moult is always suspended before all primaries have been moulted, retaining a few (1–5) outermost juvenile primaries and often also some median secondaries. These retained juvenile feathers are the most important characters for ageing birds in the second plumage.

PLUMAGES Normally three age-classes can be identified: juvenile, first-adult (second plumage) and adult. Most adults can also be sexed by plumage. Great care is needed when sexing second-plumage birds (first-adults), as males of this age-class have not yet acquired all the features typical of older adult males, and many resemble adult females. Conversely, some adult females (probably old birds) may show a very adult male-like plumage, with multiple tail-bands and heavily spotted underwing-coverts.

Juvenile

Easily aged by very pale underwings, with a dusky, narrow and mostly diffuse dark trailing edge, almost unbarred remiges (cf. Long-legged Buzzard), plain underwing-coverts and a dark, round carpal patch. They also differ from older birds by their slimmer look, explained by narrower wings and longer tail. The barring of the secondaries is rather faint and mostly confined to the distal part of the feathers, leaving the inner parts plain white. From below the tail shows just a diffuse subterminal band, while from above the distal part of the uppertail can be extensively brown, but never black as in adults, at times with some orange admixed. The inner uppertail is always white, with white uppertail-coverts showing brown drop-shaped shaft-streaks. The underbody is typically pale, with a solid dark brown belly-patch, while the head and breast are thinly streaked (more heavily marked in males?). Mantle and upperwings are brown, often with distinctly paler orange feather margins to the median coverts, while the secondaries are uniform and appear noticeably dark. The primaries show dark fingers but the bases of the upperwing primaries are variably white, creating a wide pale or white patch on the upper hand in some, not dissimilar to some juvenile Long-legged Buzzards. The iris is rather pale, greyish or greyish-brown and the black pupil is mostly clearly visible even in field situations. In direct head-on views the leading edge of the wing is strikingly pale in juveniles.

Sexes are not known to differ in juveniles, but presumed males have shown more heavily streaked breasts and multiple tail-bands above, while presumed females are overall lighter and less strongly marked and the tail shows just one broad subterminal band from above. Further studies are needed.

By late spring (May–Jun) the upperparts can be extremely faded due to wear, and distant birds could easily be mistaken for Long-legged Buzzards, but note differences in bill and leg structure, and barring of the underwings.

First-adult

Always reliably aged by retained outer 2–4 juvenile primaries, which stand out as comparatively shorter, browner and more sharply pointed than their moulted neighbours. Some juvenile median secondaries are also often retained, being clearly shorter and more faded brown and noticeably different from the new ones with a prominent black subterminal band. Iris is somewhat darker, medium greyish-brown, but the black pupil is still obvious in close views. Sexing birds of this age-class can be difficult and is not always possible.

Males of this age usually have a heavily streaked breast and a rather solid black belly-patch, hence resembling the adult female. Also, the uppertail usually shows only 1–2 additional black bands, again not dissimilar to many females. However, the underwings are already rather heavily marked, coverts as well as the remiges, and the mantle and scapulars are more pied, as in older males, but many are difficult or impossible to separate from certain adult females.

Females are similar to older females, except for the retained juvenile outer primaries, and possibly also some median secondaries.

Definitive adult

Broad black tips to all secondaries, and all fingered primaries are comparable in length and colour. The black subterminal tail-band is distinct and coloured through the feathers, being clearly visible also from below.

Males appear more variegated below than adult females, often with a dark face, throat and breast contrasting with a paler and more loosely barred belly area, from a distance looking light-bodied with a dark hood. The underwings are strongly marked, with heavily spotted coverts and strongly barred remiges, while the carpal patch is broken and thus less uniform compared to females. The uppertail shows multiple sparse black tail-bands, sometimes over the entire length of the tail, with a broad black subterminal band near the tip, but seen from below only the subterminal band is usually coloured through. The mantle appears distinctly pied, in black and grey, and the upperwing secondaries and primaries are clearly grey, with prominent dark barring. The mantle and scapulars are more clearly patterned and the upperwings are greyer compared to the browner and more uniformly coloured females.

Females are on average lighter below than adult males, more similar to the diagnostic juveniles, with a light but dark-streaked breast-shield contrasting with a solid, blackish belly-patch. If the dark belly-patch is not complete, it tends to be spotted in females, rather than barred as in adult males. The underwing-coverts are rather sparsely and finely streaked making the dark carpal patch stand out more clearly. The remiges are also less strongly barred compared to males, with fewer and finer bands. The tail-banding, which is best judged from above, shows one broad black subterminal band (sometimes with an additional 1–2 adjacent narrow bands) often set in a wider distal area of brown and orange, the latter area always sharply defined from the pure white inner parts of the tail. From below the tail looks white with just the black subterminal band showing through, while the brown outer parts and possible additional tail-bands are visible only when backlit. The upperparts of adult females are browner and less distinctly marked compared to adult males and, for instance, the upperwings look rather uniform and brown in females compared to the clearly greyer and barred remiges of males. *Some, probably old females, may show an entirely barred uppertail, as in typical adult males. These birds are also often darker and more heavily marked below than average females and are thus not possible to tell from adult males on plumage in the field.*

SEXING Sexing birds in first-adult plumage always requires great care, while many adult birds in final adult plumage are fairly straightforward to sex by plumage. However, as there seems to be quite an overlap in characters between adult males and females, only the most typical plumages are safe to sex: birds with a pale head and breast contrasting with a solid dark belly, light and poorly marked underwing-coverts, a solid round carpal-patch and just one single dark subterminal band to the tail should be safe to sex as females. Birds with a dark hood contrasting against a lighter, barred breast and belly, heavily marked underwings with a broken carpal patch and with multiple tail-bands across the entire uppertail, should mostly be males, but see above under Plumages.

CONFUSION RISKS Often easy to identify, but certain heavily marked adult males may recall Common Buzzard from below, but the tail is always diagnostic, and the entire bird appears black-and-white, lacking the pure brown colours of Common Buzzard. Juveniles in the late 2nd cy spring may be heavily worn and bleached and may therefore approach the look of a Long-legged Buzzard, but details of tail and underwing barring, as well as the fine bill and short legs differ.

NOTE The iris colour also changes with age in Rough-legged Buzzard. The iris is pale grey to pale greyish-brown in juveniles, with the black pupil showing clearly. In old adults the iris is very dark brown and the pupil cannot be discerned, even at close range. In first-adults the iris is still rather pale, almost as pale as in juveniles, and the pupil can be seen even in field situations. One year later many birds, now in their third year of life, still show a paler brown iris compared to older adults. Note, however, that individual variation in eye colour development may obscure this general pattern.

Rough-legged Buzzard is known to have hybridised successfully with Common Buzzard in Norway in 2005 and in Finland in 2006. The Finnish hybrid young recalled pale juvenile Common Buzzards, but their legs were feathered to more than halfway down the tarsus and the remiges were incompletely barred below, leaving the basal parts of the secondaries largely unbarred, mirroring the situation in Rough-legged Buzzard. Further, the inner uppertail was extensively white, while the brown distal part showed dense and fine barring similar to Common Buzzard.

References
Forsman, D. & Lämsä, E. 2007. Successful interbreeding between Common Buzzard and Rough-legged Buzzard in Finland. *Linnut* 42: 36–37 (in Finnish).
Gjershaug, J. O., Forset, O. A., Woldvik, K. & Espmark, Y. 2006. Hybridisation between Common Buzzard *Buteo buteo* and Rough-legged Buzzard *B. lagopus* in Norway. *Bull. Brit. Orn. Club* 126: 73–80.

661. Rough-legged Buzzard *Buteo lagopus*, juvenile, migrant. This is a most typical juvenile, with plain underwing-coverts, solid dark carpal patch and belly, and light shining through the white inner tail in this backlit bird. Finland, 14.10.2014 (DF)

662. Rough-legged Buzzard *Buteo lagopus*, juvenile, recently fledged. Another typical juvenile. Note how the barring of the secondaries is confined to the tips and compare with more completely barred Long-legged Buzzard, and how the diagnostic tail-pattern is difficult to make out on the underside. Norway, 20.8.2011 (DF)

663. Rough-legged Buzzard *Buteo lagopus*, juvenile, migrant. Some juveniles may have a rather dark trailing edge and subterminal tail-band, but note light iris and poorly marked underwing-coverts and remiges. Sweden, 10.10.2014 (DF)

664. Rough-legged Buzzard *Buteo lagopus*, juvenile, recently fledged (same as **662**). Note deep yellowish colour of fresh juveniles. Juveniles are clearly slimmer and longer-winged than, for example, Common Buzzards. Norway, 20.8.2011 (DF)

665. Rough-legged Buzzard *Buteo lagopus*, juvenile, migrant. Finland 15.1.2012 (DF)

666. Rough-legged Buzzard *Buteo lagopus*, juvenile, wintering. Shows typical juvenile *Buteo* contrast to upperwing between lighter forewing and darker secondaries and greater coverts. The width of the dark tail-band varies indivually and covers nearly half of the tail here. In this bird the primary panel is not very well defined, but can be largely white in others. Finland 31.1.2005 (DF)

667. Rough-legged Buzzard *Buteo lagopus*, 2nd cy female (left) with fresh juvenile, on the breeding grounds. Note above all the slimmer structure of the juvenile and the distinct black tail-tip and trailing edge of the wing of the adult. The female is best aged by its retained brown juvenile secondaries. Norway, 20.8.2011 (DF)

668. Rough-legged Buzzard *Buteo lagopus*, 2ⁿᵈ cy (female?) on breeding grounds. Aged by retained juvenile secondaries (s4 and c. ss8–9) and two outermost primaries. Norway, 20.8.2011 (DF)

669. Rough-legged Buzzard *Buteo lagopus*, 2ⁿᵈ cy migrant. After suspended moult still identified by retained juvenile outer primaries (pp8–10) and secondaries (s4 and ss7–9). Finland. 12.10.2005 (DF)

670. Rough-legged Buzzard *Buteo lagopus*, adult male, breeding. Only adult males can look black-and-white. Note also dark hood contrasting against a lighter belly, the multiple-barred tail and the heavily marked underwing-coverts. Norway, 5.7.2006 (DF)

671. Rough-legged Buzzard *Buteo lagopus*, (old?) adult male, breeding. Typical male plumage with dark hood but sparsely marked belly and heavily marked tail, remiges and underwing-coverts. Finland, 15.5.2014 (DF)

672. Rough-legged Buzzard *Buteo lagopus*, adult male, breeding. Compare with **671**. Norway, 28.5.2014 (DF)

673. Rough-legged Buzzard *Buteo lagopus*, adult male, breeding. Note strongly mottled appearance, barred thighs and multiple tail-bands. Finland, 6.6.2011 (DF)

674. Rough-legged Buzzard *Buteo lagopus*, adult male, breeding (same as **673**). Note pied mantle and scapulars, clearly barred remiges and multiple tail-bands. The upperwing looks fairly uniform from a distance, with a darker trailing edge. Finland, 6.6.2011 (DF)

675. Rough-legged Buzzard *Buteo lagopus*, adult male, breeding. Note heavily marked underwing-coverts, a typical adult male feature. Finland 14.5.2014 (DF)

349

676. Rough-legged Buzzard *Buteo lagopus*, adult male, breeding (same as 675). Note grey and clearly barred flight feathers and pied scapulars, both typical of adult male. Finland 14.5.2014 (DF)

677. Rough-legged Buzzard *Buteo lagopus*, adult female, migrant. Adult female resembles juvenile in having rather light underwing-coverts, a solid belly-patch and a single tail-band, but note the distinct black trailing edge to the wings and tail. Finland, 15.4.2011 (DF)

678. Rough-legged Buzzard *Buteo lagopus*, adult female, breeding. Note the lighter underwing compared to adult male. Finland, 14.5.2014 (DF)

679. Rough-legged Buzzard *Buteo lagopus*, adult female, breeding. Single tail-band, streaked throat/upper breast but solid flank patch and light underwings with solid dark carpal patch are typical traits of adult female. Finland, 15.5.2014 (DF)

680. Rough-legged Buzzard *Buteo lagopus*, adult female, breeding. This female has a rather male-like plumage and proves that sexing adults is not always straightforward (sexed by behaviour at the nest and through comparison with mate). Norway, 2.6.2014 (DF)

681. Rough-legged Buzzard *Buteo lagopus*, adult female, breeding (paired to **673/674**). This is a massive female, and would probably be misidentified as a male by plumage. The rather dark, mottled and pied appearance with many tail-bands suggest a male, but note that thighs are not barred. Females like this occur from time to time and are probably older birds. Finland, 6.6.2011 (DF)

682. Rough-legged Buzzard *Buteo lagopus*, adult female, breeding (paired to **670**). Compared to adult males, adult females are browner and more uniform above, lacking the adult male's pied mantle and often greyish upperwings. Note the tail-barring, which is rather typical of an adult female. Norway, 5.7.2006 (DF)

351

GOLDEN EAGLE
Aquila chrysaetos

VARIATION Two subspecies in the region: nominate *chrysaetos* over most of the range, with smaller and darker *homeyeri* in most of the Iberian Peninsula, N Africa, the Middle East and Arabia. Within the nominate subspecies the birds of N Europe are bigger and moult more slowly, taking longer to achieve adult plumage, the immatures are migratory and the adults often show rich orange markings on the breast and underwing-coverts, while C European birds are overall darker, show less orange-brown colour below and, being resident, also moult faster. The southern *homeyeri* is smaller and darker with sleek proportions, narrower wings and tail, and a slim body compared to northern birds.

DISTRIBUTION Widely distributed, mostly keeping to remote mountains, but in the taiga zone of the north also in lowland areas. Patchy distribution covers most of Europe, with major gaps in lowlands and heavily populated areas. In N Africa and the Middle East typically inhabits deserts.

BEHAVIOUR Mostly seen on the wing, either riding on updrafts along mountain ridges, or soaring high in the sky. When hunting may descend from mountains to nearby lowlands and steppes, where it sometimes quarters the ground like a huge harrier. Perches prominently on a rock, or in the top of a dominant tree, often for a considerable length of time.

SPECIES IDENTIFICATION Usually not difficult to identify thanks to its unique silhouette, with wings being broadest at the carpal with narrower base and hand, and a rather long, full tail. The diagnostic white wing-patches of immatures may occasionally be lacking in southern populations, but the inner tail is always white, with a variably broad black terminal band. Adults appear dark from a distance, but the upperparts show grey inner tail and remiges with broad dark tips, while the upperwing-coverts are partly lighter, forming a pale diagonal patch across the arm. Active flight is majestic, with slow and graceful wing-beats interspersed by glides. Soars on wings lifted in a marked dihedral, which is a diagnostic feature among our eagles (save for Verreaux's Eagle). Glides with arms lifted, while outer wing remains more level. Wings can be strongly arched in fast descending glides, but

regardless of wing posture the fingered primaries are always well splayed and bent strongly upwards.

MOULT Complete moult follows the breeding cycle, spanning from Mar–Apr to Sep–Oct, commencing earlier in southern populations. The first moult starts later, as late as Jun in Scandinavian birds. Not capable of replacing all flight feathers in one season. In adults at least two successive moults are needed to replace the entire plumage, while younger birds of the Nordic populations may take as many as four moults before the last juvenile primary is replaced in the 5th cy summer. The resident southerly populations moult more rapidly, replacing more feathers per moulting season compared to the northern populations, but further studies are needed.

PLUMAGES Acquisition of adult plumage takes about 6–7 years, but some Nordic birds of known age have still shown extensive white in their wings and tail at an age of nine years. In normal field conditions usually three main plumage types can be recognised: juvenile, immature (second to at least fourth/fifth plumage) and adult. With the help of detailed photographs the second, third and sometimes even the fourth plumage can be recognised with certainty based on moult pattern, at least in northern birds, while the ageing of subsequent immature and subadult stages is made impossible by considerable individual variation in plumage and moulting progress.

Juvenile

Best aged by uniformly dark brown upperwing-coverts, fading to a somewhat lighter brown by the 2nd cy spring, but lacking the mottled appearance of later plumages. Also the evenly indented trailing edge of the wing is a feature of juveniles. Individually variable white patches to the inner primaries, visible from both below and above (always smaller above), with white extending well onto the bases of the secondaries in some. Remiges never distinctly barred, but some individuals may show a tendency towards diffuse banding. Tail feathers white with individually variable black subterminal band and white feather-tips. Some southern juveniles may lack the white wing-patches completely, but the white inner tail remains diagnostic. Body plumage uniformly dark brown, save for tawny undertail-coverts and white bases to the uppertail-coverts. Crown and nape tawny-brown when fresh, fading to more golden when worn. Iris is dark brown.

Second plumage

Similar to juvenile, but told by generally worn and faded upperwing-coverts with a few fresh, dark coverts admixed. Always reliably aged by fresh inner primaries which, however, can be extremely difficult to notice, as these may be almost identical in colour and pattern to their predecessors. The central tail feathers are moulted and fresh in most birds, often with a greyish subterminal band, which contrasts with the remaining juvenile feathers with black tips. Mantle mostly moulted, creating a darker saddle, contrasting against paler upperwing-coverts. Most of the underbody and underwing-coverts are still retained juvenile feathers, uniformly dark brown. Iris dark brown.

Third plumage

After the second plumage immatures are very difficult to age as the plumages of later age-classes resemble one another greatly, and photographs or extremely favourable conditions are required. Upperwing-coverts are largely fresh and therefore darker compared to the previous plumage, and the white areas in the wings and tail usually start to show some dark mottling, often forming irregular, broad bands. Nordic birds can be safely aged by the number of retained juvenile remiges, with the outer 4–5 primaries and a group of some 4–5 median secondaries being still juvenile feathers and standing out as faded brown in colour. The new secondaries are broad with blunt tips compared to the narrow and more pointed juvenile secondaries. In Nordic birds this is often the plumage with the most extensive white underwing patches. Breast and underwing-coverts have been largely replaced and start to show the rusty-orange colouring connected with older birds, but the orange colour is often missing in more southerly populations. As a rule, at this stage birds of southern populations show fewer or no retained juvenile feathers in the wings, but details require further research.

Fourth plumage

Some birds of the Nordic population can still be aged by their last retained juvenile remiges, most often p10 and s9, while other individuals have already replaced them, making an exact ageing impossible. Southern populations are likely to have lost all of their juvenile feathers by now and are therefore not possible to age. Birds from northern populations still appear surprisingly immature in plumage, with a largely white base to the tail and broad white underwing patches, but the primary patch is often divided in two by darker adult-type feathers at approximately pp3–5.

Subadult

In Nordic birds the consecutive plumages still show broad white areas in the underwings and tail, while more southerly populations are darker, turning more adult-like, but because of individual differences ageing is no longer reliable. Differs from adult by retained irregular white markings on primaries, secondaries and tail. Birds in this plumage can be anything from 5–10 years old, but it appears that some Nordic adults retain white markings for life.

Adult

Appears largely dark from a distance, with paler diagonal upperwing-coverts patch, golden nape and neck contrasting with darker face, greyish inner tail with variable dark banding and with a broad black subterminal band. The primaries and secondaries are grey, both above and below, with irregular darker banding and broadly dark tips. Northern birds are noticeably pale and faded above and many show bright orange-rufous streaking on the breast and lesser underwing-coverts. Birds from the mountains of C and S Europe are darker, being more uniform above and lacking the extensive rufous coloration below. Told from similar-looking adult Imperial Eagle by different silhouette, greyish cast to the remiges above and by brown rather than blackish body plumage.

SEXING Males are smaller than females, but the difference is rarely of use in the field. Sometimes sexing can be quite challenging, even when the birds of a pair are seen together.

CONFUSION RISKS The Golden Eagle has a unique silhouette and a habit of holding its wings lifted in a shallow V when soaring, and once familiar with these, other species can normally be ruled out by these characters alone. In brief views it can be mistaken for any other large, dark eagle, most notably the Imperial Eagle, with which it shares many plumage characters. In particular the dark subadult plumage of Imperial can be quite similar to an older Golden Eagle, but the silhouette of Imperial shows more rectangular and parallel-edged wings, and a proportionately shorter tail and a longer neck.

683. Golden Eagle *Aquila chrysaetos*, immature flanked by two juvenile Bearded Vultures. Spain, 23.11.2010 (DF)

684. Golden Eagle *Aquila chrysaetos*, juvenile, autumn migrant. Juveniles are dark below save for white areas of the underwings and tail, and partly chestnut undertail-coverts. Finland, 13.10.2011 (DF)

685. Golden Eagle *Aquila chrysaetos*, juvenile, autumn migrant. The extent of white varies between birds, as does the width of the tail-band, but neither has any correlation to age. Finland, 13.10.2011 (DF)

686. Golden Eagle *Aquila chrysaetos*, juvenile, autumn migrant. Juveniles have uniformly brown upperwing-coverts, unlike any subsequent plumage. Finland, 13.10.2011 (DF)

687. Golden Eagle *Aquila chrysaetos*, second plumage. The second plumage still consists of mostly juvenile feathers, albeit slightly browner and faded compared to a fresh juvenile. The best ageing character, the new inner primaries (four in this bird), are often very difficult to make out. Finland, 3.1.2011 (DF)

688. Golden Eagle *Aquila chrysaetos*, second plumage, autumn migrant. Identical to juvenile but plumage faded and less uniform. Birds from more southerly populations can be more advanced than this northern bird, with only two new inner primaries in each wing. Finland, 3.10.2012 (DF)

689. Golden Eagle *Aquila chrysaetos*, second plumage, ssp. *homeyeri*. By early autumn, southern *homeyeri* have moulted much more extensively and replaced the inner five primaries and several secondaries. They also show a tendency for more clearly barred fresh inner primaries compared to northern birds. Spain, 9.9.2012 (DF)

690. Golden Eagle *Aquila chrysaetos*, second plumage, ssp. *homeyeri* (same as **689**). Note sharp contrast between faded median and dark greater coverts, typical of a worn juvenile upperwing. Spain, 9.9.2012 (DF)

691. Golden Eagle *Aquila chrysaetos*, second plumage (same as **687**). In northern birds the upperwing-coverts are still mostly juvenile and very faded, while southern birds have moulted a fair proportion and may look darker. Finland, 3.1.2011 (DF)

692. Golden Eagle *Aquila chrysaetos*, adult male, breeding. Underwings are softly banded with a broad dark trailing edge, while the underwing-coverts and breast may be extensively orange-brown, particularly in northern birds. Note the full crop. Finland, 15.5.2014 (DF)

693. Golden Eagle *Aquila chrysaetos*, adult, presumed male. This resident bird from the Pyrenees of N Spain shows the typical silhouette, with broad wings tapering towards the body and a longish tail, while the underwings are marked with broken, irregular and muted barring. Spain, 23.11.2010 (DF)

694. Golden Eagle *Aquila chrysaetos*, adult female, breeding. Adults from C Europe are already darker than birds from the north. Slovakia, 20.7.2011 (DF)

695. Golden Eagle *Aquila chrysaetos*, adult, (same as **693**). Adults are variably worn above, lighter or darker greyish-brown, but usually retain a lighter area across the median upperwing-coverts and a darker trailing edge. Spain, 23.11.2010 (DF)

696. Golden Eagle *Aquila chrysaetos*, adult male, breeding. The flight feathers are grey above with broad blackish tips, and the same pattern is found in the tail. Slovakia, 20.7.2011 (DF)

EASTERN IMPERIAL EAGLE
Aquila heliaca

VARIATION Monotypic. Previously considered conspecific with Spanish Imperial Eagle, which is now widely accepted as a separate species.

DISTRIBUTION The vast range extends from eastern C Europe and W Russia to C Asia, Siberia, Mongolia and China. Northern populations are partly migratory, wintering in S Asia, the Indian subcontinent, Arabia and E Africa, while others, particularly adults, winter in their respective breeding areas, conditions permitting.

BEHAVIOUR A bird of open spaces, mostly seen on the wing or perched on a prominent perch, a treetop or a pylon. Frequently also perches on the ground. When hunting, glides slowly against the wind scanning the ground below, stopping completely every now and then, before taking the final stoop, mostly aimed at medium-sized rodents or birds.

SPECIES IDENTIFICATION A long- and rather narrow-winged large eagle, with a well-protruding neck and tail. Because of a great variety of plumages Eastern Imperial Eagles can be mistaken for a number of species. Juveniles and immatures are most likely to be confused with younger Steppe Eagles, some of which may appear quite similar, especially immatures lacking the diagnostic white underwing band. From a distance juveniles could also be be mistaken for a light morph Tawny Eagle or the pale *fulvescens* form of Greater Spotted Eagle. Younger Eastern Imperial Eagles are told from all of these by their distinctly streaked breast and underwing-coverts, which is a diagnostic feature. The pied plumage of older immatures is usually straightforward to identify, as no other eagle shows this kind of boldly mixed plumage (except for Spanish Imperial Eagle). Subadults can be tricky, as the plumage is often quite similar to an adult Golden Eagle, but the silhouette is different, with rectangular and parallel-edged wings and a shorter tail, and the wings are held more flat when gliding. Older immature and adult Imperial Eagles (but not juveniles) tend to keep their tails folded when soaring, while Golden Eagles regularly fan them.

By far most difficult to tell from Spanish Imperial Eagle, and certain non-juvenile birds may appear practically identical. Immatures can be very similar, particularly birds in their third plumage, with some Easterns having already lost their streaking, but also the following pied plumages can appear almost identical. Even certain adults, e.g. Spanish birds lacking the diagnostic white leading edge to the wing, could cause serious identification problems.

MOULT Similar to other large eagles, with moult taking place mostly during breeding, from Mar/Apr–Oct/Nov, with first-timers starting somewhat later (May). Occasionally some remiges are also moulted during winter. The first three, rather juvenile-like plumages can be aged by details of moult, i.e. the number and position of retained juvenile flight feathers.

PLUMAGES Three main plumage types are easily recognised: the pale immature plumages (first to third plumages), the pied plumages (fourth and fifth/sixth plumages) and the dark adult plumage. More exact ageing requires good views, but usually the first to third plumages can be aged accurately, while later plumages should be referred to as plumage-types.

Juvenile

Although varying in ground colour from yellowish-buff to more sandy-brown, always identified by its regularly streaked breast and underwing-coverts, and its intact, broad white trailing edge of the wings. The pale head and the pale rear underbody, including the trousers, contrast sharply against the streaked and darker breast. The flight feathers are dark, save for the lighter inner three primaries, which form a distinct window, which, however, can be matched by some Steppe Eagles. The greater underwing-coverts are mostly grey, varying from light to dark grey, but usually do not contrast markedly with the rest of the underwing.

Upperparts are generally sandy-brown, with blackish secondaries and outer hand, while the inner hand shows a distinct pale window. The trailing edge of the wing is broadly white, as are the tips of the greater and median coverts, forming additional distinct white wing-bands. Not only are the uppertail-coverts whitish, but also the rump and lower back, a diagnostic character compared with Steppe Eagle, but often difficult to confirm as this tract is mostly covered by the scapulars. In closer views the upperparts can be seen to be strongly spotted, with a creamy spot to practically every feather. In head-on views approaching birds show distinctly pale carpals, a diagnostic feature compared with, for example, Steppe Eagle.

Second plumage

Being superficially similar to the juvenile plumage, the second plumage can be difficult to tell apart from a distance. The most important difference is the lack of a uniform, broad white trailing edge to the secondaries, as the birds show broad white tips only on the new moulted inner primaries. The broad white upperwing band of the juvenile greater coverts is also missing, and only traces remain. In closer views the best ageing character is the group of freshly moulted inner primaries, with the innermost three showing broadly white tips standing out from the otherwise worn trailing edge. The upperwing, scapulars and mantle appear rather dull and brownish, as they consist of mostly worn juvenile feathers.

Third plumage

From a distance still rather juvenile-like and closer views are needed for ageing. Some birds are still streaked like juveniles, but the majority are more uniformly yellowish-buff on the body and underwing-coverts, and they show, unlike birds in the second plumage, two moult fronts in the primaries, one in the outermost and one in the inner primaries. Most individuals also show the first black feathers on the throat, breast and underwing-coverts, but individual differences are great and some lack dark feathers altogether. Upperwings, mantle and scapulars are lighter coloured compared to previous plumages, as most of the juvenile plumage has by now been replaced by lighter grey and tawny feathers. New remiges and rectrices are still largely dark and uniform, similar to juvenile remiges, but the most recently moulted feathers may already show contrasting blackish tips.

Fourth plumage-type

The first of the pied plumages, in which the amount of black feathers on the body has become notable. Normally the head and upper breast have become rather adult-like, while the lower breast, belly and trousers appear largely pale. Underwing-coverts still appear pale, but dark feathers occur in variable proportions. Many of the new secondaries show dark tips and distinct dark barring, but uniformly dark secondaries from earlier generations still remain. Upperparts appear very blotchy consisting of a mix of grey and tawny feathers. Birds in this plumage can be almost identical to similar-aged Spanish Imperials, but Eastern Imperials tend to show a light forehead and crown (dark in Spanish), light rear underbody (black spotting in Spanish) while the new adult-type tail feathers are coarsely barred inside the black subterminal band (almost plain grey in Spanish; Malmhagen & Larsson 2013).

Fifth plumage-type

The second pied plumage-type, in which the amount of dark feathers has increased and pale body feathers remain mostly in the axillaries and lower belly. The underwing-coverts show a higher number of dark than pale feathers and by now most of the flight feathers show broadly dark tips with prominent dark barring further in. The head looks already adult, but the crown and nape are a rich tawny-brown, rather than golden-yellow.

Adult

A very dark, almost black eagle, with a striking Golden Eagle-like head, a barred grey tail with a broad black subterminal band and individually variable white markings on the upper scapulars (may be missing). Except for the head and tail the upperparts look very dark and uniform, but some may show a barely lighter brown area in the upperwing-coverts. The underwings show a contrast between black coverts and more glossy and grey remiges and a broad, black trailing edge. Younger adults show barred secondaries and inner primaries, but with increasing age the pattern dissolves, leading to almost uniformly dark remiges in older birds. Adults and older immatures usually keep their tails folded when soaring, whereas juveniles often soar with a fanned tail.

SEXING Although females are bigger and more heavily built than males, this is usually not possible to assess reliably in the field.

CONFUSION RISKS Most similar to Spanish Imperial Eagle, but ranges are not overlapping. However, non-breeding immatures might cross into one another's ranges, in which case identification would be extremely difficult. See under Spanish Imperial Eagle, and under Species Identification and Plumages above.

References

Malmhagen, B. & Larsson, H. 2013. Fältbestämning av subadult östlig och spansk kejsarörn. *Vår Fågelvärld* 3: 47–49 (in Swedish).

697. Eastern Imperial Eagle *Aquila heliaca*, juvenile, with juvenile Lesser Spotted Eagle. Juveniles and young immatures are easy to tell by pale, streaked plumage and dark secondaries contrasting with pale hand. Egypt, 9.10.2010 (DF)

698. Eastern Imperial Eagle *Aquila heliaca*, juvenile. Aged as a juvenile by uniform white tips to the wing and tail feathers, while the streaked breast and underwing-coverts contrasting with plain buff rear underbody and thighs identify it to species. Oman, 7.11.2013 (DF)

699. Eastern Imperial Eagle *Aquila heliaca*, juvenile. This is a typical juvenile from the paler end of variation. Oman, 7.12.2010 (DF)

700. Eastern Imperial Eagle *Aquila heliaca*, juvenile. This is a more heavily marked juvenile. Oman, 9.12.2010 (DF)

701. Eastern Imperial Eagle *Aquila heliaca*, juvenile. A rather standard-looking autumn juvenile. Oman, 9.12.2010 (DF)

702. Eastern Imperial Eagle *Aquila heliaca*, juvenile. Oman, 7.12.2010 (DF)

703. Eastern Imperial Eagle *Aquila heliaca*, juvenile. Upperparts are typically light sandy with pale streaks contrasting with dark flight feathers and tail. Note whitish inner primaries, uppertail-coverts and rump, and pale head. Oman, 10.2.2014 (DF)

704. Eastern Imperial Eagle *Aquila heliaca*, juvenile. Typical upperparts showing extensively pale lower back and rump, and diagnostic wing markings. Oman, 7.11.2013 (DF)

705. Eastern Imperial Eagle *Aquila heliaca*, second plumage. Despite extensive body moult this bird still looks juvenile. Note mix of moulted and retained juvenile remiges and lack of uniform trailing edge to the wings. Oman, 7.12.2010 (DF)

706. Eastern Imperial Eagle *Aquila heliaca*, second plumage. Note retained, worn outer primaries and mix of moulted and juvenile secondaries in this juvenile-like bird. Oman, 7.12.2010 (DF)

707. Eastern Imperial Eagle *Aquila heliaca*, second plumage. By and large like a juvenile, but remiges are a mix of fresh feathers and retained juvenile feathers. Oman, 10.2.2014 (DF)

708. Eastern Imperial Eagle *Aquila heliaca*, second plumage. This bird is less advanced, with five new primaries, a few moulted secondaries, a recently moulted breast and moulting underwing-coverts. Oman, 12.12.2012 (DF)

709. Eastern Imperial Eagle *Aquila heliaca*, second plumage. The new upperwing-coverts are not as distinctly streaked as the juvenile plumage while the greater coverts are largely worn juvenile feathers. Oman, 11.12.2010 (DF)

710. Eastern Imperial Eagle *Aquila heliaca*, second plumage (same as **707**). Note pale back and rump and clear signs of wing-moult, with faded brown juvenile feathers standing out. Oman, 10.2.2014 (DF)

711. Eastern Imperial Eagle *Aquila heliaca*, third plumage. From this age-class onwards, individual variation increases. Some birds are still rather juvenile-like, as here, but note two moult fronts in the primaries (at p3/4 and p9), unlike in the second plumage, and only one retained juvenile flight feather, the outermost primary. Oman, 12.12.2012 (DF)

712. Eastern Imperial Eagle *Aquila heliaca*, third plumage. Others are less juvenile-like, with rather plain underwing-coverts and irregular black streaks turning up on the upper breast, neck and underwings. Oman, 19.11.2014 (DF)

713. Eastern Imperial Eagle *Aquila heliaca*, less advanced fourth plumage-type. In this plumage adult-type flight feathers start to appear. Oman, 9.12.2010 (DF)

714. Eastern Imperial Eagle *Aquila heliaca*, fourth plumage-type. This plumage-type shows more dark markings to the body and underwings while many of the flight feathers are of adult-type with broad black tips. Birds of this plumage-type may be either advanced third plumage birds or fresh fourth plumage birds. Oman, 7.12.2010 (DF)

715. Eastern Imperial Eagle *Aquila heliaca*, fourth plumage-type. Thanks to an increasing number of black body feathers, birds become easier to identify as Eastern Imperial Eagles by this age. Oman, 15.2.2013 (DF)

716. Eastern Imperial Eagle *Aquila heliaca*, fifth plumage-type. Flight feathers are now largely of adult-type, but there are still many yellowish-buff feathers among the underwing-coverts. Oman, 9.12.2012 (DF)

717. Eastern Imperial Eagle *Aquila heliaca*, fifth plumage-type (same as **716**). Largely dark above, but upperwings still show pale coverts. Oman, 9.12.2012 (DF)

718. Eastern Imperial Eagle *Aquila heliaca*, fifth plumage-type. Oman, 11.12.2010 (DF)

719. Eastern Imperial Eagle *Aquila heliaca*, subadult plumage-type. By and large like adult, but underwing-coverts still show a mix of lighter feathers. Oman, 8.2.2014 (DF)

720. Eastern Imperial Eagle *Aquila heliaca*, subadult plumage-type (same as **719**). Upperwing-coverts can be very similar to older Golden Eagles in this plumage, but note darker remiges above and more regularly barred tail in Imperial. Oman, 8.2.2014 (DF)

721. Eastern Imperial Eagle *Aquila heliaca*, adult, with juvenile Steppe Eagle. Adults are nearly all black, with slight contrast between black coverts and greyer remiges. Oman, 12.12.2012 (DF)

722. Eastern Imperial Eagle *Aquila heliaca*, adult. With increasing age the remiges become more uniform as the barring dissolves to be retained in the inner primaries only. Oman, 18.2.2013 (DF)

SPANISH IMPERIAL EAGLE
Aquila adalberti

VARIATION Monotypic. Previously considered conspecific with Eastern Imperial Eagle *A. heliaca*.

DISTRIBUTION Confined to the Iberian Peninsula, with *c.* 280 breeding pairs in Spain and 7–9 pairs in Portugal (2012). Has in the past bred also in N Africa, where now extinct. Juveniles and immatures disperse, regularly reaching W Africa, rarely also W Europe.

BEHAVIOUR A bird of semi-open woodland and hills. Mostly seen soaring, or perched prominently on treetops or pylons. When hunting hangs against the wind in mid-air, sometimes high up in the sky, from where it makes impressive stoops onto ground prey, such as rabbits and partridges.

SPECIES IDENTIFICATION The dark adults are most similar to Eastern Imperial Eagle, but could also be mistaken for Golden Eagle. Slightly shorter-winged and comparatively longer-tailed than Eastern Imperial Eagle, with a white leading edge to the wings in most individuals, *which may be completely lacking in some.* Compared to Golden Eagle the wings are more parallel-edged, the head more protruding and the tail shorter. Light-coloured immatures and subadults are very similar to similar plumages of Eastern Imperial, and are not always separable. Washed-out juveniles may also be confused with the pale morph of Tawny Eagle (ranges overlapping in winter in NW Africa) or the '*fulvescens*' morph of Greater Spotted Eagle.

MOULT Moults during the breeding season, from Apr to Oct. The second plumage, and often also the third are possible to age by proportion of freshly moulted feathers and retained juvenile feathers.

PLUMAGES Normally four age-classes can be reliably identified by plumage and state of moult: juvenile, second plumage, third plumage and adult. Further subadult stages can be recognised, but only as plumage-types, not as exactly aged plumages. It is worth noting that Spanish Imperial Eagles appear to moult into adult plumage faster than Eastern Imperial Eagles, acquiring the final plumage 1–2 years sooner.

Juvenile

Overall rather uniformly rich tawny with dark flight feathers, but body plumage fades to more creamy-buff during first winter. Head merges with breast, which mostly shows no or only faint streaking, although more heavily streaked birds, similar to Eastern Imperial, are known, but in particular the underwing-coverts appear plain and unstreaked, thus differing from the much more extensively and distinctly streaked juvenile Eastern Imperial Eagle. The inner primaries are distinctly paler, forming an obvious pale window to the underwing. Remiges usually nearly uniformly dark below, lacking distinct barring, but showing a prominent pale trailing edge to the wings and tail. The upper greater coverts also have broad pale tips, forming a broad band on the upperwing, but on average the light bands are narrower than in juveniles of Eastern Imperial, and the upperparts lack the streaking of the latter.

Second plumage

Very similar to a worn and faded juvenile, but always recognised by freshly moulted inner primaries. The white trailing edge of the wing has by now worn narrow, with the broad pale tips of the new inner primaries and any fresh secondaries eye-catching. The median coverts of the upperwing have been partly moulted, showing a mixture of faded buff and more greyish feathers contrasting with the paler head. Some birds show rather distinct streaking to the upper breast, and can therefore appear similar to Eastern Imperial Eagles of comparable age. Birds are known to have paired and intended to breed in this plumage (3rd cy spring).

Third plumage

By and large still rather juvenile-like, but the upperparts are becoming darker and browner and the dark streaking below is becoming more contrasting, with some already showing a dark throat, often starting with a dark malar streak, while more advanced birds may show an adult-type head. The underwing-coverts are still largely straw-coloured, but some dark feathers may appear. The outer primaries and some median secondaries are still juvenile, by now worn and faded brown in colour. Also the tail is mostly juvenile-like and uniformly dark greyish-brown, but the first dark-tipped adult-type feathers may now appear.

Fourth plumage-type

The first of the pied plumages, in which dark feathers start to appear in greater number on the underbody and underwing-coverts, and the head appears quite adult-like. Unlike in Eastern Imperial the forehead and crown are dark, the dark mottling extends to the rear underbody and the grey parts of the new adult-type tail feathers are more uniform and lack prominent barring. Save for dark remiges above, the upperparts are still largely a mixture of tawny and grey, but dark feathers appear among the lesser and median upperwing-coverts and scapulars.

Fifth plumage-type

The second pied plumage, in which the dark feathers of the underbody are in the majority over the buffish feathers, and the bird can appear quite dark from a distance. Some may be fairly similar in plumage to an adult Golden Eagle, with a noticeable paler area in the median upperwing-coverts.

Adult

Appears very dark from a distance, with faded yellow crown, nape and hindneck, and a pale grey inner tail. Best told from adult Golden Eagle and adult Eastern Imperial Eagle by the white leading edge to the arm, which, however, varies in extent between different individuals and *can be missing altogether* in some. Upperwing usually uniformly dark brown, lacking greyish flight-feather panel and the faded brown patch across the median coverts of adult Golden Eagle, but depending on condition of plumage may show a lighter brown area in the upperwing-coverts, rather similar to but less contrasting than in Golden Eagle. Underwing-coverts nearly black, contrasting with greyer and glossier remiges, the latter showing some barring to the outer secondaries and inner primaries (barring more clear in young adults but practically missing in old birds), but overall the underwing is much darker and more uniform than in adult Golden Eagle. Younger adults may show a rather distinct dark trailing edge to the underwing, but with increasing age the barring disappears and the remiges become darker with fewer markings.

SEXING The sexes are not known to differ in plumage. In direct comparison the female is heavier and bigger than the male, with a deeper bill, but without direct comparison size cannot be used for sexing.

CONFUSION RISKS Adult distinguished from adult Golden Eagle by different silhouette, and from adults of both Golden and Eastern Imperial Eagles by its prominent white leading edge to the arm (can

be missing). Juveniles and immatures can be very similar to similarly-aged Eastern Imperial Eagles, and also to the pale morphs of Tawny and Greater Spotted Eagles, and great care is needed to separate between these.

NOTE In 2007–2008 in C Spain an adult female Spanish Imperial Eagle paired with a male *Aquila* hybrid of unknown affinities, raising chicks, but the results of the genetic analyses have yet to be published.

References

Gonzales, J. M. & Margalida, A. (eds.) 2008. Conservation biology of the Spanish Imperial Eagle *(Aquila adalberti)*. Organismo Autónomo Parques Nacionales. Ministerio de Medio Ambiente y Medio Marino y Rural. Madrid.

Malmhagen, B. & Larsson, H. 2013. Fältbestämning av subadult östlig och spansk kejsarörn. *Vår Fågelvärld* 3: 47–49 (in Swedish).

723. Spanish Imperial Eagle *Aquila adalberti*, fresh juvenile. Note plain secondaries and underwing-coverts and distinct translucent window in inner primaries and compare with juveniles of Eastern Imperial, Tawny and *fulvescens* form of Greater Spotted Eagle. The amount of breast streaking varies individually. Spain, 31.10.2009 (DF)

724. Spanish Imperial Eagle *Aquila adalberti*, juvenile, 2ⁿᵈ cy spring. Plumage is lighter due to fading, but pale trailing edge of wings is still intact. Note some new streaked feathers in the crop area. Spain, 22.4.2010 (DF)

725. Spanish Imperial Eagle *Aquila adalberti*, immature, presumed fourth plumage, with Western Black Kite. Typical pied plumage of an older immature. Spain, 26.6.2015 (DF)

726. Spanish Imperial Eagle *Aquila adalberti*, adult. Note typical, almost black body and underwing-coverts. The amount of white along the leading edge of the wing varies and may be lacking altogether. Spain, 16.2.2011 (DF)

727. Spanish Imperial Eagle *Aquila adalberti*, adult (same as **726**). Spain, 16.2.2011 (DF)

728. Spanish Imperial Eagle *Aquila adalberti*, adult (same as **726**). Usually darker and more uniform on upperwing than adult Golden Eagle. Note rather plain grey tail with broad black subterminal band. Spain, 16.2.2011 (DF)

Separating Steppe Eagle and Lesser Spotted Eagle

Lesser Spotted Eagle and Steppe Eagle can be very similar in flight in certain plumages, especially when seen from a distance. In particular the subadult plumages can be almost identical in normal field conditions (fourth/fifth plumage Steppe and third plumage and older Lesser Spotted). The juvenile plumage is distinct in both species and does not normally give any difficulties. The white trailing edge and tail-tip of Steppe is much broader than in any juvenile Lesser Spotted, and the white line in the upperwing of Steppe (tips of the greater coverts) is solid, not formed by spots. Rarely, some juvenile Lesser Spotted show a fairly broad white line to the greater underwing-coverts, which can be reminiscent of a juvenile Steppe. On average, the underwing flight feathers are darker in Lesser Spotted, lighter grey and distinctly banded in Steppe, but the variation in Steppe is considerable and anything up to nearly black remiges can occur. In close views the focus should be on the underwing barring, which is diagnostic (always sparse in Steppe and dense in Lesser Spotted). Care should be taken with immature Steppe lacking the white underwing band (missing in fewer than 10% of birds) and juvenile Lesser Spotted with broader than average white greater underwing-coverts band.

The experienced observer will notice a difference in structure between Lesser Spotted and Steppe Eagles. Steppe is longer-winged, with a longer arm in particular, and the fingered primaries are more deeply splayed (longer fingers) and more clearly curved upwards in flight compared to Lesser Spotted. On the other hand the wingtip is more rounded in Lesser Spotted compared to the almost vulture-like square tip of Steppe. Seen gliding head-on Steppe shows a rather level arm and a drooping hand, while Lesser Spotted has a smoother and more arched, less angled head-on silhouette. Lesser Spotted will appear shorter-winged with a comparatively large and rounded head compared to the long-winged and slimmer-headed Steppe Eagle.

STEPPE EAGLE
Aquila nipalensis

VARIATION Monotypic. Until recently two subspecies were recognised, with on average smaller *orientalis* in the western part of range, and larger *nipalensis* in the east, but the species is now regarded as monotypic with clinal variation.

DISTRIBUTION Breeds from E Europe, where now rare and numbers dwindling, to C Asia, E Russia, Mongolia and China. Mostly migratory, wintering in S Asia, the Middle East and Africa.

BEHAVIOUR A bird of open or sparsely wooded areas, penetrating into higher mountains. Hunts mainly from the air, hanging motionless in the wind with drooping head, scanning the ground for small and medium-sized mammals. Also hunts on foot and kleptoparasitises other raptors.

SPECIES IDENTIFICATION Medium-sized to large eagle, and depending on plumage, could be mistaken for various other *Aquila* species. The most common type of juvenile and immature plumages is easily recognised by a broad whitish band on the greater coverts of the underwing, but some lack this band and are then easily mistaken for, for example, Lesser Spotted Eagle, while pale immatures could be confused with Tawny or immature Imperial Eagles. Adult is very similar to dark morph adult Tawny, with some birds almost inseparable, but proportions slightly different with longer wings and shorter tail in Steppe. Dark adults could also be easily mistaken for adult Greater Spotted and Imperial Eagles. Most birds recognised by diagnostic, sparse and heavy underwing barring, but some birds (very rare) have uniformly dark remiges.

MOULT Most of the remex moult takes place during the breeding season, but additional feathers are also moulted in the winter quarters. Immatures of second and third plumage can be safely aged by moult pattern, as in many other large raptors. For details, see under Plumages.

PLUMAGES Normally five age-related plumage-types can be separated in the field: juvenile, second and third plumages, fourth plumage-type and adult. The main changes in the plumage take place during the breeding season, with additional odd flight feathers replaced in winter.

Juvenile

Plumage either fresh (autumn), or evenly worn (spring), with a broad white trailing edge to the wings and tail. Colour varies from dark brown to ashy grey-brown to milky coffee-coloured birds, while flight feathers are dark above but more greyish and distinctly but sparsely barred below. By spring and first summer (2nd cy) the pale feather tips may wear off completely, but the plumage still appears uniform consisting largely of juvenile feathers of similar age. Most birds feature largely white greater underwing-coverts, while others show darker centres or rarely have entirely greyish greater coverts (*c.* 5% of the birds). In normal white-banded individuals, the underwing band is also a feature of the two subsequent plumages.

Second plumage

By and large very similar to the juvenile, but up to half of the primaries have been replaced and differ by being darker, the innermost also by their broadly white tips. Some secondaries have also been replaced and are clearly longer than the juvenile feathers, protruding from the trailing edge of the wing. Most of the body plumage is worn and faded, consisting of juvenile feathers, but the head is moulted in many birds and may appear as a darker hood. However, the extent of the body moult varies considerably between individuals.

Third plumage

From a distance similar to the previous plumages, owing to the broad and white underwing band retained by most birds. However, the silhouette is by now more compact, with broader and more rectangular wings. In closer views the median and outer primaries and the majority of the secondaries have clearly been replaced, giving the trailing edge of the wing a ragged contour typical of this age-class. The most recently moulted secondaries show a broadly dark tip, the beginning of the adult's subterminal band.

Fourth plumage-type

By now the uniform white underwing band is mostly gone, but partly white greater underwing-coverts are still found. The general plumage colour is a warm sandy-brown and the barring of the underwing has become more prominent, including broad, dark tips to many secondaries, approaching the pattern of adult birds. The undertail-coverts are still predominantly pale.

Adult

The ground colour is a dull dark brown in most, while others are paler and more sandy-brown. The upperwing usually lacks a strong contrast between paler coverts and darker remiges, and the inner hand shows a variable pale or grey area, which is sometimes clearly barred. The darkest birds are very similar to adult Greater Spotted Eagles on their upsersides, and great care is needed to separate between the two. However, the tail of adult Steppe is grey with heavy dark barring and sometimes similar-looking barring is also seen on the secondaries from above (distinct barring is always lacking in adult Greater Spotted). The central back shows a white patch in most, best seen on soaring birds. Seen from below the underwing-coverts are either dark or medium brown, often with a clearly darker carpal patch, while the remiges are grey with sparse and coarse barring. In silhouette the trailing edges differ between adults of Steppe and Greater Spotted: *in Steppe the feather tips are nicely rounded, giving the trailing edge a regularly toothed shape, while in Greater Spotted the secondaries have blunter, almost square tips, and the outline of the trailing edge is straighter.*

SEXING Males are smaller and comparatively broader-winged than females, rendering them a more compact, almost Greater Spotted Eagle-like silhouette, particularly obvious in adults. Still, sexing is not recommended unless direct comparison is possible, for example when the birds of a breeding pair are seen together.

CONFUSION RISKS Depending on plumage various confusion risks occur. Dark adults can be very similar to older Greater Spotted Eagles, with almost identical silhouettes in some, while fourth-plumage birds could easily be mistaken for an immature Lesser Spotted Eagle. Juveniles and younger immatures lacking the white underwing band could be mistaken for a Lesser Spotted or Tawny Eagle of similar age. In all cases the pattern of the secondaries remains the most important key feature for separation.

729. Steppe Eagle *Aquila nipalensis*, juvenile. A classic juvenile with the diagnostic broad white underwing-band. Aged by uniform plumage with intact white trailing edge to the wings and tail. Oman, 9.12.2010 (DF)

730. Steppe Eagle *Aquila nipalensis*, juvenile. Typical juvenile, although with less well marked remiges. Oman, 7.12.2010 (DF)

731. Steppe Eagle *Aquila nipalensis*, juvenile. This juvenile has almost no white underwing-band, and also plain remiges, which both may cause confusion with, for instance, Tawny Eagle. A small percentage of juveniles look like this. Oman, 9.12.2010 (DF)

732. Steppe Eagle *Aquila nipalensis*, juvenile. Another juvenile with a partial wing-band. Similar birds could be mistaken for Lesser Spotted Eagle, but note different wing-formula, underwing barring and structure, and compare upperparts. Oman, 9.12.2010 (DF)

733. Steppe Eagle *Aquila nipalensis*, juvenile. Typical juvenile with broad wing-bands, boldly barred flight feathers and tail, and white uppertail-coverts. Note plain body plumage. Oman, 2.11.2004 (DF)

734. Steppe Eagle *Aquila nipalensis*, juvenile. A slightly darker, yet typical juvenile. Oman, 2.11.2004 (DF)

735. Steppe Eagle *Aquila nipalensis*, juvenile, 2nd cy spring. A worn juvenile from late winter, with part of the white feather tips worn off, while the head has already been moulted. Oman, 10.2.2014 (DF)

736. Steppe Eagle *Aquila nipalensis*, second plumage. Similar to juvenile, but aged by moulted inner four primaries, as well as s1 and s5, and lack of uniform white trailing edge to the wings. Oman, 7.12.2010 (DF)

737. Steppe Eagle *Aquila nipalensis*, second plumage. Note distinct difference in colour and condition between the moulted inner five and retained outer five primaries. Oman, 7.12.2010 (DF)

738. Steppe Eagle *Aquila nipalensis*, second plumage. This bird shows a slightly more advanced moult, with a higher number of replaced flight feathers and also largely moulted body plumage. Oman, 7.12.2010 (DF)

739. Steppe Eagle *Aquila nipalensis*, second plumage. A bird showing both an odd plumage and an extensive wing-moult, with more than half of all remiges and tail feathers replaced in first moult. Oman, 7.11.2013 (DF)

740. Steppe Eagle *Aquila nipalensis*, second plumage (upper), with juvenile. A very pale bird, which might cause confusion with Tawny Eagle, but note white greater coverts and distinctly barred remiges. Oman, 7.12.2010 (DF)

741. Steppe Eagle *Aquila nipalensis*, second plumage, 3rd cy spring. In most respects a rather typical second-plumage bird. Note less well marked flight feathers. Oman, 7.2.2013 (DF)

742. Steppe Eagle *Aquila nipalensis*, second plumage, 3rd cy spring. This bird shows an extremely advanced moult, with only four outer primaries being retained juvenile feathers, while all other flight feathers, the entire tail and all body plumage has been replaced. Oman, 6.2.2014 (DF)

743. Steppe Eagle *Aquila nipalensis*, second plumage, 2nd cy autumn migrant. Most of the body plumage has been moulted, but the outer four primaries, most of the scondaries and half of the upperwing greater coverts are still retained juvenile feathers. Israel, 6.11.2009 (DF)

744. Steppe Eagle *Aquila nipalensis*, third plumage, 4th cy spring. Many birds are still quite juvenile-like in their general appearance, often with a wide white underwing-band, but the primaries show two moult fronts, one in the outer primaries and a second in the inner primaries. Oman, 7.2.2013 (DF)

745. Steppe Eagle *Aquila nipalensis*, third plumage. By and large very similar to second plumage, but note fresh outer and inner primaries, with older and browner feathers in between. Oman, 9.12.2010 (DF)

746. Steppe Eagle *Aquila nipalensis*, third plumage, 3rd cy autumn. Note the moult pattern in the primaries and compare with birds in second plumage. Birds in third plumage generally show a tidy underwing, with fresh underwing-coverts, including the mid-wing band. Oman, 9.12.2010 (DF)

747. Steppe Eagle *Aquila nipalensis*, fourth plumage-type, 4th cy autumn. Birds of this age are extremely variable and may still have an immature look. However, the primary moult does not fit the third plumage pattern, with some, as here, showing three simultaneous moult fronts. This is also the plumage when the first remiges with broad dark tips start to appear in numbers. Oman, 6.12.2012 (DF)

748. Steppe Eagle *Aquila nipalensis*, fourth plumage-type, 4th cy autumn. Note adult-type body and strong barring of the flight feathers, and compare moult pattern with birds in third plumage. Oman, 9.12.2010 (DF)

749. Steppe Eagle *Aquila nipalensis*, fourth plumage-type, 5th cy spring. Rather similar to many birds in third plumage, but note different primary moult pattern and high percentage of black-tipped remiges. Oman, 7.2.2013 (DF)

750. Steppe Eagle *Aquila nipalensis*, subadult plumage. Differs from full adult by a few retained partly white greater underwing-coverts and some secondaries with poorly defined dark tips. Oman, 7.12.2010 (DF)

751. Steppe Eagle *Aquila nipalensis*, subadult/young adult. Despite the overall adult appearance with strongly marked flight feathers, dark trailing edge to the wings and a blotchy brown body plumage, some secondaries with poorly marked dark tips identify this bird as a younger adult. Oman, 11.12.2010 (DF)

752. Steppe Eagle *Aquila nipalensis*, subadult/young adult. Differs from full adult by some white-tipped greater coverts and a few brown and faded older flight feathers lacking the adult-type barring. Oman, 7.12.2010 (DF)

753. Steppe Eagle *Aquila nipalensis*, adult, autumn migrant. When gliding Steppe Eagles keep their wings angled, with level arm but drooping hand; compare this with the deeply arched posture of Greater Spotted Eagle. Israel, 10.11.2009 (DF)

754. Steppe Eagle *Aquila nipalensis*, fully adult plumage, autumn migrant. Some adults are very dark, as here, and could easily be mistaken for Greater Spotted Eagle were it not for the diagnostic sparse and broad underwing barring. Israel, 10.11.2009 (DF)

755. Steppe Eagle *Aquila nipalensis*, fully adult plumage, autumn migrant. A typical adult, even showing the individually variable ochre crown/nape patch. Israel, 10.11.2009 (DF)

756. Steppe Eagle *Aquila nipalensis*, fully adult plumage. Some adults show light bases to their outer primaries, thus being confusingly similar to adult Greater Spotted Eagle, but note diagnostic underwing barring and 'serrated' trailing edge. Oman, 11.12.2010 (DF)

757. Steppe Eagle *Aquila nipalensis*, fully adult plumage. Other adults are much lighter brown, with a darker body contrasting with lighter underwing-coverts. Note typical underwing barring and dark trailing edge. Oman, 6.2.2014 (DF)

758. Steppe Eagle *Aquila nipalensis*, fully adult plumage. Light brown adults could be mistaken for Lesser Spotted Eagle, for example, but note diagnostic underwing barring and broader and squarer wingtips. Oman, 6.2.2014 (DF)

759. Steppe Eagle *Aquila nipalensis*, fully adult plumage, autumn migrant. Note dark upperparts with greyish, heavily barred flight feathers and tail, and white back. Israel, 10.11.2009 (DF)

TAWNY EAGLE
Aquila rapax

VARIATION Three subspecies, with ssp. *belisarius* of N Africa reaching the Sahel region and Ethiopia in the south. Occurs extralimitally as nominate ssp. *rapax* in sub-Saharan Africa, south of *belisarius*, and ssp. *vindhiana* in the dry lowland areas of Pakistan and India.

DISTRIBUTION Rare in our region, with former breeding population of Souss Valley in Morocco now thought to be extinct. Occurs as a vagrant to the Middle East (Israel, Egypt and Arabia) from NE Africa and still regularly seen in S Morocco, although some recent claims from NW Africa refer to migrant immature Spanish Imperial Eagles.

BEHAVIOUR Prefers open land, perching prominently either on treetops or pylons, but also on the ground. Hunts on the wing like other large eagles, 'standing' still against the wind, but is also specialised in kleptoparasitism, stealing prey from other birds, such as kites and corvids. Often notably confiding towards man.

SPECIES IDENTIFICATION A large, nondescript, mostly brown eagle of average proportions and therefore not easy to identify with certainty. Most similar to Steppe Eagle, from which it was only recently split, but differs by having shorter wings and longer tail, often also by showing a pale underwing window of paler inner primaries. The underwing barring is mostly different, as *Tawny shows rather faint and denser barring, comparing with Steppe's sparse and distinct bars,* and Tawny also lacks the prominent black trailing edge of adult Steppe. However, some adult Tawny Eagles (old birds?) can be almost as prominently barred, including a dark trailing edge, as adult Steppe, and these birds can be truly difficult to identify. The length of the gape in relation to the eye, a classic feature of older literature, has been given far too much importance, being a character difficult to use, as the alignment of eye and gape changes as the bird turns its head.

Juveniles are usually easier to recognise. They never show the broad white underwing band featured by most immature Steppes, and the underwing barring is either missing or usually rather dense and diffuse, thus quite the opposite to young Steppe. However, juvenile and immature Steppe Eagles which lack the white underwing band (*c.* 5% of all individuals) and the underwing barring could easily be misidentified as Tawnies. In fact, several claims of Tawny Eagles from Arabia have been faded juvenile Steppe Eagles lacking the white wing-band.

MOULT As in other large eagles primary moult is protracted, with birds being almost three years of age before the last juvenile remiges are replaced. Details of moult are poorly known and more studies are needed. Possibly the timing of moult in different populations varies according to the local breeding season.

PLUMAGES A highly variable eagle, with a pale and a dark plumage morph in each age-class. Plumage differences between juveniles and adults are slight, particularly in light morph birds, but juveniles are always recognised by their intact, pale and S-curved trailing edge to the wing. Three age-classes can be identified: juvenile, immature and adult. A rare streaked plumage variant could easily be mistaken for other *Aquila* species, notably Greater Spotted and Eastern Imperial Eagles (see under Confusion risks). The streakiness appears more frequently in populations further south in Africa.

Juvenile

Occurs in two different colour forms, a pale and a brown type, but juveniles of both can be aged by uniform plumage lacking signs of moult, by a uniform pale trailing edge to the wings and tail, and by the S-curved trailing edge to the wings. The underwing barring is the most important single character when separating the species from other eagles: *it is rather faint and also rather dense and hardly visible from any great distance.* Usually a rather nondescript bird, except for pale lines on the upperwing, formed by pale tips to the greater and sometimes also the median coverts. Some juveniles show a tendency towards streakiness on the underwing-coverts and body, although less conspicuously marked compared to streaky adults. Iris is dark brown in juveniles.

Immature

Because of the harsh environment where these birds live, the plumage of juveniles soon wears and fades in colour owing to the excessive exposure to sun and sand. During the first year of life birds bleach from the warm yellowish-buff of a fresh juvenile to almost creamy-white. Birds moulting from juvenile to second plumage can be aged by their retained juvenile flight feathers, often showing some retained white on the tips. At this stage many pale morph birds appear hooded with a brown head and neck, which contrasts strongly with the rest of the plumage, and soon more brown feathers start to appear all over the body. Because of the lack of distinctive plumage features immatures cannot be aged with any certainty once the last juvenile remiges have been shed. Until then birds are aged by details of primary moult following the principles given for other large *Aquila* eagles.

Adult

Rather similar to the juvenile, but always told by less uniform plumage, showing a mixture of worn and fresher feathers. The silhouette is different, as adults show more rectangular wings with parallel edges, and the wings lack the pale trailing edge. Although the underwing barring remains rather fine and indistinct in most, some birds develop a darker subterminal band to the remiges, and in extreme cases the underwing pattern may appear quite similar to older Steppe Eagles, but the subterminal band tends to be narrower, less distinct and somewhat irregular compared to the average adult Steppe Eagle. Also, the tail-barring is finer and denser in Tawny compared to the sparse and heavy barring of adult Steppe Eagles. The iris colour of adults is variable, but is often light.

SEXING Usually not possible to sex in the field, but when a pair is seen together the female appears heavier with a deeper and longer bill.

CONFUSION RISKS A rather nondescript brown eagle, which is often confused with other *Aquila* species. Brown birds are rather similar to certain plumages of both Lesser Spotted and Steppe Eagles, but flight proportions and underwing barring differs. Some very dark Tawny Eagles with lighter streaking/ spotting on the breast and wing-coverts can be mistaken for Greater Spotted Eagle, although the silhouette and underwing barring is different. Lighter birds with darker streaking could be mistaken for a juvenile or immature Eastern or Spanish Imperial Eagle, while some pale birds can be very similar to a *fulvescens*-type Greater Spotted Eagle or to a faded juvenile Spanish Imperial Eagle, the latter occurring as a rare winter visitor in W Africa.

760. Tawny Eagle *Aquila rapax*, fresh juvenile, with Fan-tailed Raven. Poorly marked remiges identify this bird as Tawny, and the uniform plumage and complete white trailing edge on the wings as a fresh juvenile. Ethiopia, 13.2.2012 (DF)

761. Tawny Eagle *Aquila rapax*, fresh juvenile. A typical milky coffee-coloured juvenile. Note faintly barred remiges and compare with other *Aquila* eagles. Ethiopia, 10.3.2014 (DF)

762. Tawny Eagle *Aquila rapax*, fresh juvenile (same as **761**), chasing Pied Crow. Upperparts similar to young Steppe Eagle, for example, but band on greater coverts narrower, no band on median coverts, and remiges uniform. Note dark iris in juveniles. Ethiopia, 10.3.2014 (DF)

763. Tawny Eagle *Aquila rapax*, worn juvenile, about one year old. Note worn trailing edge of wings and tail, but plumage is still uniform, and all feathers are of the same generation. Ethiopia, 24.2.2014 (DF)

764. Tawny Eagle *Aquila rapax*, worn juvenile, about one year old, starting its moult to second plumage. Plumage still mostly juvenile, but head, neck and leading edge of wing moulted, as well as the inner primary. Ethiopia, 10.11.2011 (DF)

765. Tawny Eagle *Aquila rapax*, worn juvenile, about one year old. Note darker moulted head on this otherwise typical worn juvenile, robbing a Yellow-billed Kite of its prey. Ethiopia, 18.11.2011 (DF)

766. Tawny Eagle *Aquila rapax*, adult female (paired to **771**). A lighter adult aged by moulting remiges and lack of white tips to flight feathers and tail. Note indistinct barring, diagnostic of Tawny Eagles, and very pale iris. Ethiopia, 12.2.2012 (DF)

767. Tawny Eagle *Aquila rapax*, adult female (same as **766**). Rather similar to juvenile, but note ongoing moult, with remiges and coverts of varying condition. Ethiopia, 12.2.2012 (DF)

768. Tawny Eagle *Aquila rapax*, adult, presumed male. Many adults show variably blackish breast with lighter streaking, which is diagnostic for Tawny Eagle. Note also poorly marked secondaries and bright iris. Ethiopia, 22.2.2014 (DF)

769. Tawny Eagle *Aquila rapax*, adult, presumed male (same as **768**). Easily confused with other brown eagles, such as Lesser Spotted or Steppe, if just seen from above. However, note lack of white uppertail-coverts as well as type and extent of remex- and tail-barring, and compare. Ethiopia, 22.2.2014 (DF)

770. Tawny Eagle *Aquila rapax*, adult. This slightly darker brown bird also has more distinctly barred remiges which, however, differ from the barring of other *Aquila* species, with the dark bands much wider than the light, wider than in any other species. Ethiopia, 18.11.2011 (DF)

771. Tawny Eagle *Aquila rapax*, adult male, paired to **767**. A darker bird, rather similar in plumage to an older Steppe Eagle, but told by darker and less well-marked remiges and by different proportions with shorter and less rectangular wings. Ethiopia, 12.2.2012 (DF)

772. Tawny Eagle *Aquila rapax*, adult. A dark bird, which could be mistaken for a Greater Spotted Eagle, but note diagnostic underwing barring, with dense and bold bars, and light iris. Ethiopia, 6.1.2010 (DF)

773. Tawny Eagle *Aquila rapax*, adult. This rather average-looking adult shows Steppe Eagle-like barring to its remiges with broad black tips. Birds like this can be difficult to identify and care is needed to avoid misidentifications. Ethiopia, 5.2.2012 (DF)

774. Tawny Eagle *Aquila rapax*, adult. A darker adult showing more typical, yet somewhat Steppe Eagle-like barring to its remiges, lacking the broad black tips. Note lack of white uppertail-coverts. Ethiopia, 13.11.2011 (DF)

775. Tawny Eagle *Aquila rapax*, adult (same as **772**), mobbed by Fan-tailed Raven. Dark adults often show pale spots on the upperparts and are therefore often misidentified as Greater Spotted Eagles. Note, however, coarsely barred remiges and tail, and orange nape-patch, more similar to adult Steppe Eagle. Ethiopia, 6.1.2010 (DF)

Separating Greater Spotted Eagle and Lesser Spotted Eagle

These two eagles are quite distinct and relatively easy to identify if seen well. The structure (including wing-formula and the shape of the wingtip) differs, as does the plumage colour of average juveniles and older immatures. Lesser Spotted Eagles appear big-headed but fine-billed, while Greater Spotted Eagles have a flatter head and a heavier bill, and a truly eagle-like appearance. The greatest challenge is to separate the subadult and adult plumages of the two species, which are often lacking distinct species-specific plumage characters. An even greater challenge is provided by the nowadays frequently occurring hybrids, which are treated separately (see p. 418).

Juvenile Greater Spotted Eagles typically show a black body plumage with pale greyish underwing remiges creating a strong contrast against the *blackish coverts* of the underwing. Juvenile Lesser Spotted is more variable, but the underwing remiges are darker, appearing as dark or darker than the *brown underwing-coverts*. In Lesser Spotted there is also a colour difference between the coverts and remiges, the coverts being clearly brown while the remiges are dark grey.

In the following immature plumages Greater Spotted retains the pale remiges for at least two more plumages, retaining the same strong underwing contrast as in juveniles, although the underwing-coverts gradually turn from black to dark brown, while in Lesser Spotted the coverts become paler and the remiges darker, hence the contrast becomes even more clear with age.

On the upperparts most non-juvenile Lesser Spotted show a sharp contrast between straw-coloured lesser and median upperwing-coverts and clearly darker greater coverts and remiges, while in Greater Spotted this contrast is very gradual and the lighter forewing remains medium brown or darker.

The iris colour remains a reliable feature when seen well. In Lesser Spotted the iris has a pale yellowish colour in all but juveniles and first-year birds, while in Greater Spotted it remains dark brown throughout life, although some adults show a lighter, medium brown iris, but never the bright amber or almost pure yellow of adult Lesser Spotted.

Controversial individuals, showing characters of both species, should always be scrutinised carefully, as these may be hybrids. Signs of a solid yellowish nape-patch and clearly brown underwing- or upperwing-coverts in an otherwise perfect juvenile Greater Spotted are strong indications of the bird being a hybrid, as are pale grey remiges below with a fine Greater Spotted-like barring in an otherwise good looking juvenile Lesser Spotted Eagle. Note also iris colour and type of underwing barring, both of which may provide helpful clues.

GREATER SPOTTED EAGLE
Aquila clanga

Other names: Sometimes placed in the genus *Clanga*, together with the other spotted eagles.

VARIATION Monotypic. Closely related to Lesser Spotted Eagle, and large-scale hybridisation occurs in areas of sympatry resulting in fertile offspring. A rare and diagnostic pale colour morph '*fulvescens*' occurs (see below).

DISTRIBUTION From Eastern Europe in a wide zone across temperate regions of Asia to Far East Russia and the Pacific Ocean.

BEHAVIOUR A rather sluggish eagle, perching rather low in trees, often partly hidden. When hunting on the wing hangs against the wind on motionless wings scanning the ground below. In most places found close to water, preferring edges of wetlands, where it also hunts on foot, exploring meadows and shorelines.

SPECIES IDENTIFICATION A generally dark, medium-sized eagle, but because of its darkness, broad wings and compact build appears larger than it actually is. Adults can be very similar to adult Steppe Eagle in particular, but also to Lesser Spotted Eagle, and great care is needed to separate these species from one another. Both *spotted eagles are noticeably long-legged and* *in flight their yellow feet reach further out on the undertail compared to, for example, Steppe or Imperial Eagles.* When gliding, holds its wings in a diagnostic, deeply arched manner with slightly raised arm and prominently drooping hand, with fingers strongly bent upwards.

Juveniles and immatures are best told by very dark underwing-coverts contrasting with lighter grey and finely barred remiges, but plumage variation is extensive, and birds with partly pale body plumage are not rare. The diagnostic prominent white comma at the carpal, which is a good species pointer, becomes clearer once the juvenile outer primaries have been replaced (usually in early 3rd cy summer).

MOULT Moults both during the breeding season and in winter. Because of two moulting seasons per year the spotted eagles are capable of replacing more flight feathers per year than the larger eagles. Adults may in fact be capable of replacing nearly all their flight feathers in the course of 12 months. Normally moulting commences in spring between Mar and May, in first-summers in May–Jun, and continues until Oct. Moult is again resumed upon arrival on the winter quarters, and adult birds especially are moulting heavily in Dec–Jan. The moulting pattern is of importance for the ageing of immatures.

PLUMAGES Usually 4–5 age-classes can be separated based on plumage, but older immatures are not safe to age due to great individual variation. Juveniles and second-plumage birds are usually easily identified, as are full adults, but birds in third and fourth plumages are often best left as subadults.

Juvenile

A beautiful and easily identified eagle with a striking black plumage, marked with individually variable creamy spotting. From below characterised by black wing-coverts contrasting with much paler, almost silvery remiges, while the wingtip is dark and the three inner primaries form a pale window. Head and underbody are blackish, with a variable amount of creamy streaking, while the lower belly and undertail-coverts are creamy. The border between the dark forebody and pale rear usually runs across the belly, but in some individuals the pale area may reach the upper breast. One of the most important identification features against other *Aquila* species is the diagnostic barring of the secondaries: *the barring is fine and well spaced (dark bars are much finer than the grey spaces in between) and covers only the middle section of the feather,*

leaving the base and distal parts plain. Juveniles with all-dark secondaries, lacking any kind of barring, occur rarely, with the exception of juvenile '*fulvescens*', which always has uniformly dark remiges. Another important detail to note is the shape of the whitish crescent at the base of the outer primaries: in Greater Spotted it has the shape of a half-moon at the very bases of the outer three primaries. In juveniles there is often another band of lighter markings at the base of the greater primary coverts, much as in Lesser Spotted; the crescent is thus less diagnostic in juveniles compared to later plumages.

The upperparts are strikingly marked with pale drop-shaped tips to the upperwing-coverts forming regular lines. The size of the drops and the number of wing bands varies considerably from one bird to another, but compared to juvenile Lesser Spotted the ground colour is blackish throughout in autumn juveniles, lacking the distinct contrast between brown coverts and darker remiges of fresh juvenile Lesser Spotted. The uppertail-coverts are partly whitish, forming a prominent U across the tail-base, and the tip of the tail as well as the trailing edge of the wings are broadly white, well visible both from above and below. The inner three primaries are paler than the rest, forming a pale window, which from a distance is enhanced by the white tips of the upper primary coverts and the white shafts of the primaries, to form a wide white flash on the upper inner hand.

The plumage variation in juveniles is quite extensive. Some birds show rather small spots on the upperwing and appear very dark-bodied below, while others show big spots in the upperwing-coverts, sometimes overlapping to create a large pale area across the median upperwing-coverts; the underbody and underwing-coverts can be largely pale tawny or sometimes heavily streaked. Regardless of type and colour of body plumage the underwing barring remains a valid identification character.

By spring the blackish plumage has faded to dark brown and many of the drop-marks have almost worn away. It is worth remembering that first-summer birds are more easily mistaken for Lesser Spotted Eagles, as the upperwing contrast between remiges and coverts has become much more distinct; however, the characters of the underwing remain.

Second plumage

By and large very similar to a juvenile and best told by various signs of moult. Body plumage always comprises a mixture of worn juvenile feathers and freshly moulted feathers, which is best seen in the upperwing-coverts, where the fresh feathers show bright creamy spots while the juvenile feathers are abraded with worn-off spots. The best ageing character is the stage of the primary moult, which shows only one moult front. By the end of the 2nd cy summer the birds have moulted the inner 4–6 primaries, which are clearly darker and also somewhat longer than the replaced juvenile feathers, thus clearly protruding from the trailing edge of the wing. Although moult is often resumed in the winter quarters most birds retain a few juvenile outermost primaries until the next spring (3rd cy). The underparts still appear rather juvenile, with a strong underwing contrast and a mostly dark body plumage, and even the newly moulted secondaries still show the diagnostic fine barring, similar to the juvenile.

Third plumage

By now the bird has changed colour and become more brown, but there are still signs of immaturity in the plumage. The new greater upperwing-coverts still show pale markings, but they are no longer distinct as before, but appear more like pale but diffuse feather-tips. The body plumage still shows some pale streaking and the undertail-coverts are variably pale. The new secondaries show broader barring than in second plumage and a dark subterminal band is developing, the pattern thus approaching that of similar aged Lesser Spotted.

After this plumage there is still at least one more immature plumage before the final adult plumage is acquired. This plumage is very similar to the adult, but retains the odd pale markings to the tips of the underwing and undertail-coverts in particular, while many of the secondaries are still barred. Birds like this are best referred to as subadults.

Adult

Usually a dark brown and nondescript, broad-winged and plump-looking eagle, with a rounded head and a short and thickset tail. Very old birds can be paler brown and thus almost identical in coloration to some adult Lesser Spotted Eagles, but the upperparts lack the strong contrast of the latter, as the dark remiges are only marginally darker than the brown coverts, with a gradual and diffuse transition between the two. The uppertail-coverts are whitish while the faint pale area formed by the white shafts of the inner primaries is far less obvious than in adult Lesser

Spotted. The best species character is found in the underwing, which looks generally very uniform and dark, but shows a strong white crescent at the base of the outer primaries. This clear and striking crescent is usually lacking in Lesser Spotted and Steppe Eagles, in which the crescents are either greyer (in Steppe) or narrower and longer as in Lesser Spotted, which further normally shows not just one but two parallel crescents. In adults the remiges are uniformly dark, lacking any kind of barring, but the underwing-coverts may vary from dark to a lighter brown, often with darker carpal area. Note that some adult Steppe Eagles can be very similar, even showing a whitish carpal crescent, but they are told by their greyer and heavily barred remiges, more serrated (and dark) trailing edge of the wing, and their shorter legs, reaching only the base of the tail on a flying bird.

Importantly, the iris colour remains dark or medium brown in Greater Spotted throughout its life, which is a good character compared to Lesser, in which it turns amber or pale yellowish during the first two years of life, even changing to bright yellow in some older birds.

Identification of the '*fulvescens*' form

The distinct pale form of Greater Spotted Eagle, called '*fulvescens*', is poorly known among birders. It is rare in Europe, increasing in numbers further east, but even there it comprises only a few per cent of the population. Being a light-coloured eagle, its main confusion risks would be blond Tawny Eagles and immatures of Eastern and Spanish Imperial Eagles, or possibly also some very pale immature Lesser Spotted Eagles. Light morph Booted Eagles and juvenile Bonelli's Eagles are very different in general structure and also in plumage, and are thus unlikely to cause confusion.

In '*fulvescens*', as in ordinary Greater Spotted, the silhouette of the bird is compact, with broad wings and a comparatively short tail, but it is important to note that the shape is less compact in juveniles than in adults. The juveniles are very striking birds with a uniformly yellowish-buff body plumage and blackish tail and remiges. Important underwing features to look for are the white crescent at the base of the outer primaries, diagnostic of Greater Spotted, and the dark greater coverts, forming a conspicuous band between the creamy coverts and the plain, dark remiges. Also, the faintly darker face and throat found in most juveniles is a good feature against other light-coloured *Aquila* eagles. The remiges are uniformly dark in '*fulvescens*', lacking the diagnostic barring of normally coloured juveniles, but the inner three primaries are slightly paler than the rest, while the fingers are the darkest. Both spotted eagle species are well-known for being long-legged, and *in flight the yellow toes reach to the mid-tail, which is much further out than in other Aquila eagles*, an important identification detail.

The distinct plumage described above is typical of juveniles and second-plumage birds. Older *fulvescens* birds are browner and irregularly streaked, and far less striking in appearance compared to the juveniles. Some are foxy-red in colour, others just lighter brown, but as a rule the plumage appears untidy, particularly on the upperparts. With their rather nondescript appearance these subadults are even more likely to be mistaken for other *Aquila* species, but the all-dark and unbarred flight feathers, the white carpal crescent below and the dark greater coverts band usually help to clinch the identification. Ageing of immatures is best done by details of remex moult, applying the details given above for normally coloured birds.

SEXING Greater Spotted Eagles cannot normally be sexed in the field, but males are smaller, appear bigger-headed and narrower-winged compared to the massive and compact-looking females.

CONFUSION RISKS Very similar in shape to adult Steppe and Lesser Spotted Eagles, and plumage details are required for a positive identification. Type of barring of the remiges, detailed notes on the carpal comma, wing-formula and general coloration are all essential characters when dealing with these species.

In particular certain dark adult Steppe Eagles can be confusingly similar in general structure and silhouette and can be told from an average adult Greater Spotted only in the best of viewing conditions when, for example, the underwing barring or other important plumage characters can be confirmed.

NOTE Hybridises commonly with Lesser Spotted Eagle in E Europe. As a result genetically pure Greater Spotted are becoming rare, and the species may face a rapid extinction in areas of sympatry. F1-generation hybrid juveniles and immatures can also be identified in the field, if seen well, while later generation back-crosses (the hybrids are fertile) become increasingly difficult to identify.

References

Lontkowski, J. & Maciorowski, G. 2010. Identification of juvenile Greater Spotted Eagle, Lesser Spotted Eagle and hybrids. *Dutch Birding* 32: 384–397.

Maciorowski, G., Lontkowski, J. & Mizera, T. 2014. *The Spotted Eagle – Vanishing Bird of the Marshes.* Poznan.

Väli, U. & Lõhmus, A. 2004: Nestling characteristics and identification of the Lesser Spotted Eagle *Aquila pomarina*, Greater Spotted Eagle *A. clanga* and their hybrids. *J. Ornithol.* 145: 256–263.

Väli, U., Dombrovski, V., Treinys, R., Bergmanis, U., Daróczi S. J., Dravecky, M., Ivanovsky, V., Lontkowski, J., Maciorowski, G., Meyburg, B.-U., Mizera, T., Zeitz, R. & Ellegren, H. 2010. Widespread hybridisation between the Greater Spotted Eagle *Aquila clanga* and the Lesser Spotted Eagle *Aquila pomarina* (Aves: Accipitriformes) in Europe. *Biol. J. Linnean Soc.* 100: 725–736.

776. Greater Spotted Eagle *Aquila clanga*. When gliding, Greater Spotted Eagles keep their wings deeply arched, a reliable identification feature if used with care. Oman, 12.12.2012 (DF)

777. Greater Spotted Eagle *Aquila clanga*, juvenile. A classic juvenile with blackish body plumage contrasting with grey remiges. Note the diagnostic finely barred secondaries with the barring ending well before the tips of the feathers and compare this with other juvenile *Aquila* eagles. Oman, 12.12.2012 (DF)

778. Greater Spotted Eagle *Aquila clanga*, juvenile. A more heavily streaked individual, but with characteristic underwing. Note typical bright white carpal crescent. Oman, 7.12.2012 (DF)

779. Greater Spotted Eagle *Aquila clanga*, juvenile. This spring bird is slightly browner from wear, but shows the typical juvenile underwing. Note almost plain flight feathers. Oman, 7.2.2013 (DF)

780. Greater Spotted Eagle *Aquila clanga*, juvenile. Many juveniles are extensively streaked on the underbody, with ochre streaking reaching up to the upper breast. Oman, 12.12.2012 (DF)

781. Greater Spotted Eagle *Aquila clanga*, juvenile (same as **787**). Odd plumages with largely light underparts are not rare in juveniles. Oman, 16.11.2014 (DF)

782. Greater Spotted Eagle *Aquila clanga*, juvenile. Another case of plumage aberration. Oman, 7.11.2013 (DF)

783. Greater Spotted Eagle *Aquila clanga*, juvenile. The spotting of the upperparts varies individually a lot. Note the general darkness of the upperparts and compare with juvenile Lesser Spotted Eagle. Oman, 11.11.2013 (DF)

784. Greater Spotted Eagle *Aquila clanga*, juvenile. A more extensively spotted juvenile. Oman, 9.11.2013 (DF)

785. Greater Spotted Eagle *Aquila clanga*, juvenile. An individual with even larger spots. Oman, 8.11.2013 (DF)

786. Greater Spotted Eagle *Aquila clanga*, juvenile. By spring the upperparts have faded browner, and less spotted birds can look quite similar to fresh juvenile Lesser Spotted Eagle. However, by this time Lesser Spotted would be extremely worn and faded, lacking spots and wing-bands and looking generally much paler above than this bird. Oman, 10.2.2014 (DF)

787. Greater Spotted Eagle *Aquila clanga*, juvenile (same as **781**). Variably pale individuals are heavily spotted above, with spots almost entirely covering the ground colour. Oman, 16.11.2014 (DF)

788. Greater Spotted Eagle *Aquila clanga*, juvenile (same as **782**). Oman, 7.11.2013 (DF)

789. Greater Spotted Eagle *Aquila clanga*, second plumage, with Brown-necked Raven. Rather juvenile-like, but aged by moulted inner and retained juvenile outer primaries. Note more square wing-shape compared to juvenile. Oman, 3.11.2004 (DF)

790. Greater Spotted Eagle *Aquila clanga*, second plumage. This rather streaked bird looks brown, but note typical moult pattern and the diagnostic fine barring of the flight feathers. Oman, 3.11.2004 (DF)

791. Greater Spotted Eagle *Aquila clanga*, second plumage. Most second-plumage birds still appear rather juvenile-like, save for moulted remiges. Oman, 12.11.2013 (DF)

792. Greater Spotted Eagle *Aquila clanga*, second plumage. Birds are typically rather brown above in this plumage, with individually varying amounts of spotting depending on the state of moult and individual variation. Oman, 23.11.2014 (DF)

403

793. Greater Spotted Eagle *Aquila clanga*, third plumage, 4th cy spring. Largely similar to second plumage, but note two moult fronts in the primaries, with fresh outer and inner primaries. Note also white carpal flashes as well as finely barred remiges, both typical of Greater Spotted Eagles. Oman, 15.2.2013 (DF)

794. Greater Spotted Eagle *Aquila clanga*, third plumage, 3rd cy autumn. Typically barred underwing remiges but poorly defined carpal crescents in this bird. Oman, 11.12.2010 (DF)

795. Greater Spotted Eagle *Aquila clanga*, subadult, possibly fourth plumage, in transition from immature to adult plumage. The freshly moulted secondaries are uniformly dark, as are the vent and undertail-coverts, but the greater coverts still have light fringes and the old secondaries are still barred. Oman, 9.12.2010 (DF)

796. Greater Spotted Eagle *Aquila clanga*, subadult/young adult. Similar to adult apart from quite extensively barred remiges. Oman, 2.11.2004 (DF)

797. Greater Spotted Eagle *Aquila clanga*, subadult plumage. This subadult, showing a classic broad-winged Greater Spotted Eagle silhouette, still retains pale undertail-coverts and pale-fringed greater coverts, while most flight feathers are turning plain. Oman, 11.2.2013 (DF)

405

798. Greater Spotted Eagle *Aquila clanga*, adult. Note overall dull brown plumage with dark flight feathers. Oman, 10.2.2014 (DF)

799. Greater Spotted Eagle *Aquila clanga*, adult. Note typical adult silhouette with broad and rectangular wings and a short tail. Adults appear very uniform below, except for white carpal flashes and yellow feet. Oman, 10.12.2012 (DF)

800. Greater Spotted Eagle *Aquila clanga*, adult. This adult is moulting its body plumage heavily, with missing feathers explaining the white markings in the underwing-coverts. Oman, 3.11.2004 (DF)

801. Greater Spotted Eagle *Aquila clanga*, adult. Some adults appear truly uniform below. Oman, 12.12.2012 (DF)

802. Greater Spotted Eagle *Aquila clanga*, juvenile plumage of the form *'fulvescens'*. Note uniform yellowish-cream body plumage with a dark band on the greater coverts and uniformly dark flight feathers, the latter typical of *'fulvescens'*. Also note white bases to the outer primaries. Oman, 11.12.2010 (DF)

803. Greater Spotted Eagle *Aquila clanga*, juvenile *'fulvescens'* (same as **802**). Upperparts appear almost black-and-white and could be confused with faded juveniles of Tawny or Spanish Imperial Eagle, but note compact proportions and typically browner face of *'fulvescens'*. Oman, 11.12.2010 (DF)

804. Greater Spotted Eagle *Aquila clanga*, immature *'fulvescens'* moulting from juvenile to second plumage in 2nd cy autumn. Note plain flight feathers, white carpal crescent and dark greater coverts (partly missing due to moult), and compare with other largely pale *Aquila* plumages. Oman, 12.11.2013 (DF)

805. Greater Spotted Eagle *Aquila clanga*, second-plumage immature *'fulvescens'* in 2nd cy autumn. Similar to juvenile, but broader-winged owing to new inner primaries. Note white carpal crescent and dark band on the greater coverts as well as stocky proportions. Oman, 5.11.2004 (DF)

806. Greater Spotted Eagle *Aquila clanga*, adult *'fulvescens'*. With age *'fulvescens'* birds turn more foxy in colour with darker spots and streaking. Note typical uniform flight feathers, white carpal comma and dark greater coverts. Oman, 7.11.2013 (DF)

LESSER SPOTTED EAGLE
Aquila pomarina

Other names: Sometimes placed in the genus *Clanga*, together with the other spotted eagles.

VARIATION Monotypic. The form *hastata*, from the Indian subcontinent and adjacent areas, was formerly regarded as a subspecies of Lesser Spotted Eagle, but has recently been accorded full species rank as Indian Spotted Eagle *Aquila hastata*. Appearing somewhat intermediate in appearance between Greater and Lesser Spotted Eagle, combining features from both and thus looking much like many Greater x Lesser hybrids/intergrades, the possibility of it being a hybrid species should be further investigated.

DISTRIBUTION A typically European species ranging from E Europe to Russia W of the Ural Mts. Lesser Spotted Eagle is a long-distance migrant, following the eastern flyway through the Near East to its wintering grounds in southern Africa.

BEHAVIOUR A rather sluggish eagle. Frequently seen hunting on the wing, standing against the wind with lowered head as it scans the ground for prey. Also perches for long times when hunting, often partly hidden in the lower part of the canopy, or may stalk prey on foot, walking through fields and meadows.

SPECIES IDENTIFICATION By and large very similar to Greater Spotted and Steppe Eagles, and good viewing conditions and great care is always needed to separate these species. Head-on silhouette when gliding is rather similar to Black Kite and compared to Greater Spotted Eagle the wings are more moderately and gently bowed and the wingtips are less strongly upcurved and also kept more closed (showing a narrower 'broom' of splayed fingers). Seen head-on Steppe Eagles have longer wings and a comparatively smaller body in relation to their wingspan, the wing plane looks more angled with a flat arm and drooping hand, and the broom of the splayed fingers is wider. Also the wing-formula can be of help: in Lesser Spotted the seventh finger (p4) is on average shorter and more rounded than in the other species, and some juveniles especially may show only six clearly fingered primaries, in which case it is diagnostic. Juveniles are usually easily identified by their distinctly bicoloured upperparts, with dark remiges and greater coverts contrasting with chocolate-brown median and lesser coverts, with lines of pale spots forming distinct wing-bands. Older immatures and adults are more variable and could easily be confused with either Greater Spotted or Steppe Eagles depending on plumage. Birds with retained spotting can always be told from Steppe Eagles, as they lack spotting altogether, and from Greater Spotted Eagle by details of underwing (barring of remiges and type of carpal comma), and by generally paler, more faded brown plumage and paler iris. More uniformly coloured older immatures and adults are best told by using a suite of different underwing characters, like the type of carpal crescent and underwing barring as well as iris colour. A feature shared by both spotted eagles and separating them from other *Aquila* eagles are the long legs: *in the two spotted eagles the yellow feet reach to the mid-tail in flight, while in Steppe, Tawny and Imperial they only reach the base of the tail.*

MOULT Similar to Greater Spotted Eagle, moults part of its plumage in summer in the breeding areas and part in the winter quarters, with breaks for the migration seasons. Owing to the two annual moulting seasons Lesser Spotted is capable of replacing most of its plumage in the course of one year, although individual differences appear to be notable.

Juveniles begin to moult during their first summer, at the age of one year, and replace between three and six inner primaries before the autumn migration. The extent of the body moult varies individually and while some birds are still mostly in a worn juvenile plumage by the 2nd cy autumn, others have moulted most of the head, breast and mantle. Ageing immatures is best done by combining plumage features with details of moult in the primaries and secondaries.

PLUMAGES Normally four different plumage types can be separated: juvenile, second plumage, third plumage and adult plumage. Because of prolonged moulting seasons, spanning over both breeding and non-breeding seasons, the plumage is more or less in a constant change.

Juvenile

The colour of the body plumage is chocolate-brown in most, but varies from a warm sandy-brown to dark brown, with the darkest birds appearing almost as dark as a juvenile Greater Spotted from a distance. This is the only *Aquila* to show an ochre nape-patch in juvenile plumage, but the size of the patch varies considerably from almost non-existent to covering the whole hindcrown and nape. The nape-patch is surprisingly difficult to see from most angles in flight, but shows well in head-on views on circling birds.

One of the best identification characters is the type of barring of the underwing. The barring is rather dense and coarse, with dark bars being equal in width to the pale spaces in between. Owing to the dense barring, the underwing remiges appear clearly darker than in, for example, juveniles of Steppe or Greater Spotted Eagles, and the white tips to the secondaries and rectrices are always sharply defined against the dark barring (cf. Greater Spotted). Iris is dark greyish-brown from the start, turning paler in some by the following (2nd cy) spring.

Second plumage

Seen from a distance, some sparsely moulted birds are superficially very similar to an adult, with faded and evenly worn upperwing-coverts, and care is needed to avoid making this mistake. More advanced birds may show largely fresh and spotted upperwing-coverts, being thus overall darker and more similar to the third plumage. Underparts are largely worn and faded, sandy-brown, often with freshly moulted darker head and upper breast. If the head has been moulted, then the fresh orange-yellow nape-patch will stand out against the darker neck, whereas a faded and worn head may lack it.

Best aged by moult pattern of the primaries, with freshly moulted darker inner primaries which contrast with faded and shorter retained juvenile outer primaries. Secondaries are still mostly juvenile feathers, although a few may have been replaced, mostly s1 and s5, sometimes ss1–2 and ss5–6. Undertail-coverts in most birds are retained juvenile feathers and therefore uniformly off-white, while more advanced birds may show fresh buffish feathers with some darker markings. The iris is, as a rule, already paler brown than in juveniles, and noticeably pale yellowish-brown in some.

Third plumage

This plumage is characterised by huge individual variation, both when it comes to the extent of moult, as to the amount of spots in the plumage. By and large

rather similar to the second plumage, but as a rule the plumage is largely fresh, and therefore darker, with lots of spotted feathers on the upperwing-coverts. The underparts are variably spotted, often with large pale creamy-buff blotches on the breast and/or underwing-coverts. Some birds can be almost covered in creamy spots, and may therefore look confusingly pale from a distance.

Best told from the previous plumage by showing two simultaneous moult-fronts in the primaries, with fresh outermost and innermost primaries, but with median primaries appearing worn and faded. The trailing edge of the arm is typically uneven and ragged, caused by secondaries of different age and length. The new secondaries still show the diagnostic Lesser Spotted barring, with the most recently replaced starting to develop a broad, dark subterminal band, similar to Steppe Eagle although less distinct. Undertail-coverts are pale with darker mottling, but the amount of mottling varies individually. Iris is already pale.

During the successive 1–2 moults the plumage gradually loses the spotting and the undertail-coverts turn browner. At the same time the remiges will turn darker as the new feathers lack barring.

The definitive adult plumage is probably not attained until the fifth or sixth plumage.

Adult

A rather nondescript brown eagle, which varies in darkness depending on the stage of body moult: largely worn birds are faded and more sandy-brown, while freshly moulted birds are darker. As a rule the adult Lesser Spotted is lighter brown than adult Greater Spotted, with a more marked contrast to the upperwing between lighter forewing and darker flight feathers, but some individuals may be difficult, and identification has become further complicated by the extensive hybridisation between the species.

Very similar to adult Greater Spotted also in structure, although Lesser has a marginally finer bill and more rounded, rather than square-cut wingtips, but the reliability of both these characters is obscured by differences between males and females in both species.

The most reliable differences between adult Greater and Lesser Spotted Eagles are the yellowish iris colour of adult Lesser, as well as its sharper upperwing contrast and on average shorter and more rounded seventh finger (p4), although this latter character shows a lot of variation. Adult Lesser also shows a more distinct pale patch on the upperwing, at the base of the inner primaries, compared to Greater, in which the patch is usually formed by pale feather shafts only and therefore is much more diffuse. On the underwing the pale carpal crescent is of importance when separating the adults: adult Lesser Spotted normally shows two rather long crescents, one at the base of the outer primaries and another at the base of the greater coverts. Adult Greater Spotted normally shows only one shorter, half-moon shaped white crescent at the base of the outermost primaries. Because of the extensive hybridisation between the species this formerly useful character has lost some of its reliability, and today various intermediate types of carpal crescents can be seen.

SEXING Lesser Spotted Eagles are as a rule not possible to sex in the field, although males are smaller with a proportionately larger head, finer bill and narrower wings.

CONFUSION RISKS Easy to confuse with other largely brown *Aquila* eagles, particularly Greater Spotted and Steppe, with risk of confusion varying and depending on the age and plumage of the bird. Although slight differences in structure and silhouette may aid the experienced, these eagles should always be identified using a suite of the more reliable characters, such as type of underwing barring, type of carpal crescent and iris colour.

NOTE Hybridises regularly with Greater Spotted Eagle, which is also the main reason for the latter's critical situation in Europe (see under Greater Spotted Eagle). This large-scale hybridisation has made the identification of older, non-juvenile spotted eagles an almost impossible task.

References

Lontkowski, J. & Maciorowski, G. 2010. Identification of juvenile Greater Spotted Eagle, Lesser Spotted Eagle and hybrids. *Dutch Birding* 32: 384–397.

Väli, U. & Lõhmus, A. 2004. Nestling characteristics and identification of the Lesser Spotted Eagle *Aquila pomarina*, Greater Spotted Eagle *A. clanga* and their hybrids. *J. Ornithol.* 145: 256–263.

Väli, U., Dombrovski, V., Treinys, R., Bergmanis, U., Daróczi S.J., Dravecky, M., Ivanovsky, V., Lontkowski, J., Maciorowski, G., Meyburg, B.-U., Mizera, T., Zeitz, R. & Ellegren, H. 2010. Widespread hybridisation between the Greater Spotted Eagle *Aquila clanga* and the Lesser Spotted Eagle *Aquila pomarina* (Aves: Accipitriformes) in Europe. *Biol. J. Linnean Soc.* 100: 725–736.

807. Lesser Spotted Eagle *Aquila pomarina*, fresh juvenile, with juvenile Steppe Buzzard above. A typical juvenile with uniform plumage, lighter brown underwing-coverts than body, and rather dark and densely barred remiges. Note very short and rounded seventh finger, counting in from the outermost primary, typical of Lesser Spotted Eagle. Egypt, 9.10.2010 (DF)

808. Lesser Spotted Eagle *Aquila pomarina*, fresh juvenile. The pale markings of the greater underwing-coverts sometimes form a conspicuous underwing band. Egypt, 10.10.2010 (DF)

809. Lesser Spotted Eagle *Aquila pomarina*, fresh juvenile. The brown plumage and the dark and densely barred remiges differ clearly from any juvenile Greater Spotted Eagle. Egypt, 11.10.2010 (DF)

810. Lesser Spotted Eagle *Aquila pomarina*, fresh juvenile. Many juveniles are variably streaked yellow on the upper breast. Egypt, 11.10.2010 (DF)

811. Lesser Spotted Eagle *Aquila pomarina*, fresh juvenile. Note how the brown upperwing-coverts contrast sharply against dark greater coverts and remiges, which is typical for fresh juveniles. The number of wing-bands varies, but this is a rather average individual. The diagnostic ochre nape patch can just be seen against the bird's left shoulder. Egypt, 10.10.2010 (DF)

812. Lesser Spotted Eagle *Aquila pomarina*, second plumage, 2ⁿᵈ cy autumn. The second plumage is a mix of retained juvenile and moulted feathers. The inner five primaries and some secondaries are new and dark, as is most of the head, neck and underbody except for the underwing-coverts. The eyes have already turned lighter compared to juveniles. Egypt, 9.10.2010 (DF)

813. Lesser Spotted Eagle *Aquila pomarina*, second plumage, 2ⁿᵈ cy autumn. This bird has moulted its head and neck and half of its primaries and some secondaries, but the overall impression is that of a worn juvenile. Note short and rounded seventh finger. Egypt, 9.10.2010 (DF)

814. Lesser Spotted Eagle *Aquila pomarina*, second plumage, 2ⁿᵈ cy autumn. Another rather average bird with newly moulted head and breast, while most of the body plumage is still comprised of worn juvenile feathers. Note typical wing-barring and compare with other *Aquila* species. Egypt, 10.10.2010 (DF)

815. Lesser Spotted Eagle *Aquila pomarina*, second plumage, 2nd cy autumn. Egypt, 11.10.2010 (DF)

816. Lesser Spotted Eagle *Aquila pomarina*, second plumage, 2nd cy autumn. Upperparts vary a lot between birds, depending on extent of moult. This bird wears a worn and faded juvenile plumage, and therefore looks very much like an adult, showing the typical upperwing contrast of older Lesser Spotted Eagles. Egypt, 10.10.2010 (DF)

817. Lesser Spotted Eagle *Aquila pomarina*, second plumage, 2nd cy autumn. This immature has moulted a fresh head, neck and mantle/scapulars as well as some upperwing-coverts with bright pale spot-marks. It also shows the pale nape-patch and the primary moult typical of its age-class. Egypt, 9.10.2010 (DF)

414

818. Lesser Spotted Eagle *Aquila pomarina*, third plumage, 3rd cy autumn. This bird still retains its juvenile p10, while all other remiges have been moulted, some, like pp1–3, already for the second time. Plumage generally much more variegated from ongoing patchy body moult, usually with lots of spotted feathers appearing both above and below. Egypt, 11.10.2010 (DF)

819. Lesser Spotted Eagle *Aquila pomarina*, third plumage, 3rd cy autumn. Another typically heavily spotted third-plumage bird, with retained juvenile pp9–10 and ss8–9. Israel, 5.10.2009 (DF)

820. Lesser Spotted Eagle *Aquila pomarina*, presumed third plumage, 3rd cy autumn. Although this bird has replaced all its juvenile remiges, the typically spotted and variegated plumage indicates a third-plumage bird. Note rather light iris. Israel, 6.10.2009 (DF)

821. Lesser Spotted Eagle *Aquila pomarina*, third plumage, 3rd cy summer. Immatures, particularly third-plumage birds, with variable light feather tracts are seen regularly, but birds as light as this must be considered exceptional. Estonia, 17.8.2012 (Ülo Väli)

822. Lesser Spotted Eagle *Aquila pomarina*, adult plumage. Adults are rather uniform and nondescript, but the retained typical barring of the remiges indicates a younger age, not yet a full adult. Israel, 5.10.2009 (DF)

823. Lesser Spotted Eagle *Aquila pomarina*, final adult plumage, presumed male. Adults are uniformly light brown, with darkish greater underwing-coverts and remiges. The almost uniformly dark remiges and the bright yellow iris are typical of fully adult birds. Israel, 8.10.2009 (DF)

824. Lesser Spotted Eagle *Aquila pomarina*, adult plumage, presumed female. The light-spotted greater underwing-coverts and the barred remiges indicate a younger adult. Israel, 6.10.2009 (DF)

825. Lesser Spotted Eagle *Aquila pomarina*, final adult plumage, breeding female. Most adults are sandy-brown with a distinctly darker band on the greater underwing-coverts. Note short and rounded seventh finger, counting inwards. Estonia, 4.5.2015 (DF)

826. Lesser Spotted Eagle *Aquila pomarina*, final adult plumage (same as **825**), Typical upperparts of a non-juvenile Lesser Spotted Eagle, with lighter brown median upperwing-coverts contrasting with darker remiges and mantle/scapulars. Estonia, 4.5.2015 (DF)

Identification of hybrid spotted eagles

The two species of spotted eagle frequently hybridise where they co-exist, and this problem has been studied in depth in the Baltic States, particularly in Estonia and in Poland (see References above under Greater and Lesser Spotted Eagles).

The hybridisation has probably been going on for a longer time, but has increased recently due to changes in the environment, when logging of swampy old-growth forests, the prime habitat of Greater Spotted Eagle, have forced the birds to move into agricultural land, which has been the traditional habitat of Lesser Spotted Eagle.

The hybrid pairs, mostly a male Lesser and a female Greater Spotted Eagle, are capable of reproducing successfully. Their hybrid chicks are fertile and are themselves capable of successfully raising young. Over a few generations this leads, at least in theory, to a huge variation in hybrid offspring, with not just 50–50 hybrids, but all possible back-cross intergrades ranging from nearly pure Lessers to nearly pure Greaters.

Even if the 50–50 hybrid juveniles are often easy to identify as hybrids, showing a mix of characters from both species, including some intermediate features, many of the features are lost as the birds grow older and the plumage becomes more uniform. The second generation back-cross offspring (intergrades) would already be much harder to identify, as they would most likely feature 75% Lesser and 25% Greater genes (providing that the 50–50 hybrid has paired with a pure Lesser Spotted Eagle, which statistically is the most likely option). Or, the genetic composition could be something totally different, if the 50–50 hybrid is paired to a Greater Spotted Eagle, or perhaps to another hybrid with an unknown mix of Greater and Lesser genes. This extensive variation of genetic possibilities makes the identification of spotted eagles very difficult today, and in many cases impossible. In most cases intergrades will pass as one or the other species, without anyone knowing that they are in fact carrying genetic material from two species. Only in clear cases can hybrid juveniles and perhaps second/third-plumage birds be identified, and with luck also some intergrades, but the majority of birds will pass undetected.

Again, in order to be able to spot a hybrid, a good knowledge of both species is desirable and preferably good photographic documentation is required. Features to focus on are the barring of the flight feathers and the overall appearance of the plumage, the latter best judged on fresh autumn juveniles. Many hybrids look like Greater Spotted but they are browner, not blackish, they have an ochre nape patch and the underwing barring is heavier, more similar to that of Lesser Spotted. Conversely, others (later intergrades?) may look like juvenile Lesser Spotted, but the remiges are lighter grey with finer Greater-type barring. For a more in-depth review of the hybrid problem, Lontkowski & Maciorowski (2010) can be recommended.

In the subsequent immature plumages the underwing barring is still important, but now the colour of the iris also gives an indication of ancestry: a Greater Spotted-like bird with a pale iris and an ochre nape patch is most probably of hybrid origin. As the birds get older the remiges become plain and this character is lost, but instead the colour of the iris becomes more evident.

Apart from plumage the structure between the two species also differs, but assessing the subtle differences is even more difficult. Greater Spotted has a sleek, more eagle-like head, a heavier bill and more rectangular wings, while Lesser Spotted has a big head but a relatively small bill and the wings are more rounded at the tip.

827. Presumed hybrid Greater x Lesser Spotted Eagle *Aquila clanga x pomarina*, autumn juvenile. This juvenile shows brown underwing-coverts and the tail pattern of Lesser, while the barring of the remiges bears resemblance to Greater. The wing-formula, particularly the length and shape of p4, is intermediate. Oman, 7.11.2004 (DF)

828. Presumed hybrid/intergrade Greater x Lesser Spotted Eagle *Aquila clanga x pomarina*, 2ⁿᵈ cy autumn. This bird was widely twitched as a Greater and only later found to be a presumed hybrid. Note short and rounded p4, while body plumage is too dark for a normal 2ⁿᵈ cy Lesser. Barring of remiges intermediate and wingtip rounded for a Greater. Iris was already light in this immature. Finland, 19.9.2007 (DF)

829. Presumed hybrid/intergrade Greater x Lesser Spotted Eagle *Aquila clanga x pomarina*, 2ⁿᵈ cy autumn. This migrant was identified as a Greater as it travelled along the Finnish south coast. Note overall dark plumage, indicating Greater, while short and rounded p4 and rounded wingtip suggest Lesser. Barring of underwing intermediate, typical of hybrids. Finland, 1.10.2011 (DF)

830. Confirmed hybrid Lesser x Greater Spotted Eagle *Aquila pomarina x clanga*, 3rd cy spring. This bird was identified by its rings as the offspring of a male Lesser Spotted and female Greater Spotted Eagle, hatched in Estonia in 2012. Aged also from photographs as a 2nd-summer by its retained outer two juvenile primaries and juvenile s9 on the left wing. Note light brown body plumage and light iris and compare underwing barring and wing-formula with corresponding plumages of Lesser and Greater Spotted Eagles. Finland, 6.6.2014 (Jaakko Esama)

831. Presumed hybrid/intergrade Greater x Lesser Spotted Eagle *Aquila clanga x pomarina*, presumed 3rd cy autumn. Most features point towards a Lesser Spotted, but note heavier bill and more eagle-like head suggestive of Greater Spotted and typically intermediate underwing barring. Israel, 5.10.2009 (DF)

VERREAUX'S EAGLE
Aquila verreauxii

Other names: Black Eagle (used in South Africa)

VARIATION Monotypic.

DISTRIBUTION An African species with a wide sub-Saharan distribution. Isolated populations in S Arabia, SE Egypt and Sinai, in the Tibesti Mts. of N Chad and possibly also still in Jordan. A rare and rarely seen bird in our region.

BEHAVIOUR A bird of ragged mountains and steep cliffs, where it feeds on medium-sized mammals, specialising in hyraxes. Mostly seen in flight, either when soaring at great height, or gliding along steep escarpments.

SPECIES IDENTIFICATION Adults are unmistakable thanks to the diagnostic shape and plumage. The wings have a peculiar shape, being very broad at the wrist, with long outer secondaries and strongly bulging trailing edge, but tapering markedly towards the tip and the base, while the tail is fairly long for a large eagle. The more rarely seen juveniles and immatures are slightly less extreme in proportions, but still readily recognisable. Juveniles can also be easily identified by their distinct plumage, which is unique among large raptors.

Verreaux's Eagle soars on wings lifted in a rather deep V, with the long fingers steeply curved upwards, producing a most diagnostic silhouette.

MOULT Moult of the flight feathers is protracted, as in other large raptors, and replacing the juvenile plumage takes more than one year to complete.

PLUMAGES The adult and juvenile plumages are distinct, supplemented by another 2–3 transitional immature stages, during which the bird gradually moults into definitive adult plumage.

Juvenile

A most distinctive plumage, which, together with the diagnostic silhouette and the habit of soaring on raised wings in a deep V, makes the bird rather unmistakable. The underparts are defined by dark throat and breast contrasting with a pale rear body, while the underwings show a diagnostic broad pale primary window, larger than in any other large eagle.

The upperside is characterised by a pale rump and lower back, and the large primary window is also well visible from above.

Immature

Identified by retained juvenile feathers in the wings and tail, which differ from the newly replaced dark feathers of adult-type. The proportion of new black flight feathers gives an indication of the bird's age, following the principles familiar from other large eagles.

Adult

The distinctive black-and-white plumage makes the adult unmistakable. Plumage all black, except for the extensive white primary windows both above and below, and a clean white rump with attached white 'suspenders' framing the mantle. The yellow feet stand out against the black plumage, as does the yellow base of the bill and the swollen bare yellow skin around the eye.

SEXING Females are bigger than males, but the size difference is rarely of use in the field, except when the pair is seen together.

CONFUSION RISKS The only species with a remotely similar silhouette is the Golden Eagle, which is sympatric in some of the areas listed above for Verreaux's Eagle. Golden Eagles show a slimmer silhouette, with narrower wings and a proportionately longer tail, and the plumage is different in all age-classes.

832. Verreaux's Eagle *Aquila verreauxii*, fresh juvenile. Diagnostic silhouette with strongly bulging mid-wing tapering both ways. The underwing shows darker secondaries and a lighter hand with fine barring ending well short of the feather tips, while the underwing-coverts are dark, speckled with distinct pale tips. South Africa, 28.10.2007 (Johann Knobel)

833. Verreaux's Eagle *Aquila verreauxii*, fresh juvenile (same as **832**). The entire plumage is adorned with light feather tips and edges which will soon wear off, making the plumage look much darker and more uniform. South Africa, 28.10.2007 (Johann Knobel)

834. Verreaux's Eagle *Aquila verreauxii*, adult, breeding. Unmistakable black eagle with large primary-flashes and bright yellow, swollen skin around the eyes and base of the bill. Ethiopia, 21.2.2006 (DF)

835. Verreaux's Eagle *Aquila verreauxii*, breeding adult (same as **834**). Both shape and plumage are unique, and make the adult birds unmistakable. Ethiopia, 21.2.2006 (DF)

836. Verreaux's Eagle *Aquila verreauxii*, adult. The wing-shape of the adults is even more extreme than in juveniles. The white markings on the upperside are visible over long distances making it easy to identify, even from afar. Ethiopia, 3.1.2010 (DF)

837. Verreaux's Eagle *Aquila verreauxii*, adult pair. Females (left) are bigger and broader-winged than males, but the plumage looks the same. Ethiopia, 1.3.2014 (DF)

BONELLI'S EAGLE
Aquila fasciata

Other names: Until recently *Hieraaetus fasciatus*

VARIATION Nominate *fasciata* in the region; isolated ssp. *renschi* of the Lesser Sundas in Indonesia may prove to be specifically distinct.

DISTRIBUTION In Europe a typically Mediterranean species, but range extends to the mountains of N Africa, the Middle East and India. In Djibouti and Somaliland meets with closely related African Hawk Eagle *Aquila/Hieraaetus spilogaster*. Photographic evidence from this area (Nik Borrow and Callan Cohen *in litt.*) suggests that the two apparently hybridise, as shown by birds with intermediate features (personal interpretation).

BEHAVIOUR Outside the breeding season adult pairs are often seen hunting together, with the more agile and smaller male following close behind the larger female. A bold and fierce raptor capable of tremendous stoops, like Booted Eagle or even Peregrine Falcon, when hunting Rock Doves in flight. Its appearance also causes panic among smaller raptors, and harriers, kites and kestrels always rise high in the air to escape attacks. Sometimes perches for considerable time on rocks and other suitable vantage points.

SPECIES IDENTIFICATION Often identified by diagnostic silhouette alone, with rather broad mid-wing tapering towards the body and wingtip, not that dissimilar to adult honey-buzzards or a massive Northern Goshawk, but with broader wings with six fingers at the tip. Tail is relatively long for an eagle, adding to the similarity with honey-buzzards. Adults are unmistakable due to a combination of unique plumage and silhouette, but juveniles and second-plumage birds are more similar to other raptor species and have been mistaken for both Long-legged Buzzard and the rufous form of Booted Eagle, although silhouette and plumage details differ.

MOULT Complete moult during the breeding season, but does not moult all the flight feathers in one season. Second-plumage birds can be identified by retained juvenile remiges.

PLUMAGES Usually three plumage-types can be recognised: juvenile, second plumage and adult. Even after attaining the first-adult plumage the appearance still changes for some years before the final adult plumage is acquired.

The type of barring on the remiges and rectrices is important for ageing, as is coloration of the underbody, underwing-coverts and head.

Juvenile

Quite distinctive and the plumage variation is rather limited. The underparts vary from rich tawny to a lighter buff, depending on degree of fading, with the plumage becoming paler as the season progresses. The remiges are pale greyish below with fine dark barring, the wingtips being darkest while the inner hand is paler than the secondaries. Underwing-coverts are coloured as the body, with dark greater coverts creating an individually variable wing-band. Upperparts appear rather uniform, greyish-brown, with a slight contrast between the darker remiges and browner coverts, a paler window in the inner primaries and a greyish, finely barred tail. A small white patch on the upper back is visible on most birds.

Second plumage

In many respects intermediate between juvenile and adult plumages. The body plumage is superficially similar to the adult, but the markings on the breast are less distinct, often on a buffish or even light brownish, not a clean white, background. Most birds show a broad dark band to the midwing, along the rear edge of the underwing-coverts, broader and more obvious than in any other plumage, being highlighted by the paler forewing and the greyish remiges, while the brown thighs, as in the adult, make the rear flanks look dark. Best aged by a mixture of retained juvenile and new adult type remiges, the latter being finely barred but showing a broad black tip, forming a distinct buzzard-like dark trailing edge to the underwing. After moult is completed in late autumn some outer juvenile primaries and often also some median juvenile secondaries, lacking the black tips, are retained, this being the single most reliable ageing character.

Adult

Clean white below on the underbody with brown thighs, which appear as a dark V on the rear underbody. The underwing looks generally dark with a variably whitish forewing in some, while the median and greater underwing-coverts are black. The remiges are black distally, with greyer and barred or marbled bases, most clearly marked on the long primaries. The upperparts appear rather uniformly dark greyish-brown, with the upperwing-coverts slightly paler and browner than the darker remiges, and with a variable white patch on the mantle, while the tail is a lighter grey with a broad black subterminal band and some finer barring further in.

Young adults (presumed third plumage) have grey and finely barred underwing remiges with a broad subterminal band, while old birds show almost uniformly black remiges with only some paler marbling visible proximally; the older the bird the darker the underwing flight feathers and the less barring remains. The head also changes, from being mostly brown with a whiter throat in younger adults, to nearly white with a contrasting dark crown and ear-covert patch in older adults. With increasing age the amount and size of the spots on the underbody also decreases, and truly old birds can appear almost immaculate white below.

SEXING Males are considerably slimmer and also smaller than females, which is easy to appreciate when the two are seen side by side. However, size cannot be safely used on single birds, but males appear comparatively larger-headed, smaller-billed, shorter- and more rounded-winged and longer-tailed than females. Recent studies show that adult males have on average whiter trousers and undertail-coverts compared to adult females, which are darker with more brown showing in respective feather tracts (Garcia, V. *et al.* 2013).

CONFUSION RISKS Adults are usually easy to identify when seen well, but juveniles and immatures are sometimes mistaken for Long-legged Buzzard or Booted Eagle, although silhouette and details of underwing are diagnostic.

References

García, V., Moreno-Opo, R. & Tintó, A. 2013. Sex differentiation of Bonelli's Eagle *Aquila fasciata* in Western Europe using morphometrics and plumage colour patterns. *Ardeola* 60: 261–277.

838. Bonelli's Eagle *Aquila fasciata*, juvenile. Note typical shape with broad mid-wing tapering towards the body and a longish tail. Underwings appear fairly light with fine dark barring and slightly darker fingers. Spain, 17.9.2012 (DF)

839. Bonelli's Eagle *Aquila fasciata*, juvenile (same as **838**). Spain, 17.9.2012 (DF)

840. Bonelli's Eagle *Aquila fasciata*, juvenile. The darkness and length of the dark mid-wing band varies individually and is practically missing in many. Note light underwings with darker fingers only. Spain, 25.10.2009 (DF)

841. Bonelli's Eagle *Aquila fasciata*, juvenile. The upperparts appear rather uniformly light greyish-brown, with a slight contrast between lighter coverts and darker secondaries. Note finely barred flight feathers and tail, and a small white speck on the middle of the back. Oman, 9.12.2010 (DF)

427

842. Bonelli's Eagle *Aquila fasciata*, juvenile female, 2nd cy spring. This juvenile has just started its first moult; it has lost its central tail feathers and is growing its inner primaries. Israel, 22.3.2012 (DF)

843. Bonelli's Eagle *Aquila fasciata*, second plumage. Immatures moulting for the first time are variable, but are easily identified by retained juvenile flight feathers lacking the dark tips. In this bird two outermost primaries and about half of its visible secondaries are retained juvenile feathers. Oman, 28.11.2014 (DF)

844. Bonelli's Eagle *Aquila fasciata*, second plumage. A slightly more advanced immature male, with two retained juvenile secondaries and one primary in each wing. Note light sandy overall impression with variable dark streaking. Oman, 11.12.2012 (DF)

845. Bonelli's Eagle *Aquila fasciata*, second plumage. In second plumage the difference between the lighter primaries and darker secondaries becomes more obvious, as does the dark trailing edge of the wings and tail. Oman, 16.11.2014 (DF)

846. Bonelli's Eagle *Aquila fasciata*, second plumage. This bird has replaced its entire plumage, but can still be recognised by its immature, sandy-coloured body and rather light underwings with a distinct dark trailing edge and distinctly barred secondaries. Oman, 10.12.2012 (DF)

847. Bonelli's Eagle *Aquila fasciata*, second plumage female. This bird, paired to an adult male, is still showing some retained juvenile remiges, although the body plumage already looks adult, with a conspicuous white mantle-patch. Israel, 25.3.2010 (DF)

848. Bonelli's Eagle *Aquila fasciata*, adult. Clean white body with variable black streaking identifies an adult, but the rather clearly barred secondaries and the dirty-looking forewings indicate a younger bird, not yet a full adult. India, 12.12.2007 (DF)

849. Bonelli's Eagle *Aquila fasciata*, adult male, breeding. The almost uniformly dark secondaries and the pure white colour of the forewing indicate an older individual in fully adult plumage. Oman, 7.2.2014 (DF)

850. Bonelli's Eagle *Aquila fasciata*, adult female, breeding. The underbody streaking gets finer with age and can almost disappear in some, particularly in old males. Oman, 9.12.2012 (DF)

851. Bonelli's Eagle *Aquila fasciata*, adult. The upperparts are fairly uniform, with a slight contrast between coverts and remiges, while the white mantle-patch and the grey tail with a broad subterminal band are more diagnostic. India, 12.12.2007 (DF)

852. Bonelli's Eagle *Aquila fasciata*, adult male, breeding (same as **849**). Uniformly greyish-brown upperparts contrast with diagnostic white-speckled mantle and grey, black-banded tail. Oman, 7.2.2014 (DF)

853. Bonelli's Eagle *Aquila fasciata*, adult female, breeding (same as **850**). Oman, 9.12.2012 (DF)

431

BOOTED EAGLE
Hieraaetus pennatus

Other names: Sometimes named as *Aquila pennata*

VARIATION Monotypic.

DISTRIBUTION Widespread and fairly common in France and in the Iberian Peninsula, with the range extending through E Europe to C Asia and S Siberia. Winters mostly in S Asia and sub-Saharan Africa, mainly in the Sahel, but also further south on the continent; recently also wintering in increasing numbers in S Spain.

BEHAVIOUR Primarily an aerial hunter, patrolling high over the ground looking for prey underneath. Often seen hanging motionless against the wind with drooping neck and lowered legs. Known for its breathtaking Peregrine-like stoops, sometimes diving straight through the canopy.

SPECIES IDENTIFICATION Small, buzzard-sized eagle, with broad wingtip (six fingers), relatively short wings and a fairly long tail. Pale morph birds rather straightforward to identify, but dark birds often confused with Black Kite or Marsh Harrier (see p. 232). Upperparts diagnostic, with pale diagonal area across upperwing-coverts, whitish uppertail-coverts, pale scapular ovals and white 'headlights' at the base of the forewing; only certain juvenile honey-buzzards can show a similar set of characters, thus looking similar from a distance, but proportions differ. The remiges of the underwing appear dark, except for a lighter window to the inner primaries, and the sparse and heavy barring of the secondaries is often difficult to discern. Head colour diagnostic, with crown and nape always lighter and face darker, not unlike Golden Eagle. Feathered legs leave only yellow toes visible. In active flapping flight wing-beats are rather fast, and the wing also retains its diagnostic arched shape with strongly up-curved primary tips when soaring (cf. Marsh Harrier and Black Kite).

MOULT Annual moult starts on the breeding grounds, and after a pause for the autumn migration, is resumed again in the winter quarters. As a rule all flight feathers are replaced annually, but some secondaries may be left unmoulted. Immatures can be aged by retained juvenile flight feathers until the 2nd cy autumn.

PLUMAGES Occurs in two distinct colour morphs, a pale and a dark morph, with the proportion of light morph birds decreasing from W Europe eastwards. The light brown or tawny variants of the dark morph have sometimes been separated as an intermediate or rufous morph. Juveniles and older birds can easily be told from each other with close views, but more precise ageing of non-juveniles usually requires photographic documentation, as details of moult allow further separation between second-plumage birds and adults.

Juvenile

In autumn shows a uniform and pristine plumage, *with an even white trailing edge to the wings and tail-tip, and no signs of wing-moult.* The wings are narrower than in adults, with a more clearly S-curved trailing edge. The iris colour is dark brown.

Pale morph juveniles are whitish below, with contrasting dark wing feathers. The underbody, sometimes including the underwing-coverts, is often stained with a rich tawny colour, unlike light morph adults, and as a rule, juveniles are less streaked on the underbody than adults. The head is often rich rufous-brown in juveniles. The diagnostic markings of the upperparts are more contrasting in pale birds than in dark ones.

Dark morph juveniles vary from almost blackish-brown to a lighter tawny colour on the underbody and underwing-coverts, while the flight feathers and tail are similar to pale juveniles. The upperparts markings are less contrasting than in pale morph juveniles.

By spring the plumage has changed only through wear, and the pale tips and edges of the upperwing-coverts especially are mostly worn off. The white secondary tips are mostly lost, but the uniform trailing edge is still quite obvious. Some dark birds fade considerably, to become more tawny-brown in colour, with a darker band through the median and greater underwing-coverts.

Transitional plumage (2nd cy summer–autumn)

Immatures can still be aged up to the time of the autumn migration by the group of 3–5 retained juvenile outer primaries, which are abraded, faded brown, shorter and very pointed compared to the fresh inner fingers, and with broad white tips to the inner three primaries. Most of the secondaries are still juvenile, retaining the S-shaped trailing edge to the arm, but lacking the white tips of a fresh juvenile, while any newly moulted secondaries (normally s1 and s5) stand out as being longer with a broad white tip. Once the last juvenile wing feathers have been replaced, usually during the second winter, birds can no longer be separated from adults. Some individuals still carry a few juvenile secondaries, mostly s4 and ss8–9, when returning from Africa in the 3rd cy spring.

Adult

Appears rather similar to juvenile, but easily told by signs of moult, always showing worn and fresh secondaries and primaries side by side and the *pale trailing edge of the wing and the pale tip of the tail is irregular* in adults, due to feathers of different age and wear. Adults also show more clearly barred tail feathers (although the outermost pair is mostly plain), and in the light morph the adults are more clearly streaked or barred on the body compared to juveniles. The iris is lighter and more brightly coloured compared to juveniles, orange or chestnut.

SEXING Females are on average heavier than males, but the size difference is mostly impossible to assess in the field, even when both birds of a pair are seen together.

CONFUSION RISKS The pale morph is usually straightforward to identify by diagnostic plumage, both above and below. Dark morph birds are often confused with Black Kite and Marsh Harrier in particular, which from a distance may appear similar in plumage, but not in silhouette. Rufous morph birds with a dark underwing band may appear similar in plumage to a young Bonelli's Eagle, but differ in size, different silhouette and by darker flight feathers. The upperparts markings are diagnostic in all colour morphs, and exclude any other raptor species, except some odd dark juvenile Honey-buzzards, which may show paler uppertail and upperwing-coverts, but have a different flight silhouette.

854. Booted Eagle *Hieraaetus pennatus*, juvenile light morph. Distinct contrast between whitish body plumage and dark remiges with lighter inner primaries. Uniform white trailing edge is diagnostic of juvenile. Spain, 10.9.2012 (DF)

855. Booted Eagle *Hieraaetus pennatus*, juvenile light morph. Juveniles tend to have a strongly rufous head and a plain or sparsely marked body. Spain, 14.9.2011 (DF)

856. Booted Eagle *Hieraaetus pennatus*, juvenile light morph. Others are more marked on the underwing-coverts and body. Spain, 16.9.2011 (DF)

857. Booted Eagle *Hieraaetus pennatus*, juvenile light morph. Note uniform trailing edge of juvenile. Spain, 10.9.2012 (DF)

858. Booted Eagle *Hieraaetus pennatus*, juvenile dark morph. The darkness of the plumage varies individually, but juveniles are always aged by their uniform plumage. Note white 'headlights' at the base of the wings. Egypt, 9.10.2010 (DF)

859. Booted Eagle *Hieraaetus pennatus*, juvenile dark morph. A typical juvenile with uniform translucent trailing edge and tail-tip. Egypt, 10.10.2010 (DF)

860. Booted Eagle *Hieraaetus pennatus*, juvenile dark morph. The lightest birds of the dark morph have also been called the intermediate or rufous morph. They could be mistaken for juvenile Bonelli's Eagle, but note the more heavily barred flight feathers and darker secondaries of Booted. Spain, 16.9.2011 (DF)

861. Booted Eagle *Hieraaetus pennatus*, juvenile light morph. The upperside is diagnostic with large light patches, one diagonally across the upperwing-coverts and another in the scapulars; note also pale uppertail-coverts and white 'headlights'. Aged by uniform plumage. Spain, 10.9.2012 (DF)

862. Booted Eagle *Hieraaetus pennatus*, juvenile light morph. A fresh autumn juvenile migrant. Cameroon, 17.10.2010 (Ralph Buij)

863. Booted Eagle *Hieraaetus pennatus*, juvenile dark morph (same as **858**). Dark birds show the same markings above as light birds, but the pale areas are slightly less contrasting. Egypt, 9.10.2010 (DF)

864. Booted Eagle *Hieraaetus pennatus*, juvenile dark morph, 2nd cy spring. By spring the light feather tips of the wings and tail are largely lost due to wear, but the trailing edge and tail-tip are still even and uniform. Note fresh innermost primary in both wings. Israel, 18.4.2009 (DF)

865. Booted Eagle *Hieraaetus pennatus*, immature dark morph, 2ⁿᵈ cy autumn. By their second autumn, immatures have replaced about half of their juvenile primaries but only some juvenile secondaries. Note the four faded and worn outer juvenile primaries and the brownish juvenile secondaries, which are clearly narrower than the few new ones. Juvenile tail feathers are on average plainer, contrasting with the newly moulted and strongly barred feathers. Oman, 11.12.2010 (DF)

866. Booted Eagle *Hieraaetus pennatus*, adult light morph. Some 3ʳᵈ cy birds in spring can still be aged if they have retained any juvenile remiges. This bird shows at least one short and narrow juvenile secondary (s8) in its right wing, with both p10 and some tail feathers also being juvenile feathers. Spain, 4.5.2011 (DF)

867. Booted Eagle *Hieraaetus pennatus*, adult light morph. This 3ʳᵈ cy spring bird can be aged by its retained juvenile p10, s4 and ss8–9, the latter being clearly shorter and narrower than their moulted neighbours. Note diagnostic upperparts. Spain, 4.5.2011 (DF)

868. Booted Eagle *Hieraaetus pennatus*, adult light morph. Adults do not differ much from juveniles, but are on average more marked below, the light trailing edge is broken up and the iris is paler, more amber-coloured. Spain, 15.9.2012 (DF)

869. Booted Eagle *Hieraaetus pennatus*, adult light morph. Some adults can be quite strongly marked below on the body and underwing-coverts. Spain, 4.9.2013 (DF)

870. Booted Eagle *Hieraaetus pennatus*, adult dark morph. Differs from dark morph juveniles by signs of moult in the wings and tail, a more distinctly barred tail, and brighter iris colour. Oman, 7.12.2012 (DF)

871. Booted Eagle *Hieraaetus pennatus*, adult dark morph. A lighter-coloured bird of the dark morph, potentially confusable with an immature/subadult Bonelli's Eagle, but note different underwing markings with heavier barring and the lack of Bonelli's' prominent dark trailing edge. Israel, 25.3.2013 (DF)

872. Booted Eagle *Hieraaetus pennatus*, adult light morph. Note diagnostic brightly marked upperparts, but condition of the plumage varies individually depending on the state of moult. This bird shows two simultaneous moult-waves in its primaries. Spain, 4.9.2013 (DF)

873. Booted Eagle *Hieraaetus pennatus*, adult light morph. Compared to **872** this bird's plumage is in far better condition. Note the amber iris. Spain, 15.9.2012 (DF)

441

WAHLBERG'S EAGLE
Hieraaetus wahlbergi

Other names: Sometimes named as *Aquila wahlbergi*

VARIATION Monotypic

DISTRIBUTION Extensive range covers most of sub-Saharan Africa. Prefers semi-open savanna and woodland avoiding dense forest. Some tropical populations may be resident, but South African breeding birds migrate to Nigeria, Cameroon, Chad and Sudan for the austral winter. Recorded once in the Western Palearctic, a juvenile in rather fresh plumage which was seen and photographed near Ras Shuqeir, north of Hurghada, Egypt, on 3 May 2013.

BEHAVIOUR Hunting behaviour very similar to Booted Eagle, often seen hanging against the wind scanning the ground below.

SPECIES IDENTIFICATION A rather uniform and nondescript eagle with size and proportions most similar to Booted Eagle, the most likely pitfall in a European context. Silhouette differs from Booted Eagle by comparatively longer and narrower wings and a longer tail, often also a longer neck; the silhouette is even more extreme in adults, while juveniles are more similar to Booted in proportions. Occurs in a common dark morph and a much rarer pale morph (less than 10%?). Juveniles also occur in a sandy variant, but these birds turn dark during the first moult. Dark morph birds are overall dull brown lacking all the diagnostic upperparts markings of Booted, including whitish uppertail-coverts. The face and throat are often somewhat darker, making the yellow gape and cere quite prominent. Underparts may appear equally uniform, save for a lighter diffuse carpal crescent at the base of the outer primaries but lacking Booted's paler inner primaries. Pale morph birds have an off-white body, including the underwing-coverts and, importantly, also the head, while the upperparts are more uniform compared to the conspicuous plumage of pale morph Booted. The small crest on top of the head, often erect on perched birds, cannot be seen in flight. The iris colour is dark in Wahlberg's Eagle (cf. adult Booted Eagle).

MOULT Not well understood and more studies are needed, but moulting process probably similar to Booted Eagle. Flight feather moult resumes with the start of the breeding season, which in turn depends on the geographical region, and great variation in timing can be expected between birds from different parts of Africa. Also, resident and migratory populations may well have developed different moulting strategies.

PLUMAGES Adults and juveniles are rather similar in plumage and are best told from each other by different silhouette and signs of wing-moult in non-juveniles.

Juvenile

Best aged by uniform plumage, lacking signs of wing-moult, by clearly S-shaped trailing edge to the wing and by pale tips to the tail and secondaries. Unlike dark morph adults, juveniles often show a clear contrast between browner upperwing-coverts and darker remiges, and the lighter tips of the greater coverts form a narrow upperwing-band. Pale morph juveniles may show extensive whitish tips and fringes to the mantle, scapulars and upperwing-coverts. Juveniles often circle with the tail fanned, adding to the similarity with Booted Eagle.

Second plumage

Possible to age until replacement of the last juvenile flight feathers, which are lost approximately at the age of two years from hatching. Retained juvenile secondaries are shorter than neighbouring adult feathers and the outer primaries are faded and pointed. The juvenile secondaries also lack the broad dark subterminal band of the adult feathers. Before the onset of the actual wing moult, birds moult their head and neck feathers, which in dark birds stand out as clearly darker than the pale and faded body.

Adult

Told from the juvenile by more rectangular wing-shape and by habit of keeping tail mostly closed in flight, which together create a rather diagnostic silhouette. Dark birds are rather uniformly brown above, sometimes with a faded patch across the median upperwing-coverts (may be reminiscent of Booted Eagle), and lacking the more clearcut contrast between coverts and flight feathers found in juveniles. From below plumage is rather similar to the juvenile, but barring of the flight feathers is often more distinctive with a broad dark subterminal band to the tail and wings.

SEXING Birds cannot normally be sexed in the field.

CONFUSION RISKS Wahlberg's Eagle is most similar to Booted Eagle in size and silhouette, but because of similar uniformly brown plumage, it could also be mistaken for a Tawny Eagle, which is a much larger bird with broader wings and a comparatively shorter tail.

874. Wahlberg's Eagle *Hieraaetus wahlbergi*, fresh juvenile. A small and rather nondescript lighter or darker brown eagle, between Tawny and Booted in shape. Note pale trailing edge diagnostic of juvenile. Juveniles frequently fan their tails when soaring, unlike adults. The Gambia, 8.2.2008 (DF)

875. Wahlberg's Eagle *Hieraaetus wahlbergi*, fresh juvenile (same as **874**). Uniform plumage with evenly marked upperparts and pale tips to tail and flight feathers identifies a juvenile. Note lack of pale rump and compare with Booted and other similar eagles. The Gambia, 8.2.2008 (DF)

876. Wahlberg's Eagle *Hieraaetus wahlbergi*, worn juvenile, about one year old. This bird has grown a fresh, darker head and some inner primaries. The Gambia, 4.12.2013 (DF)

877. Wahlberg's Eagle *Hieraaetus wahlbergi*, worn juvenile, about one year old (same as **876**). Note heavily worn and faded upperwing-coverts, but a light band on the greater coverts, typical of juveniles, is still visible. The Gambia, 4.12.2013 (DF)

878. Wahlberg's Eagle *Hieraaetus wahlbergi,* adult. Typical silhouette with rectangular wings, extended neck and closed tail. The Gambia, 29.11.2006 (DF)

879. Wahlberg's Eagle *Hieraaetus wahlbergi,* dark adult. Note broad dark trailing edge to the wings and a darker subterminal tail-band, both typical features of adults. Uganda, 20.12.2009 (DF)

880. Wahlberg's Eagle *Hieraaetus wahlbergi,* dark adult. The Gambia, 8.2.2008 (DF)

881. Wahlberg's Eagle *Hieraaetus wahlbergi,* adult. Plumage similar to other brown eagles, but note typical proportions and lack of pale markings on the upperparts. The Gambia, 16.4.2007 (DF)

Differences in flight proportions between juvenile and adult falcons

As in many other raptors, the proportions of the silhouette in falcons differs between juveniles and adults. The difference is distinct in most species, but appears to be taken to the extreme in specialised aerial hunters, such as Sooty and Eleonora's Falcons, Eurasian Hobby and Peregrine, while it is almost non-existent in the Common and Lesser Kestrels, and surprisingly also in Merlin.

In most cases the juveniles show a 'generalist' falcon silhouette, and this shape appears to be shared by most species. In other words, just given the silhouette of a juvenile Red-footed, Eurasian Hobby, Lanner, Eleonora's or Sooty Falcon, it would be impossible to say which species it belongs to, although the silhouette of the adults may be highly diagnostic, as is the case with, for example, Eleonora's Falcon, Eurasian Hobby or Peregrine.

The difference in the silhouette is explained by the different lengths of the feathers of the wings and tail in adults and juveniles. For instance in Eurasian Hobby, Red-footed, Eleonora's and Sooty Falcons the adult primaries are considerably longer than those of the juveniles, which is best seen in moulting immatures, showing a mixture of juvenile and adult primaries.

Also the length of the tail changes from juvenile to adult, but the difference varies from species to species. In Eleonora's Falcon, for example, the new adult-type tail feathers can be up to 3–4 cm longer than the juvenile ones, giving the bird a totally different outline compared to the juvenile, while in other species, like Eurasian Hobby or Sooty Falcon, the difference is much less, and the tail retains almost its former proportions.

Not only does the silhouette vary between juveniles and adults, but also between males and females in adult birds, but again with major differences between species. The general rule is that males are slimmer, with narrower wings, a difference which is clearly seen in, for example, Eurasian Hobby, and Eleonora's and Sooty Falcons. On the other hand, in species like the kestrels, and Red-footed and Amur Falcons, this difference is too small for the human eye to detect.

It is quite apparent that the different silhouette is an adaptation to the ecology and lifestyle of the species. The juveniles have a generalist silhouette which is like a 'survival kit', adequate for most things, including successful hunting and migration, while in adults the proportions have been honed to perfection, to cater for the special requirements of each species.

Identification of kestrels

The problem of separating Eurasian and Lesser Kestrels still remains one of the greatest identification challenges among Western Palearctic raptors. Sometimes even the brightly coloured males can be difficult to identify, be it due to distance or poor light, but the true problem is separating the females and juveniles of the two species.

When the two species are seen together, as is often the case on migration, it is possible to pick out the two by differences in size and wing-action, the Lesser having a faster wing-rate when hovering. But this obvious difference soon disappears, once direct comparison is lost. Since the wing-beat frequency during hovering varies with wind force, one needs to be able to compare both species in the same wind conditions, preferably side by side, to see the difference.

For reasons mentioned above, the plumage characters remain the only reliable way of telling

the two apart. Although there are differences in the barring of the upperparts between Lesser and Common Kestrels, these differences are usually impossible to see in flight. Even the striking grey coverts-band on the upperwing of adult male Lessers is often surprisingly difficult to see and varies in extent between different males.

For females and juveniles the head markings and the underwing pattern, especially that of the primaries, are the most important features to note. In Lesser Kestrel the side of the head is pale, lacking prominent features, except for the dark eye and a tear-drop shaped moustache, the latter often quite inconspicuous. Common Kestrels have a dark eye-line behind the eye and the pale cheek is bordered also from behind, giving the face more character. The outer primaries forming the wingtip are often unbarred in Lesser and show a *broad and diffuse greyish wingtip.* In Eurasian the primaries are always barred (but also in some Lessers) and the *dark primary-tips are narrower, darker and more sharply defined,* forming a distinct dark line along the trailing edge of the hand.

The two species also have a different wing-formula, giving the wingtip a different shape, but the details are often impossible to see in the field (for details see under Common Kestrel). The impression is that

Lesser appears to have a more rounded wingtip in flight, but the true wing-formula can only be assessed from a sharp photograph, showing the wingtip from the right angle.

Common Kestrel also appears clearly longer-tailed in most situations, but since the tail length varies in both species, partly depending on age, many birds appear intermediate making this feature less reliable.

COMMON KESTREL
Falco tinnunculus

Other names: Eurasian Kestrel

VARIATION Nominate *tinnunculus* occurs over most of Europe, the Middle East and C Asia. In Egypt, *F. t. rupicolaeformis* is deeper in colour with bolder and darker markings, for example to the underwing; in adult males the grey of the head and the rufous upperparts are darker than in *tinnunculus*, and females are darker above with broader and darker barring, both thus approaching ssp. *rufescens* of sub-Saharan Africa. Even within *tinnunculus,* birds from the southern part of the range are deeper in colour than northern, migratory populations.

Isolated island forms occur in the Canaries and on Cape Verde. In the Canary Islands *F. t. dacotiae* from Lanzarote and Fuerteventura is rather similar to southern populations of *tinnunculus*, while *canariensis,* a resident of the C and W Canary Islands is clearly darker and more heavily marked. Both forms show a broad subterminal tail-band, comparatively broader than in *tinnunculus* (up to 25% of tail length in *canariensis*), and in particular *canariensis* is distinctly shorter- and broader-winged, and shorter-tailed than the nominate, at times almost approaching Eurasian Sparrowhawk in silhouette.

In the Cape Verdes two further subspecies occur, ssp. *alexandri* on the eastern and southern islands, small and with sexes almost identical and female-like, while *neglectus* of the northwestern islands is smaller still with sexes similar, dark and saturated in colour. Both of the Cape Verdean subspecies have been given species rank by some authorities.

In Egypt, the fast developing tourist towns on both sides of the Red Sea, Sharm el Sheikh in Sinai and Hurghada on the western shore, have both

experienced an invasion of breeding kestrels into the hotel areas, where they are now numerous. These birds are definitely darker and more heavily marked than typical European birds, but whether they are Egyptian *rupicolaeformis* or expansive *tinnunculus*, or a mix of the two, awaits further studies.

DISTRIBUTION One of the commonest and most widespread raptors of the region. Prefers all kinds of open areas, from semi-deserts to high mountains and from agricultural fields and meadows to road verges, locally also in towns and villages.

BEHAVIOUR Perches openly on telegraph poles and lamp posts and is frequently seen hovering in flight when looking for prey. Besides the two kestrel species only the Red-footed and Amur Falcons hover frequently when hunting. On migration often in flocks, sometimes together with Lesser Kestrel, and flocking behaviour and habit of hunting flying insects in the air is shared by both kestrel species, and is thus not a behaviour diagnostic of Lesser Kestrel.

SPECIES IDENTIFICATION In most situations easy to identify as a kestrel by largely *rufous upperparts* with contrasting darker outer wing, rufous tail of juveniles and most adult females, and a *generally pale underwing,* but the two kestrel species are notoriously difficult to separate, and close views (or photographs) are required to clinch the identification.

The largely rufous upperparts exclude all other raptor species except for Lesser Kestrel. The kestrels also have the palest underwings of all falcons, appearing uniformly pale from a distance.

Separating the two kestrel species is arguably the most difficult identification problem among the raptors of the region. The Common Kestrel is best told from Lesser Kestrel by its wing-formula and its differently marked underwing. Common shows completely barred primaries below, save for some juvenile males, in which the barring of the outer primaries may be reduced, while many (but not all) Lessers have largely unmarked and very pale outer primaries. Also, the dark trailing edge of the hand is different: in Common the primaries have a narrow, but distinct dark tip, which contrasts sharply against the pale inner part of the feather, while in Lesser the dark trailing edge is paler, greyer and wider, and its inner edge is diffuse.

The wing-formula is distinctly different in Common and Lesser Kestrels. In Common the wing tip is formed by the four longest primaries, of which *the outermost and the fourth (counting inwards from the outermost) are equally long* (second and third are longer). In Lesser the wingtip is formed by the longest three primaries, with *the outermost clearly longer than the fourth.*

In flight Common Kestrel appears longer-tailed and the wings appear pointed, while Lesser looks more compact, with a shorter tail and a more rounded wing-shape. When the two species are seen hovering together, Lesser Kestrel flaps more rapidly, indicating a smaller size, but when birds are seen singly this difference is lost, as varying conditions, such as strength of wind, affects the flight of both species considerably. For further differences see under Lesser Kestrel.

MOULT The plumage is moulted during the breeding season, between May and Sep–Oct. The juveniles undergo a partial body moult during their first winter, but this is very limited compared to, for example, Red-footed Falcon and Lesser Kestrel, and comprises usually only parts of the mantle and rump.

PLUMAGES (nominate *tinnunculus*) Normally only two plumage-types can be identified in the field: the adult male and the female-type plumages, the latter including adult females and all juveniles. Adult males are straightforward to identify, while adult females and juveniles are quite similar. In practical fieldwork female-type birds are very difficult to age and sex, since plumage differences are slight and difficult to see. Actively moulting birds in late summer–early autumn are quite obvious, and these are at least one-year-old birds, since hatchlings of the year do not moult flight feathers. Given good views female-type birds can also be aged and sexed by details of the wings and markings of the underparts.

Juvenile

Fresh autumn juveniles feature broad pale tips to their flight feathers, and also to their upper greater primary coverts and secondary coverts, forming narrow buffish wing-bars. These bands are lacking in adult females. The underbody is streaked in juveniles with broader markings to the flanks, while adult females are distinctly spotted below rather than streaked. Unlike adult females, juveniles never show signs of wing-moult during late summer–autumn.

Given exceptionally good views many juveniles can be sexed by the pattern of the upperparts and tail: in females the dark barring is broad, with the dark bands of the mantle and upperwing as broad as the rufous spaces in between, and the tail is distinctly and completely barred. In juvenile males the upperparts are more finely barred, with the dark bars being narrower than the rufous spaces in between, and the tail is finely, often even incompletely, barred compared to the females. Many juvenile males also already show a greyish rump and uppertail-coverts, while others are rufous and thus similar to females. From a distance juvenile females look browner above, while young males appear more brightly rufous.

During the winter the dorsal plumage of the juvenile becomes quite worn, while in adults the plumage remains in a better condition. By late winter juveniles can be aged by the contrast between newly moulted, fresh mantle feathers and scapulars and worn upperwing-coverts, and from now on sexing of immatures becomes easier, males showing brick-red mantle feathers with black spots, while females show rufous feathers with dark cross-bars. By now the underbody may be entirely moulted, showing adult-type spotting rather than juvenile-type streaking, but many juveniles retain the streaked juvenile breast feathers throughout spring.

Transitional plumage (2nd cy summer–early autumn)

Males can still be aged until moult is completed by retained female-type barred feathers in the tail and upperwing. Some males can still be tentatively aged as 2nd cy, even after the completed moult, by their head colour, showing a strong rufous cast to the greyish crown and nape. It also appears that the size of the dorsal and ventral spots, and the extent of the tail-barring, is linked to age so that first-adult males show larger and more spots and more tail bars than older birds, but further studies are needed.

2nd cy summer females are practically impossible to age. With the completion of the first moult during the 2nd cy autumn it is no longer possible to separate immature and adult females.

Adult male

From above easy to recognise by brick-red mantle and inner wing contrasting with dark outer wing and grey rump and tail, the latter with a white tip and a broad black subterminal band. The rufous upperparts show a subterminal black spot to each feather, and although varying in size they are a good character compared to adult male Lesser Kestrel. The crown and nape is grey, with a dark moustache and pale cheek-patch, the entire head being thus more conspicuously marked compared to adult male Lesser Kestrel.

Adult female

Resembles a juvenile in plumage, but upperparts usually more brightly rufous with finer and black rather than brown barring. Adult females lack the broad pale tips to the greater coverts and flight feathers above, obvious in fresh juveniles, and the underparts are distinctly marked with blackish spots or tear-drops turning to bolder markings on the flanks. Active moult easily separates adult females from juveniles, with primary moult lasting from the early breeding season until late summer, with northern migratory populations not completing their moult until Sep/Oct. With increasing age adult females start to develop male-type characters, such as greyish head, reduced dark markings above, and greyish tail feathers with reduced barring, but the timing of this development requires further studies with birds of known age.

SEXING Males are marginally smaller than females, even as juveniles, but the size difference is rarely of use in the field. See also under Plumages. Adults are straightforward to sex by plumage.

CONFUSION RISKS In most situations easily told from all other raptors by rufous upperparts, and from other falcons by pale underwings, with the exception of Lesser Kestrel, which shares the same features. When silhouetted in poor light kestrels could be mistaken for any small falcon or even a small *Accipiter*. Wing-beats are rather weak and flappy in active flight, while the tail always appears long and narrow.

NOTE Some (old?) adult males may lack black spots on the upperwing-coverts almost completely, but some spots are always visible on the scapulars.

882. Common Kestrel *Falco tinnunculus*, juvenile. Note wing-formula with tip of outermost primary (p10) equalling p7 (fourth in) in length, and compare with Lesser Kestrel. Israel, 6.10.2009 (DF)

883. Common Kestrel *Falco tinnunculus*, juvenile, migrant. Note diagnostic wing-formula and distinct and narrow dark trailing edge to the hand. Finland, 23.8.2012 (DF)

884. Common Kestrel *Falco tinnunculus*, juvenile, migrant. Juveniles are brown-streaked below compared to the blacker spot-marks of adult females. The dark line behind the eye is missing in Lesser Kestrel. Finland, 2.9.2014 (DF)

885. Common Kestrel *Falco tinnunculus*, juvenile, migrant, presumed male. Juveniles are rufous-brown above, similar to adult females, but the tips of the flight feathers and greater coverts are broadly buffish in fresh autumn plumage. In juveniles, narrowly barred upperwing-coverts, incompletely barred tail and greyish rump indicate a male. Finland, 15.9.2006 (DF)

886. Common Kestrel *Falco tinnunculus*, 2nd cy female, spring migrant. Young females are darker above due to heavier dark barring. Israel, 20.4.2009 (DF)

887. Common Kestrel *Falco tinnunculus*, 2nd cy male, spring. This immature male has acquired a colourful adult-type breast with black spots, and has grown one adult-type tail feather. Egypt, 15.3.2009 (DF)

888. Common Kestrel *Falco tinnunculus*, 2nd cy male, autumn migrant. Males in first-adult plumage often have a rufous rather than a blue-grey head, like this bird, although many are difficult to age once the last juvenile feathers have been replaced. Finland, 24.9.2006 (DF)

889. Common Kestrel *Falco tinnunculus*, adult male, breeding, presumed ssp. *tinnunculus* from Sinai. Adult males are not that different from adult females when seen from below; some retain some fine barring to the tail, particularly younger adults. Egypt, 20.3.2009 (DF)

890. Common Kestrel *Falco tinnunculus*, adult male. On the upperwing the dark hand contrasts with the brick-red innerwing, both tracts lacking the dense barring of the females. Morocco, 21.1.2014 (DF)

891. Common Kestrel *Falco tinnunculus*, adult male, breeding, presumed ssp. *rupicolaeformis* from the western Red Sea coast. Adults of more southerly populations are more deeply coloured compared to birds from further north. Egypt, 16.3.2008 (DF)

892. Common Kestrel *Falco tinnunculus*, adult male, breeding, presumed ssp. *tinnunculus* from Sinai (with nestling House Sparrow robbed from nest). Sparsely spotted brick-red upperparts contrast with dark hand and grey tail, the latter with a broad black subterminal band. Egypt, 20.3.2009 (DF)

893. Common Kestrel *Falco tinnunculus*, adult female, spring migrant. Distinct dark spotting on the breast differs from the streaking in juveniles. Israel, 18.3.2010 (DF)

894. Common Kestrel *Falco tinnunculus*, adult female, breeding. Eastern and northern populations can appear very pale and faintly marked; they could potentially turn up on migration in the Middle East. Mongolia, 9.6.2012 (DF)

895. Common Kestrel *Falco tinnunculus*, adult female, presumed ssp. *rupicolaeformis* from the western Red Sea coast. Note how dark this female is compared to the previous two females. Egypt, 15.3.2009 (DF)

896. Common Kestrel *Falco tinnunculus*, adult female, spring migrant. Females are densely barred above, although immature males may be difficult to distinguish in early spring. Israel, 23.3.2010 (DF)

897. Common Kestrel *Falco tinnunculus*, adult, resident ssp. *alexandri* from Santiago, Cape Verde. A small and dark island form, in which male and female differ only slightly in plumage. Santiago, Cape Verde, 3.2.2007 (DF)

898. Common Kestrel *Falco tinnunculus*, resident ssp. *neglectus* from Santo Antão, Cape Verde. An even smaller subspecies, with male and female practically identical in plumage. Santo Antão, Cape Verde, 2.2.2007 (DF)

899. Common Kestrel *Falco tinnunculus*, adult male of endemic resident ssp. *canariensis*. A richly coloured subspecies with broadly barred underwings and broad tail-band. Structure notably compact and differs from nominate *tinnunculus* by short and broad wings, and shorter tail. La Palma, Canary Islands, 13.12.2011 (DF)

900. Common Kestrel *Falco tinnunculus*, female of endemic resident ssp. *canariensis*. A richly coloured and heavily marked subspecies. Note compact structure and boldly barred primaries and tail. La Palma, Canary Islands, 13.12.2011 (DF)

901. Common Kestrel *Falco tinnunculus*, adult male of endemic resident ssp. *dacotiae*. A far less distinctive subspecies compared to *canariensis*, not that different from the nearest continental populations in Morocco. Lanzarote, Canary Islands, 10.12.2011 (DF)

LESSER KESTREL
Falco naumanni

VARIATION Monotypic.

DISTRIBUTION Breeding range extensive, stretching from S and C Europe across E Europe and C Asia to China and Far-eastern Russia. Long-distance migrant, with birds wintering scattered across all of sub-Saharan Africa. Some Spanish populations are resident, remaining in the breeding area year-round.

BEHAVIOUR Similar to Common Kestrel and Red-footed Falcon, both of which it associates with on migration as well as when breeding. Hunts either from an elevated perch, such as telegraph poles and overhead wires, but also hovers for long periods in suitable winds. Breeds mostly in derelict buildings and hunts in the surrounding open countryside.

SPECIES IDENTIFICATION One of the most difficult species to identify because of similarity with Common Kestrel. Easily recognised as a kestrel by rufous upperparts contrasting with darker outer wings, but very difficult to tell from Common Kestrel in the field, even when plumage details can be seen. Pale claws and wing-formula differ from Common Kestrel, but both are difficult to confirm in normal field conditions. For further details see under Common Kestrel.

MOULT Juveniles undergo a partial body moult in their first winter during which most of the body plumage, and often also some tail feathers are moulted. After this moult juvenile males and females resemble respective adults, but can usually be identified by their worn juvenile upperwing-coverts and retained juvenile tail feathers, which especially in males differ greatly from new adult-type feathers. The first complete moult starts in the breeding area in the 2nd cy summer, but is suspended for the autumn migration. Moult is resumed upon arrival in the wintering area and completed on the winter quarters.

Adults begin to moult while breeding and roughly follow the same timing as given above for immatures.

PLUMAGES Adult males and 2nd cy spring/summer males are usually easily identified, while females and autumn juveniles are hard to separate and need to be seen exceptionally well.

Juvenile (1st cy autumn)

During autumn told by uniformly fresh plumage lacking signs of wing-moult, and aged much like juvenile Common Kestrel. Very difficult to tell from juvenile Common Kestrel, but underwing primaries appear whiter, often with reduced barring on the outer feathers (particularly in juvenile males), while the *darker wingtip is wider, greyer and more diffuse.* The wing-formula is diagnostic (see under Common Kestrel), but this is mostly impossible to assess in flight; however, the wings of Lesser appear more rounded in flight compared to Common Kestrel. The *face appears paler and featureless compared to Common Kestrel*, lacking the dark eye-line of the latter, but normally shows an obvious dark moustachial stripe, appearing as a dark tear-drop below the eye. *Upperparts are more finely barred than Common Kestrel's, with individual marks showing as crescents rather than transverse bars.* Colour of rump and tail vary from rufous to more greyish, young males being on average greyer than females.

Moults body plumage partially while in the wintering area, enabling sexing of winter immatures by plumage.

Transitional plumage (first-winter to 2nd cy autumn)

Immatures can be told from older birds by a mixture of worn juvenile and fresher adult-type feathers.

Male The partial body moult is individually variable, but usually most of the body plumage has been replaced by adult-type plumage, except for the upper- and underwing-coverts, which are still retained juvenile feathers. The remiges are all juvenile, and normally at least some barred juvenile-type brownish tail feathers are also retained. Males in this plumage are safest to age by their retained barred upperwing-coverts. In many individuals the primaries show barring below, differing from the cleaner-looking feathers of adult males.

Females are much more difficult to age, since the plumages of immatures and adults are rather similar and the marginal differences are hard to see in the field. However, only immature females would show a clear difference between fresh mantle and worn upperwings, and often the wingtips also appear abraded in young birds, but not in adults.

Adult plumage (from 2nd cy late autumn onwards)

Male Easily told by uniformly brick-red upperparts, with variable amount of grey in the coverts, although the latter can be surprisingly difficult to discern in flight, even in good light. The amount of spotting on the underbody and underwing-coverts varies individually, with markings probably disappearing with increasing age.

Unlike statements in older literature adult males may show a darker moustache and/or a paler cheek-patch, characters normally connected with male Common Kestrel, although both features are more frequently seen on 2nd cy males in transitional plumage.

It also appears that males in first-adult plumage (second-winter to second- (3rd cy) summer) tend to show some faint barring on the underwing remiges, while definitive adult males tend to look cleaner, with unbarred white underwing remiges and a broad greyish trailing edge and tips, but whether this difference is consistent or not needs to be verified from known-aged birds.

Female Adult females are difficult to tell from younger females, and exceptionally good conditions are required; even so, the slight differences may be visible only on perched birds. Compared to younger females the entire plumage appears uniformly fresh or worn, depending on time of year, without obvious differences in wear between, for example, the mantle and upperwings.

SEXING Adults are easy to sex by different plumages. Juveniles differ as in Common Kestrel: males are more brightly coloured above with reduced markings on the upperparts and tail, while females are more extensively marked above, with complete barring on the upperwing-coverts and tail.

CONFUSION RISKS Very similar to Common Kestrel and good views are required to tell the species apart.

NOTE In Spain an adult male was photographed showing fine dark spots on its upperwing-coverts (pers. obs.).

902. Lesser Kestrel *Falco naumanni*, juvenile, presumed male, autumn migrant. Note important wing-formula with p10 (outermost) being clearly longer than p7 (fourth in), which is always diagnostic compared with Common Kestrel. For young males note also markings of primaries, with largely unbarred 'mid-hand' contrasting with broadly greyish tips and trailing edge. Israel, 2.10.2008 (DF)

903. Lesser Kestrel *Falco naumanni*, juvenile, presumed female, resident. Young females are more similar to Common Kestrel in respect of markings on the primaries, but different wing-formula is reliable. Spain, 19.9.2011 (DF)

904. Lesser Kestrel *Falco naumanni*, juvenile, presumed female, resident. Note faint moustache and pale cheeks, but otherwise rather featureless head; compare with Common Kestrel. Spain, 19.9.2011 (DF)

905. Lesser Kestrel *Falco naumanni*, juvenile, migrant. Fly-by kestrels are very difficult to identify, but note typical facial markings of this bird. Spain, 15.9.2005 (DF)

906. Lesser Kestrel *Falco naumanni*, juvenile. Note the diagnostic pale claws of this juvenile, and typical head and primary markings. Spain, 12.9.2005 (DF)

907. Lesser Kestrel *Falco naumanni*, juvenile. Upperparts look very similar to Common Kestrel and the fine details cannot be studied in detail on a flying bird. Note wing-formula and bland face, lacking Common Kestrel's dark eye-line. Spain, 12.9.2005 (DF)

908. Lesser Kestrel *Falco naumanni*, 2ⁿᵈ cy male in transitional plumage, migrant. By spring young males look like adults, but they still retain their juvenile wings; some also retain part of their barred juvenile tail. Israel, 21.4.2009 (DF)

909. Lesser Kestrel *Falco naumanni*, 2ⁿᵈ cy male in transitional plumage, migrant. This bird has moulted its tail completely, but can still be aged with difficulty by its barred flight feathers, especially secondaries. Israel, 22.4.2009 (DF)

910. Lesser Kestrel *Falco naumanni*, 2nd cy male in transitional plumage, migrant. Transitional males are more easily aged by their juvenile, barred upperwings. Israel, 22.4.2009 (DF)

911. Lesser Kestrel *Falco naumanni*, 2nd cy male in transitional plumage, migrant. Note the difference between the moulted back and the entirely juvenile upperwing. Israel, 22.4.2009 (DF)

912. Lesser Kestrel *Falco naumanni*, 2nd cy female, migrant. Seen only from below spring females are very difficult to age. Israel, 20.4.2009 (DF)

913. Lesser Kestrel *Falco naumanni*, 2nd cy female, migrant. Israel, 22.4.2009 (DF)

914. Lesser Kestrel *Falco naumanni*, 2nd cy female, migrant. In this case the partly moulted upperwing-coverts and the juvenile-type underwing-coverts confirm this as a 2nd cy female. Israel, 22.4.2009 (DF)

915. Lesser Kestrel *Falco naumanni*, 2nd cy male, resident. This male can still be aged by its few retained, barred juvenile secondaries and the extremely worn outer primary. Spain, 19.9.2011 (DF)

916. Lesser Kestrel *Falco naumanni*, breeding adult male and female (age uncertain). Spain, 3.5.2011 (DF)

917. Lesser Kestrel *Falco naumanni*, adult male, resident. Note the markings of the primaries and compare with adult male Common Kestrel. This male is just about to complete its annual moult. Spain, 19.9.2011 (DF)

918. Lesser Kestrel *Falco naumanni*, adult male, breeding. Some adults show faintly barred flight feathers, making the ageing of such males difficult, if just seen from below. Spain, 3.5.2011 (DF)

461

919. Lesser Kestrel *Falco naumanni*, adult male, migrant. Others have plain underwings and are easily aged as full adults. Israel, 22.3.2010 (DF)

920. Lesser Kestrel *Falco naumanni*, adult male, migrant. Adult males are easily told from male Common Kestrels by their grey wing-covert panels, but sometimes these may be small or even almost lacking. Israel, 20.3.2010 (DF)

921. Lesser Kestrel *Falco naumanni*, adult male, breeding. In this male all the greater coverts and even some inner secondaries are grey. Note also the spotless scapulars and upperwing-coverts. Spain, 3.5.2011 (DF)

922. Lesser Kestrel *Falco naumanni*, female, breeding. Note wing-formula and extensively dusky wingtip, and compare with female/ juvenile Common Kestrel. Spain, 3.5.2011 (DF)

923. Lesser Kestrel *Falco naumanni*, female, breeding. Spain, 3.5.2011 (DF)

AMERICAN KESTREL
Falco sparverius

VARIATION Several subspecies recognised, but nominate *sparverius* of N America is the most likely to reach Europe.

DISTRIBUTION Widely distributed across N and S America. Extremely rare vagrant to the Azores and Britain, with just a few records.

BEHAVIOUR Perches prominently on treetops and overhead wires. Hovers when hunting. Active flight is a combination of a series of erratic, rapid flapping interspersed by short glides.

SPECIES IDENTIFICATION Small falcon, with proportions and size of a male Merlin, with relatively long tail and shortish but pointed wings. Males are unmistakable owing to colourful plumage, while females are less striking, except for the head pattern: both sexes display two vertical black lines across the face, one a moustachial mark, another parallel to this across the ear-coverts. In flight from below, shows a prominent dark trailing edge to the hand bordered proximally by a conspicuous line of white spots.

MOULT Juveniles undergo an individually variable partial body moult in Aug–Nov, during which the males will attain an adult-type body plumage, while females change less. Yearlings as well as adults undergo a complete moult, involving remiges and rectrices, from late spring to Oct/Nov.

PLUMAGES Males and females differ in all plumages. Also juvenile and adult males differ, while females appear similar.

Males have grey upperwing-coverts with dark spots contrasting with rufous mantle and scapulars. Tail uniformly bright rufous with broad black subterminal band and white tip and edges, the latter with sparse black bands. Underwings are pale with contrasting dark markings, trailing edge of hand black with band of white spots.

Juvenile male

Nape is whitish and the underparts are streaked or spotted (longitudinal markings), with markings increasing in size from the upper breast towards the rear flanks. Mantle and scapulars as a rule more intensely barred compared to adult males, but this varies. During the first autumn (Aug–Nov) juveniles replace much of their body plumage and look similar to adults after this. The underbody changes from intensely streaked/spotted to sparsely spotted and the barring of the mantle/scapulars is reduced, while the nape turns rufous. No signs of wing-moult at this age.

Adult male

Much like juvenile male, but the nape is rufous and the underparts are sparsely spotted (rounded markings) and more deeply buffish. Active moult of primaries is obvious in summer and early autumn.

Females have upperparts, including upperwing-coverts, entirely rufous with dense dark barring.

Juvenile female

Much like adult female, but streaking of underbody is softer and the nape behind the black auricular marking is whitish. No signs of wing-moult at this age.

Adult female

Upperparts and tail are rufous with dense dark barring, the tail often with a broader subterminal band. Nape and hindneck are rufous. Underbody shows distinct markings, often drop-like spots becoming wider towards the flanks. Active moult of primaries is obvious in summer and early autumn.

CONFUSION RISKS Males are unmistakable, if seen well. Females may be mistaken for Common or Lesser Kestrels, but smaller size and different head and underwing markings are useful features.

924. American Kestrel *Falco sparverius*, male. Males have an orange breast and a plain rufous tail with a broad black subterminal band and white tips (the outer tail feathers are marked). USA (Jerry Liguori)

925. American Kestrel *Falco sparverius*, male. The upperwing-coverts are blue-grey in all males. Note the 'band of white pearls' just inside the primary tips. USA (Jerry Liguori)

926. American Kestrel *Falco sparverius*, male. Adult males are unmistakable if seen properly. USA (Jerry Liguori)

927. American Kestrel *Falco sparverius*, female. Note double dark facial marks. Females have barred tails much like female/young Common Kestrels, but lack their broad black subterminal band. USA (Jerry Liguori)

928. American Kestrel *Falco sparverius*, female. Resembles Common Kestrel from above but note striking facial and primary pattern. USA (Jerry Liguori)

Identification of small, Hobby-like falcons

The group of small to medium-sized hobby-like falcons includes Eurasian Hobby, and Eleonora's, Sooty, Red-footed and Amur Falcons. All of these share, at some point in their plumage development, confusingly similar features. The juvenile and first-winter plumages are very similar for all species, and for Eurasian Hobby and Eleonora's Falcon the similarity remains until adulthood.

The larger size of Eleonora's Falcon, compared to the other species, can only be judged when it is seen side by side with one of the smaller species, but generally speaking the impression of size is of limited value for identification.

The most important characters are found in the details of the underwing, tail and head pattern, but for most species the upperparts also differ in colour and pattern. It is important to learn the pattern of the juvenile remiges and rectrices, which then can be traced in the transitional plumages of the immatures the following year. All species undergo a partial body moult during their first winter, and the extent of the moult varies not only from species to species, but also between individuals of the same species. Immature Sooty, Red-footed and Amur Falcons will look almost like adults by the time they return to the breeding grounds the following year, save for their retained juvenile remiges (sometimes also rectrices), which will act as excellent markers for another six months, until their second winter.

Young Eleonora's Falcons have moulted on average less of their body plumage compared to the previous species, but the individual variation is extensive, while Eurasian Hobbies moult the least number of feathers, often only a few feathers on the mantle and scapulars. The main identification challenge is to separate some of the less-moulted immature Eleonora's Falcons from immature or adult Eurasian Hobbies, as their body plumage can be almost identical. For these the different underwing patterns remain the key feature.

RED-FOOTED FALCON
Falco vespertinus

VARIATION Monotypic. Amur Falcon *Falco amurensis*, was formerly considered to be an eastern subspecies of Red-footed Falcon.

DISTRIBUTION A breeding migrant to the steppe zones, ranging from E Europe to C Asia. Winters mostly in the western parts of southern Africa.

BEHAVIOUR Breeds colonially in old stick nests of other birds, mostly of corvids, and is also gregarious on migration and in winter. Feeds mainly on large insects, but also takes small mammals and lizards. Often seen hovering like a kestrel when hunting, but also still-hunts from poles and overhead wires.

SPECIES IDENTIFICATION Easily confused with other small falcons, particularly the less conspicuous juveniles. Juveniles best identified by strikingly black and white barring to the underwing remiges with broad dark trailing edge, completely barred tail above, and dark facial mask and white neck-band. Adult males told from dark Eleonora's or Sooty Falcons by silvery primaries above and red thighs and vent, although latter is often difficult to discern in the field, and adult females by bright orange underbody, pale head and greyish upperparts, including uppertail, and distinctly marked underwing remiges, as in juvenile.

MOULT A typical long-distance migrant to the tropics, with moult mostly occurring in the winter quarters. Adults only start their wing-moult on the breeding grounds before moult is suspended for the autumn migration. By then males show on average fewer moulted feathers than females, while 2nd cy immatures have moulted on average more feathers than adult birds of the same sex. After migration moult is resumed and finished in the winter quarters, in good time before spring migration. Returning adults in spring appear in mint plumage, but birds showing several (3–5) worn and faded median primaries are first-adults (3rd cy spring).

Juveniles undergo an extensive body moult during their first winter, but the juvenile remiges are always retained until the following summer, some of the feathers even until the following winter. The tail can be partly or completely moulted.

PLUMAGES In autumn three age-classes can be separated, juvenile, second plumage (transitional), and adult, while in spring only 2nd cy birds and adults (sometimes also 3rd cy) can be distinguished.

Males and females always differ in plumage, except for autumn juveniles, which are similar.

Juvenile (1st cy autumn)

Head appears pale or even whitish from a distance, with a black mask through the eye and a hint of a short moustache. Crown varies from whitish to deep ochre and darker birds may approach certain juvenile Eurasian Hobbies as far as head pattern is concerned. Red-footed Falcons will, however, always show a white neck-band compared to Eurasian Hobby, making the dark face markings look detached by comparison. The upperparts appear greyish-brown, varying in shade depending on the light, but the blackish primaries will usually stand out clearly. The tail is greyish to greyish-brown above, finely barred dark both above and below (cf. juvenile Eleonora's and Sooty Falcons and Eurasian Hobby), with the pale and dark bars about equal in width (some birds show a female-like broader subterminal band), and with a broad pale tip. The underbody is pale buff with brownish streaking, often confined to the breast, leaving the belly and vent unmarked. The underwings are distinctly barred black and white (cf. Eurasian Hobby), with a broad dark trailing edge, while the pale buffish underwing-coverts are densely marked with rufous-brown (blackish barring in juvenile Amur Falcon). From a distance the underwing-coverts often appear as being slightly darker than the remiges.

Transitional plumage (2nd cy spring)

In the winter quarters juveniles undergo a partial body moult, which changes their appearance considerably compared to fresh autumn juveniles.

Males are individually variable, but by and large resemble adult males. They can always be aged by their retained juvenile underwing-coverts and wing feathers, which are barred below, not black as in older

males, and they also lack the silvery sheen of the upperwing of adult males. The head and underbody are variable and some look just like adult males, while others have an orange patch on the mid-breast and the head may show a pale cheek-patch and a darker moustache. The tail is often completely moulted by spring and shows new dark grey feathers with a broad black subterminal band. Bare parts as in adult male, deep orange.

Females are already very similar to adult females, but can be told by the retained juvenile flight feathers and barred underwing-coverts, while the worn upperwing looks dull and brown. The tail is often a mixture of juvenile and adult-type feathers, with the juvenile feathers being short and abraded and the new ones longer with a broad dark subterminal band and a pale tip. The breast shows on average more streaking compared to adult females.

Transitional plumage (2nd cy summer–autumn)

By the 2nd cy autumn both sexes wear a transitional immature plumage, comprising a mixture of juvenile, transitional and adult plumage feathers.

Males are easily recognised by their diagnostic wing pattern, with sections of black adult-type feathers amidst retained barred juvenile feathers, both in the primaries and secondaries.

Females are very similar to adult females, but can be aged given good views. As a rule immature females have moulted more primaries than adults (average of 4–6 fresh primaries in immatures compared to only 0–2 in adult females) and the new ones stand out as clearly darker but also as being clearly longer than the retained juvenile primaries. The underwing-coverts regularly retain juvenile feathers with dark barring, making the underwing-coverts look untidy, compared to the neat coverts of older females.

Adult plumage

Adult males are grey and black with reddish thighs, undertail-coverts and vent, and the remiges and rectrices lack any kind of barring. Usually the head is darker while the breast and belly are a lighter grey. Tail and underwings are black. From above the bird looks dark grey with a black tail and silvery-grey primaries, but the silvery bloom largely wears off during summer, making the upperwing look less conspicuous by autumn. 3rd cy males can sometimes be identified by their faded and colourless median upperwing primaries (see also under Moult above). Cere, orbital skin and feet are deep orange to red-orange.

Adult females are very colourful with orange underbody and underwing-coverts and a paler head with a black mask through the eye and a short black moustache. The head can vary from almost white to deep orange, but the throat and cheeks are always whitish. The upperparts are grey with fine dark barring, darkest on the upper mantle, while the tail is paler grey, finely barred in black with a broad subterminal 'kestrel-band'. The primaries are dark above, contrasting clearly with the grey inner wing (for identification of 3rd cy spring females, see under Adult males above). The flight feathers are white below with dense black barring and a broad dark trailing edge to the wing. Cere, orbital skin and feet are deep orange-yellow.

CONFUSION RISKS Juveniles are rather similar to juvenile Eurasian Hobby (also see juvenile Amur Falcon) and great care is needed to separate the two. The head patterns can be confusingly similar in flight, although Red-footed as a rule has a whiter head with more restricted dark facial markings (e.g. shorter moustache and a more complete white neck-band). In juvenile Red-footed the tail is always greyish and completely barred above, even when folded, often standing out as the palest tract above, while in Eurasian Hobby it is all dark, with rufous barring showing only from below, or partly from above when the tail is fanned. The central tail feathers are, however, always uniformly dark in Eurasian Hobby. From below juvenile Red-footed Falcons are pale buffish with sparser and often also paler brown streaking (blackish streaking on a more yellowish-ochre ground colour in juvenile Eurasian Hobby), and the underwing is distinctly patterned in black and white with a broad dark trailing edge to the wing, often leaving a pale central area on the hand, while the whole underwing looks uniformly patterned and dark in Eurasian Hobby. Juvenile Red-footed potentially can also be confused with juvenile Eleonora's or juvenile Sooty Falcon, but details of head, uppertail and underwing are different.

NOTE Has probably hybridised with Amur Falcon in Europe. A female showing mixed characters was observed in a colony of Red-footed Falcons in Hungary in 2006 (see under Amur Falcon).

929. Red-footed Falcon *Falco vespertinus*, juvenile, vagrant. Note buffish body and underwing-coverts and compare with juvenile Amur Falcon; compare underwing and tail-barring with Eurasian Hobby. Finland, 30.8.2005 (DF)

930. Red-footed Falcon *Falco vespertinus*, recently fledged juvenile. Hungary, 23.7.2011 (DF)

931. Red-footed Falcon *Falco vespertinus*, juvenile, vagrant. A darker-headed individual. Note diagnostic black-and-white underwing barring and distinct dark trailing edge of wing. Finland, 30.8.2005 (DF)

932. Red-footed Falcon *Falco vespertinus*, juvenile, vagrant (same as **929**). Upperparts are typically brownish-grey with sandy-brown tips. Colour and markings of head and tail vary individually; in this bird they look quite similar to an adult female. Finland, 30.8.2005 (DF)

933. Red-footed Falcon *Falco vespertinus*, juvenile, vagrant. Depending on the light the upperparts may appear either greyish or brownish, often with a clear contrast between lighter coverts and darker flight feathers. Finland, 30.8.2005 (DF)

934. Red-footed Falcon *Falco vespertinus*, 2nd cy male in transitional plumage, spring migrant. The body plumage has been largely moulted in winter and now resembles an adult male, but the underwing-coverts and flight feathers are retained juvenile feathers, easily recognised by their barring. A variable number of tail feathers will also have been moulted, in this case just the central pair. Greece, 26.4.2007 (DF)

935. Red-footed Falcon *Falco vespertinus*, 2nd cy male in transitional plumage, spring migrant. Greece, 26.4.2007 (DF)

936. Red-footed Falcon *Falco vespertinus*, 2nd cy male in transitional plumage, spring migrant. Immature males are easily recognised by their dull brown upperwings, lacking the silvery sheen of the adults. Greece, 26.4.2007 (DF)

937. Red-footed Falcon *Falco vespertinus*, 2nd cy male in transitional plumage, autumn migrant. By autumn half of the wing and tail feathers have been replaced by black adult-type feathers, rendering the immature males unmistakable. Israel, 2.10.2008 (DF)

938. Red-footed Falcon *Falco vespertinus*, 2nd cy female in transitional plumage, autumn migrant. Autumn females are more difficult to age, but the contrast between old juvenile and new feathers is clearer in immatures, when comparing colour and wear as well as difference in feather length. Israel, 2.10.2008 (DF)

939. Red-footed Falcon *Falco vespertinus*, 2nd cy female in transitional plumage, autumn migrant. Immature females have also moulted more primaries than adult females. Israel, 2.10.2008 (DF)

940. Red-footed Falcon *Falco vespertinus*, adult male, autumn migrant. Adult males are unmistakable with black underwings and tail, grey body, and rufous thighs, vent and undertail-coverts. Note also the bright orange bill and feet. Israel, 2.10.2008 (DF)

941. Red-footed Falcon *Falco vespertinus*, adult male, spring migrant. Greece, 26.4.2007 (DF)

942. Red-footed Falcon *Falco vespertinus*, adult male, breeding. Hungary, 22.7.2011 (DF)

943. Red-footed Falcon *Falco vespertinus*, adult male, spring migrant. The upperside of an adult male is grey with silvery outer wings and a black tail. Greece, 26.4.2007 (DF)

944. Red-footed Falcon *Falco vespertinus*, adult female, spring migrant. Females are difficult to age in spring, but adult females have less marked underwing-coverts and the primaries are in a better condition. Greece, 26.4.2007 (DF)

945. Red-footed Falcon *Falco vespertinus*, adult female, spring migrant. Greece, 26.4.2007 (DF)

946. Red-footed Falcon *Falco vespertinus*, adult female, autumn migrant. By autumn adult females have fewer freshly moulted primaries compared to 2nd cy females, in this case only one (p4). Their body and underwing-coverts are also more uniform, with finer markings. Israel, 2.10.2008 (DF)

947. Red-footed Falcon *Falco vespertinus*, adult female, spring migrant. In spring adult females are silvery-grey above, while immature females would show a dull brownish upperwing contrasting with a newly moulted grey mantle. Greece, 26.4.2007 (DF)

AMUR FALCON
Falco amurensis

Other names: Eastern Red-footed Falcon

VARIATION Monotypic. Formerly considered to be a subspecies of Red-footed Falcon.

DISTRIBUTION Breeding migrant to the E Palearctic, in Mongolia, China and E Russia, wintering in the eastern parts of southern Africa, making the longest non-stop sea-crossing across the Indian Ocean of any migratory raptor. Rare vagrant to Europe, with records from Italy, France, Britain and Sweden, mostly in spring or summer. Scarce but regular on passage along the eastern seaboard of Arabia, e.g. in Oman.

BEHAVIOUR Similar in behaviour to kestrels and Red-footed Falcon. Hunts either from a perch or hovers like a kestrel above the ground. Also hunts flying insects high up in the air in Eurasian Hobby-fashion and concentrates in numbers around emerging termites in the wintering grounds and on migration.

SPECIES IDENTIFICATION Identical in proportions and behaviour to Red-footed Falcon and Lesser Kestrel, often joining the latter on migration and in the wintering areas. Adult males are most likely to be confused with adult male Red-footed Falcons, while females resemble Eurasian Hobby. Juveniles very similar to juvenile Red-footed Falcon (and Eurasian Hobby), and separation requires good views.

MOULT Similar to Red-footed Falcon.

PLUMAGES In autumn three age-classes can be separated: juvenile, 2nd cy birds and adult; in spring only two: immature and adult.

From the first winter onwards birds are easily sexed by plumage, while fresh autumn juveniles appear indistinguishable.

Juvenile (1st cy autumn)

Head appears like Eurasian Hobby from a distance, with quite extensive individual variation ranging from light-headed to darker and more Eurasian Hobby-like birds. The extent of the dark markings, the black mask around the eye and the short moustache vary individually, as does the crown which varies from light greyish to dark, often with a distinct pale supercilium. Contrary to Eurasian Hobby, juvenile Amur Falcon shows a distinct pale neck-band, making the whole head look paler compared to Eurasian Hobby. The upperparts appear greyish-brown and surprisingly uniform, varying in shade depending on the light, and contrary to juvenile Red-footed the primaries don't stand out as clearly darker. As in juvenile Red-footed the tail is sandy-brown to greyish above, finely barred throughout (contrary to juvenile Eleonora's and Sooty Falcons and Eurasian Hobby), with more or less equally broad pale and darker bands and a broader pale tip; some juveniles show a broader dark subterminal band, which is normally a feature of adult females, but it is not black as in adult females. The underbody is *whitish with blackish streaking*, the pattern being much more contrasting compared to juvenile Red-footed, with the streaking of the breast changing to *crescents or arrowheads on the flanks,* while the belly and vent are often more deeply coloured, uniformly yellowish-buff. The underwing remiges are distinctly barred with black and white, as in Red-footed (cf. Eurasian Hobby), with a variably broad dark trailing edge, again varying quite a lot between birds, but the underwing-coverts are contrastingly *whitish with dense black barring*, compared to more buffy coverts with brown markings in juvenile Red-footed Falcon, another diagnostic difference between the species. Feet and base of bill are bright orange-yellow as in juvenile Red-footed Falcon.

Transitional plumage (2nd cy spring)

Upon reaching the wintering areas the juveniles undergo a partial body moult, which will change their appearance completely compared to fresh autumn juveniles.

Males are individually variable, with the body plumage being a mix of fresh adult-type and worn juvenile feathers. Regardless of the state of body moult they can always be aged by their retained juvenile remiges and underwing-coverts, and also by partly retained barred juvenile tail feathers, which are not grey as in older males; they also lack the silvery sheen on the upperwing of adult males. The head is normally first to acquire an adult-type look while the underbody is more variable, often with a greyish upper breast, but retaining a streaked belly and barred flanks. The tail is also partly moulted by spring and shows new uniformly grey feathers mixed in with barred juvenile rectrices. Bare parts are similar to adult male, deep orange.

Females can be very similar to adult females, but are always told by their faded and worn juvenile flight feathers, making the upperwing look dull and brown, and lacking the bluish-grey finish of older females. The tail is often a mixture of juvenile and adult-type feathers, the juvenile ones being shorter and abraded (by now lacking the pale tips) and the new ones being longer and silvery-grey with finer dark barring compared to the juvenile feathers, and with a clearly broader black subterminal band and a pale tip. As a rule the dark trailing edge of the wing is narrower compared to older females and the underwing-coverts are more strongly marked, making the whole underwing look rather light with extensive barring and a contrasting dark trailing edge.

Transitional plumage (2nd cy summer–autumn)

By the 2nd cy autumn both sexes wear a transitional immature plumage, comprising a mixture of juvenile, transitional and adult plumage feathers.

Males are easily recognised by their diagnostic wing pattern, with median primaries and median secondaries being black contrasting sharply with the retained and barred juvenile feathers. Best told from the similar plumage of male Red-footed Falcon by a more contrasting pattern to the underwing-coverts, and by grey, not black new tail feathers.

Females are very similar to adult females, but can still be aged given good views. As a rule immature females have moulted a larger number of primaries than the adults (cf. Red-footed Falcon), and the new ones stand out as clearly darker and longer compared to the retained, by now very worn, juvenile primaries. The underwing-coverts are largely faded juvenile feathers with dark barring, making the underwing-coverts look untidy, compared to the whiter and more sparsely marked, cleaner-looking coverts of adult females.

Adult plumage

Adult males are rather uniformly grey, compared to adult male Red-footed Falcon, with reddish thighs, vent and undertail-coverts, while the uniformly dark remiges and rectrices lack any kind of barring. Usually the head is somewhat darker than the body, with discernibly paler cheeks and throat, often with a clearly indicated darker moustache-mark and eye-mask, and a paler cheek-patch, while the breast and belly are lighter grey, clearly paler than the average adult male Red-footed Falcon. From below the *underwing is black with diagnostic strikingly white underwing-coverts*, while the tail is clearly grey, not black as in adult male Red-footed. The upperparts appear rather uniformly grey, with somewhat paler and more silvery-grey remiges in fresh plumage (spring and early summer), but *lacking the striking almost silvery-white outer primaries of adult male Red-footed, and the tail is grey and concolorous with the upperparts, not black and darker than the back as in adult male Red-footed Falcon.* Cere and base of bill, orbital skin and feet are deep orange. The adult male is thus overall paler, greyer and more uniform compared to adult male Red-footed, and the underwing remiges are the only black feathers in the entire plumage.

By autumn the plumage is duller grey above because of wear, but any moulted (median) primaries would stand out as clearly paler on the upperwing.

Adult females are strongly marked in black and white below and may superficially resemble Eurasian Hobby, but several details make the identification rather straighforward. The flanks are clearly barred, not streaked, the breast is irregularly spotted and streaked, often forming a darker band across the upper breast, and the vent, thighs and undertail-coverts are uniformly peachy-buff, not rusty-red. The flight feathers are white below with dense black barring, with a broad and contrasting black trailing edge to the wing, and the underwing-coverts are pure white with fine black streaks near the leading edge changing to finer bars further in. Older females may show nearly all-black remiges from below, with hardly any visible barring at all, and from a distance the underwing closely mimics that of an adult male. The head pattern also recalls Eurasian Hobby, but the small dark moustache and the more extensively white cheeks and white neck-band give the bird a friendlier look, while the deep orange cere and base of bill are diagnostic compared to Eurasian Hobby. The upperparts are pure grey, darkest on the upper mantle and finely barred throughout, while the tail is paler silvery-grey, finely barred with black and with a broad subterminal 'kestrel-band'. The upperwing appears entirely grey above with finely barred coverts and inner wing. Cere, base of bill, orbital skin and feet are deep orange-yellow.

SEXING Adults and immatures in transitional plumage can be sexed by plumage differences, but autumn juveniles are inseparable in the field.

CONFUSION RISKS Juveniles and adult females are most likely to be mistaken for Red-footed Falcon or Eurasian Hobby, but could possibly also be mistaken for other Eurasian Hobby-like small falcons, such as Eleonora's or Sooty Falcons. Differs from Eurasian Hobby by paler, two-toned underwing, with light forewing contrasting with dark trailing edge, compared to uniformly darkish underwing in Eurasian Hobby. Adult males are similar to adult male Red-footed, but generally paler grey with diagnostic white underwing-coverts.

NOTE Amur and Red-footed Falcons may have hybridised in Europe, probably as a result of Amur Falcons joining Red-footed Falcons on their spring migration. A female showing mixed characters (author's interpretation) was observed in a Red-footed Falcon colony in Hungary in 2006 (photograph in *Birding World* 19: 276).

948. Amur Falcon *Falco amurensis*, juvenile, autumn migrant. Distinguished from very similar juvenile Red-footed Falcon by more contrasting plumage, with black markings on a whiter ground colour. Oman, 8.12.2012 (DF)

949. Amur Falcon *Falco amurensis*, juvenile, autumn migrant. Note contrasting underwing with conspicuous black trailing edge. The amount of black around the eye varies individually, this bird being from the lighter end of variation. Oman, 27.11.2014 (DF)

950. Amur Falcon *Falco amurensis*, juvenile, autumn migrant. The width of the black trailing edge of the wings is variable but always distinct, while the tail looks pale with fine dark barring (cf. Eurasian Hobby). Oman, 8.12.2012 (DF)

951. Amur Falcon *Falco amurensis*, juvenile, autumn migrant (same as **949**). The upperparts are practically identical to juvenile Red-footed Falcon and the two species can be inseparable. Oman, 27.11.2014 (DF)

952. Amur Falcon *Falco amurensis*, 2nd cy male in transitional plumage, breeding. This plumage can be quite similar to Red-footed Falcon in corresponding plumage but note contrasting black-and-white underwing-coverts and grey rather than blackish new tail feathers. Young Amur males also have a tendency to show some retained streaking on the breast and a dark head with white chin and cheeks, as here. Mongolia, 4.6.2012 (Tom Lindroos)

953. Amur Falcon *Falco amurensis*, 2nd cy female in transitional plumage, breeding. Immature females still retain heavily barred juvenile underwing-coverts and flight feathers, while the head and underbody have been moulted. Mongolia, 4.6.2012 (DF)

954. Amur Falcon *Falco amurensis*, adult male, breeding. Easily recognised by gleaming white underwing-coverts. Mongolia, 4.6.2012 (DF)

955. Amur Falcon *Falco amurensis*, adult male, autumn migrant. Although superficially similar to adult male Red-footed Falcon, apart from the obvious white underwing-coverts, note lighter grey body with contrasting darker head, and grey, not black tail. Oman, 8.12.2012 (DF)

956. Amur Falcon *Falco amurensis*, adult male, autumn migrant (same as **955** and **957**). Oman, 8.12.2012 (DF)

957. Amur Falcon *Falco amurensis*, adult male, autumn migrant (same as **955** and **956**). The upperwings are more uniformly grey compared to the silvery wingtips in adult male Red-footed Falcon. Note suspended primary moult. Oman, 8.12.2012 (DF)

958. Amur Falcon *Falco amurensis*, adult female, breeding. This most diagnostic plumage almost mimics the underwing pattern of adult males, with largely black flight feathers and white, finely marked coverts. The underbody markings differ markedly from those of Eurasian Hobby, while the tail-barring is identical to adult female Red-footed Falcon. Mongolia, 4.6.2012 (DF)

959. Amur Falcon *Falco amurensis*, adult female, autumn migrant. Adult females show a peach-coloured wash to the rear flanks and vent, and distinctly barred flanks. Oman, 9.12.2012 (DF)

960. Amur Falcon *Falco amurensis*, adult female, breeding. The upperparts of females are uniformly grey with darker barring. Mongolia, 4.6.2012 (DF)

EURASIAN HOBBY
Falco subbuteo

VARIATION The nominate subspecies occurs as a breeding migrant and passage visitor throughout the region; another race, ssp. *streichi* occurs in S China.

DISTRIBUTION Breeding migrant across the Palearctic, from Europe to Japan, wintering in sub-Saharan Africa (eastern birds winter in S and SE Asia). Seen on migration mostly in Aug–Oct and Apr–May. Concentrations at favourable feeding sites in late May–early Jun refer mostly to non-breeding first-summer birds.

BEHAVIOUR Aerial, either hunting leisurely for flying insects, or in dashing pursuits for small birds, often rather high in mid-air, but also swooping low over open ground. Perches mostly on prominent dead tree tops, but also uses man-made structures.

SPECIES IDENTIFICATION The uppertail appears uniformly dark lacking any kind of barring (median tail feathers are always dark) and the upperparts lack any strong contrast. Little variation in head pattern, but juveniles may show a largely pale forecrown, sometimes rather similar to juvenile Red-footed Falcon. From a distance, and especially in low light, may appear overall dark below, with only the pale cheek-patch showing, adding to the confusion risk with, for example, male Red-footed or Sooty Falcons, or Eleonora's Falcon. Note that adults and juveniles have a different silhouette, as only adults show the characteristic 'hobby-shape' (see also under sexing of adults), while juveniles, which have shorter wings, have a more general falcon shape, which is more or less identical to the silhouette of juveniles of Red-footed, Eleonora's, Sooty, Peregrine and Lanner Falcons.

MOULT Moults mainly on African wintering grounds, but 2nd cy birds start replacing primaries while still in the breeding area, whereas adults as a rule only after arriving in Africa. Juveniles undergo a partial body moult in their first winter replacing their upperparts partially, in particular the scapulars and rump, and sometimes also parts of the underbody. The extent of this moult varies individually, but it is always less extensive than in juveniles of Red-footed Falcon and Lesser Kestrel. Unlike adults, most 2nd cy summer birds replace p4 and p5 before migrating south.

PLUMAGES In late summer and autumn three plumage-types can be separated, juvenile, transitional and adult. By spring, once moult is completed, only juveniles and adults can be recognised. Pattern of tail feathers differs between juveniles and adults.

Juvenile

In fresh plumage easy to tell by yellowish-buff, not white, underparts, lacking the bright rufous thighs, vent and undertail-coverts of older birds (but note that these tracts may be more deeply coloured than the remaining underparts). The *distinct pale tip to the tail and trailing edge of the wing*, visible from afar against a darker background, are also diagnostic ageing features compared to adults.

The upperparts are uniformly dark with narrow creamy fringes and tips to the scapulars and upperwing-coverts. The tail is dark with (mostly hidden) distinct ochre barring, except for the central pair of tail feathers which are uniformly dark. This tail-barring is normally visible only from below, but when the tail is fully fanned it can also be seen from above. In fresh autumn plumage the upperwing can show a slight contrast between the dark primaries and slightly paler innerwing, owing to the pale fringing and a greyish bloom of the fresh plumage, and the combined effect may be reminiscent of some juvenile Red-footed Falcons. This contrast, as well as the pale fringes, is lost during winter, and by spring the upperwing appears uniformly dull brown.

Transitional plumage

The 2nd cy plumage is highly variable and is a transitional stage between the juvenile and adult plumages. It is best identified by juvenile tail feathers showing the distinct pale barring. The pale tips of the tail and remiges are worn off by now, leaving spiky feather-tips. Upperparts may be largely worn and faded dark brown, or may show a variable number of moulted grey feathers among the scapulars and upperwing-coverts, but always looks somewhat tatty in close views, often with brownish upperwings contrasting with a fresh, grey mantle, creating a saddle-effect. Underparts are much as in adults, but the amount of rufous varies from practically nothing to fully adult-like. By the 2nd cy autumn the remiges become very worn and frayed at the tips, with a few newly moulted median (pp4–5) primaries (as a rule 2nd cy birds show a few moulted primaries contrary to the adults at this time).

Adult

Adults appear neat and tidy above while on the breeding grounds. The upperparts are dark with even darker primaries, the adult males being greyer and more colourful above than the duller females. Adults also have slightly elongated and rather pointed central tail feathers, which is not found in juveniles. The underparts are clean white with black streaking and the vent, undertail-coverts and thighs are rusty-red. The tail-bars are rather indistinct and diffuse compared to the juvenile pattern, with the barring becoming even fainter towards the distal parts of the tail. Adults lack the prominent pale tail-tip and trailing edge of the wing, which are diagnostic features of fresh autumn juveniles.

SEXING is only possible for adults: males are slimmer in silhouette than females, with narrower wings and a slimmer body, and the difference in structure is quite obvious when the two are seen together. Males also tend to show unstreaked or only finely streaked trousers, while females are more marked, but this difference is also partly age-related and old females may be rather poorly streaked.

CONFUSION RISKS In poor light conditions, or when distant, Eurasian Hobbies can be confused with most other small falcons. Even with good views they can appear confusingly similar to Eleonora's, Red-footed or Sooty Falcons, not to mention vagrant Amur Falcon, especially in juvenile and immature plumages.

NOTE A very rare darker morph of Eurasian Hobby has been reported, mostly from the Mediterranean basin, but also from W Europe (Faveyts 2010, with photographs). These birds are heavily streaked below, the ground colour of the breast can be rufous and the underwings and tail are dark with diffuse barring only. Birds like this can be very similar in appearance to first- or second-summer Eleonora's Falcons, and great care is needed to separate these plumages.

References

Corso, A. & Monterosso, G. 2000. Ein unbeschriebene dunkle variante des Baumfalken *Falco subbuteo* und ihre Unterscheidung vom Eleonorenfalken *F. eleonorae*. *Limicola* 14: 209–215.

Corso, A. & Monterosso, G. 2004. Further comments on dark Hobbies in southern Italy. *British Birds* 97: 411–414.

Faveyts, W. 2010. Ringvangst van een donkere Boomvalk te Brecht in mei 2009. *Natuur.oriolus* 76: 109–112.

Ristow, D. 2004. Exceptionally dark-plumaged Hobbies or normal Eleonora's Falcons? *British Birds* 97: 406–411.

961. Eurasian Hobby *Falco subbuteo*, juvenile, migrant. Rather uniformly patterned underwings are typical of the species, while distinct pale tail-barring and tail-tip are typical of juveniles. Finland, 5.9.2008 (DF)

962. Eurasian Hobby *Falco subbuteo*, juvenile, migrant. In normal lighting the underparts look rather dark save for the pale cheeks. Finland, 13.9.2013 (DF)

963. Eurasian Hobby *Falco subbuteo*, juvenile, migrant (same as **961**). Note distinct tail-barring, typical of juveniles, and the dark central tail feathers. Finland, 5.9.2008 (DF)

964. Eurasian Hobby *Falco subbuteo*, juvenile, migrant. A rather darker bird, which also shows faintly rufous trousers, all part of the individual variation. Finland, 20.9.2014 (DF)

965. Eurasian Hobby *Falco subbuteo*, juvenile, migrant. From a distance the wingtips appear darker than the rest of the upperparts. Also note the plain dark uppertail when folded. Finland, 20.9.2014 (DF)

966. Eurasian Hobby *Falco subbuteo*, juvenile, migrant. In close views note finely scalloped upperparts and unbarred tail. The crown varies from dark to rather pale, as here. Also note the plain dark uppertail when folded. Finland, 30.9.2010 (DF)

967. Eurasian Hobby *Falco subbuteo*, recently fledged juvenile. Finland, 20.8.2014 (DF)

968. Eurasian Hobby *Falco subbuteo*, 2nd cy in transitional plumage, migrant. Largely adult-like, but note distinctly barred juvenile tail feathers and the very abraded trailing edge of the wings. Note also freshly moulted median primary (p4), typical of this age-class. Finland, 6.9.2005 (DF)

485

969. Eurasian Hobby *Falco subbuteo*, 2ⁿᵈ cy in transitional plumage. Upperwings are clearly brownish, compared to more uniformly grey in adults. This bird has replaced one median primary in each wing. Finland, 9.8.2015 (DF)

970. Eurasian Hobby *Falco subbuteo*, breeding adult. Note uniformly patterned and rather darkish underwing, and rufous thighs, vent and undertail-coverts. The tail-barring is less contrasting compared to juveniles. Finland, 6.9.2005 (DF)

971. Eurasian Hobby *Falco subbuteo*, adult male, breeding (paired to **974**). Males are slimmer than females, mostly with unmarked rufous parts of the rear underbody, and their underwings tend to be more finely barred. Finland, 19.8.2014 (DF)

972. Eurasian Hobby *Falco subbuteo*, adult male, breeding. Finland, 8.8.2012 (DF)

973. Eurasian Hobby *Falco subbuteo*, adult male, breeding (mate of **976**). Finland, 31.8.2012 (DF)

974. Eurasian Hobby *Falco subbuteo*, adult female, breeding (mate of **971**). Females are more robust with broader wings and they are usually more heavily marked below, including the rufous plumage tracts of the rear underbody. Finland, 15.8.2014 (DF)

975. Eurasian Hobby *Falco subbuteo*, adult female, breeding (mate of **972**). Finland, 8.8.2012 (DF)

976. Eurasian Hobby *Falco subbuteo*, adult female, breeding (paired with **973**). Finland, 1.9.2012 (DF)

977. Eurasian Hobby *Falco subbuteo*, adult male, breeding (same as **971**). Adults are uniform above with marginally darker primaries, the males being greyer than the duller females. Finland, 1.9.2012 (DF)

ELEONORA'S FALCON
Falco eleonorae

VARIATION Monotypic.

DISTRIBUTION As a breeding bird confined to the islands of the Mediterranean and the NW corner of Africa, wintering in Madagascar, which is reached following a direct route across the African continent from the breeding colonies. Mostly non-breeding first-year birds occur as rare summer vagrants to S Europe and further north, occasionally reaching W Europe and the Nordic countries.

BEHAVIOUR A highly aerial falcon, spending much of its time on the wing hunting for insects or birds. Perches prominently on dead treetops, outside the breeding season often several together. Breeds colonially and is also typically gregarious on the wintering grounds.

SPECIES IDENTIFICATION A slim and medium-sized falcon with proportions varying depending on age and sex. Occurs in two distinct colour morphs, light and dark. Adults are normally rather straightforward to identify when seen well, but juveniles and older immatures can be confusingly similar to other smaller falcons, Eurasian Hobby in particular. The underwing markings are diagnostic in most plumages, but good light is needed for confirmation, while head pattern and underbody colour may provide further clues. Appears noticeably small-headed compared to, for example, Eurasian Hobby.

MOULT The annual complete moult takes place mostly on the wintering grounds. Some females may start their primary moult on the breeding grounds, before suspending it in late Oct in time for the autumn migration, but most adults migrate south with an old set of flight feathers. The moult is completed on the wintering grounds by Feb/Mar. Juveniles undergo a partial body moult during their first winter, but this moult does not include the remiges. The extent of the body moult varies individually, which explains why returning first-year birds are so variable. Immatures start the complete moult of the primaries during their first summer, when about one year old, and by autumn they have replaced 0–4 of their median primaries.

PLUMAGES In autumn three different age-classes can be separated: juvenile, a 2nd cy transitional plumage, and adult; in spring only two: immature and adult. Dark and light morph birds are found in all age-classes.

Juvenile (1st cy autumn)

Easy to age by scaly upperparts, broadly white-tipped tail and distinctly marked underwings, with a broad dark subterminal band. The tail is distinctly barred throughout save for the central pair, which is all-dark as in Eurasian Hobby, but the pale bands tend to be broader than the dark (cf. juvenile Eurasian Hobby), and the outermost dark band is often clearly broader than the rest, unlike in Eurasian Hobby. However, the barring of the tail varies depending on the colour-morph of the bird, dark juveniles having broader dark markings compared to lighter birds. The underparts are buffish with the dark streaking of the breast changing to broader arrowheads on the flanks, while the rear underbody always lacks rusty-red tones. The head markings also differ from Eurasian Hobby (with rare exceptions) in showing a neatly rounded pale cheek-patch, lacking the notch behind the eye of Eurasian Hobby.

Light morph juvenile About 60% of the birds belong to the light form, but the percentage varies somewhat geographically. The underwing-coverts are more sparsely patterned compared to dark morph juveniles, and may be similar to Eurasian Hobby, while the lower belly, vent and undertail-coverts are uniformly buffish lacking dark markings. The breast streaking is fine, with dark streaks covering less than 50% of the surface, but this varies depending on the general degree of pigmentation (amount of melanin).

Dark morph juvenile About 40% belong to the dark form. The underwing-coverts are darker, often contrasting with the rest of the underwing and the underparts also show dark markings on the lower belly, vent and undertail-coverts. The light ground colour is more deeply tanned throughout and the streaking is heavier compared to light morph juveniles, with the dark breast-streaks often covering more than 50% of the ground colour. A small percentage (*c.* 3–4%) of the birds are almost completely dark (homozygous dark morph birds), but they can be told from dark adults by their broadly pale tail-tip and faint light markings on the upperparts.

Transitional plumage (2nd cy)

This is a plumage comprising partly retained juvenile feathers and partly moulted adult-type feathers, and is carried from the first winter to the second. The plumage is highly variable and usually poorly described in the literature; some birds in this plumage can appear confusingly similar to Eurasian Hobby.

During their first winter juveniles start to moult the body plumage, but the extent of this partial moult varies individually. As a rule it is more extensive than in Eurasian Hobby, but the great variation ensures that practically every individual has a unique appearance. The juvenile remiges are retained in this partial moult, mostly also the tail, although some odd tail feathers may be moulted, most often the central pair. During the 2nd cy summer the immatures also start to moult their primaries and by autumn most of them show a few fresh median primaries, which stand out from afar as being both darker and longer than the neighbouring juvenile feathers.

Light morph transitional plumage Most birds have moulted the entire underbody and some may show a very Eurasian Hobby-like plumage, with rufous vent and undertail-coverts contrasting with a white breast with fine black streaks. Always told from adults and from Eurasian Hobby by their retained juvenile remiges, which show a contrasting broad dark trailing edge, and many birds (but far from all) also show contrasting darker underwing-coverts. Upperparts look tatty, with new feathers of the rump and mantle contrasting with worn and brown upperwings and tail, the uppertail in particular looking very faded brown. A good ageing feature and a reliable identification character compared to Eurasian Hobby are the worn inner secondaries and tertials of the upperwing, which always show some rusty barring, lacking in the more uniform Eurasian Hobby. The tail and uppertail-coverts also show clearly more rufous barring than Eurasian Hobby, making the entire rear body and tail stand out as more marked and more rufous compared to the duller brown wings.

Dark morph transitional plumage Similar to light morph birds of the same age, but the underbody is largely dark, often heavily mottled, not uniformly dark as in adults, and the underwing-coverts are always clearly darker than the barred remiges. Upperparts are similarly worn as in light morph birds, with some retained rufous markings on the tertials and tail. All-dark immatures are best told from similar-looking

adults by their more distinctly barred tail and bases of underwing primaries, and in spring by their partly worn plumage.

Adult (3rd cy spring and older)

In flight adults (but not juveniles) are recognised by slim proportions, with markedly long wings and tail, and they appear clearly longer-armed compared to Eurasian Hobby. When soaring with fully spread wings and tail the outer primaries are clearly bulging and the tail-base appears noticeably narrow. Adults show uniformly dark underwing-coverts (sometimes finely pale-spotted, probably in younger adults) and the remiges show at the most faint paler barring at the bases of the longest primaries, more obvious on younger adults and more distinct in females than males. Also, the tail-barring is more diffuse than in previous plumages, becoming even fainter with increasing age. Adults are very uniform above, often dark coffee-brown with a greyer mantle, wing-coverts and rump/inner tail, and more grey in males than in females and greyer still in older males. Some adults are still quite distinctly marked on the underwing-coverts, primaries and tail, and these may be birds in their first-adult plumage (3rd cy), but further studies of known-age birds are required to confirm this.

Light morph adult Always told from dark morph birds by the contrasting Eurasian Hobby-like head pattern, but lacking Eurasian Hobby's second dark notch behind the eye and also lacking the pale neck-patches of Eurasian Hobby. Rather variable below, with males tending to be more deeply and more uniformly coloured than females. The ground colour of the underbody is whitish with a rufous hue, the rustiness varying in extent from just covering the rear underbody (much as in adult Eurasian Hobby) to covering the entire underbody up to the white throat. The dark streaking of the underbody may be rather sparse and restricted to the lower breast and vent, or it may cover the ground colour of the underside completely, just leaving the throat and cheeks white.

Dark morph The entire bird is uniformly dark coffee-brown, with a slight greyish tone to the upperparts in males, with somewhat darker wingtips and outer tail above. The underwing shows a marked contrast between dark coverts and more silvery remiges, while the tail looks grey from below, sometimes with faint barring proximally. Old birds appear very uniform, including the underside of the remiges and tail, while presumably younger adults may show some markings to the underwing-coverts and belly, with the bases of the primaries and tail faintly barred.

SEXING Adult males are slimmer-bodied and narrower-winged and longer-tailed compared to adult females, but this difference is not apparent in younger birds. Cere and eye-ring are yellow in adult males, bluish in adult females. In juveniles these bare parts are pale greenish regardless of sex.

CONFUSION RISKS Although Eleonora's Falcons are often described to have a distinctive flight, they are in fact very similar to other smaller falcons. The behaviour may be distinctive at the breeding colonies, but vagrant stragglers may behave and fly like other falcon species of similar size and structure. The shape of the extremely slim adult males may be diagnostic, but adult females are far less striking and juveniles do not differ that much in silhouette from juveniles of other falcon species. In a European context the Eurasian Hobby, but also juvenile Red-footed Falcon, is the most likely confusion risk, but juvenile Lanners and Peregrines could equally be potential pitfalls, although they usually look heavier-bodied with a stronger flight.

978. Eleonora's Falcon *Falco eleonorae,* recently fledged juvenile of light morph. Superficially resembles juvenile Eurasian Hobby, but note on average darker underwing with prominent dark trailing edge and lighter, more finely streaked underbody. Cyprus, 17.10.2014 (DF)

979. Eleonora's Falcon *Falco eleonorae,* recently fledged juvenile of light morph. Open tail is lighter than in juvenile Eurasian Hobby, owing to finer dark barring, but the central tail feathers are similarly unbarred and black. Cyprus, 17.10.2014 (DF)

980. Eleonora's Falcon *Falco eleonorae,* recently fledged juvenile of light morph. Cyprus, 17.10.2014 (DF)

981. Eleonora's Falcon *Falco eleonorae*, recently fledged juvenile of dark morph. Note dark underwing-coverts, heavily barred undertail-coverts and deeper rusty-buff ground colour compared to light morph juveniles. Cyprus, 17.10.2014 (DF)

982. Eleonora's Falcon *Falco eleonorae*, recently fledged juvenile of light morph. Differs from juvenile Eurasian Hobby by more distinctly scalloped upperparts; in fresh plumage also by paler crown and lack of pale neck-spots. Cyprus, 17.10.2014 (DF)

983. Eleonora's Falcon *Falco eleonorae*, 2nd cy bird of light morph in transitional plumage. This bird is still in practically complete juvenile plumage, save for moulted head and underbody. Birds like this, with juvenile proportions but partly adult plumage can be extremely similar to Eurasian Hobby. Spain, 8.10.2007 (DF

984. Eleonora's Falcon *Falco eleonorae*, 2nd cy bird of light morph in transitional plumage. This bird, also with retained juvenile wings and tail, is easy to identify thanks to contrasting dark underwing-coverts and distinct dark trailing edge to the wings. The head and underbody have been moulted. Spain, 7.10.2007 (DF)

985. Eleonora's Falcon *Falco eleonorae*, 2nd cy bird of light morph in transitional plumage. Apart from head and body, this bird has also moulted in some adult-type primaries and secondaries, as well as about half of its lesser underwing-coverts. Differs from Eurasian Hobby by uniformly dark new primaries, while the pale bars of the juvenile tail feathers are wider than in juvenile Eurasian Hobby. Cyprus, 17.10.2014 (DF)

986. Eleonora's Falcon *Falco eleonorae*, 2nd cy bird of light morph in transitional plumage. This bird has moulted its head and body, and has some new median primaries and secondaries. Note how the moulted central tail feathers project well past the tip of the juvenile feathers, indicating the change in proportions when moulting from juvenile to adult plumage. Spain, 8.10.2007 (DF)

987. Eleonora's Falcon *Falco eleonorae*, 2ⁿᵈ cy bird of light morph in transitional plumage. Easily identified by its worn and clearly barred juvenile uppertail and the worn brownish upperwings contrasting with newly moulted grey feathers of the scapulars, mantle and back/rump. This bird has moulted p4 in both wings. Spain, 8.10.2007 (DF)

988. Eleonora's Falcon *Falco eleonorae*, 2ⁿᵈ cy bird of dark morph in transitional plumage. Told from dark morph adults by distinctly barred juvenile flight feathers and tail. Spain, 8.10.2007 (DF)

989. Eleonora's Falcon *Falco eleonorae*. This presumed 3ʳᵈ cy dark morph features a rather mottled underbody and still retains some pale markings on the underwing-coverts and tail, features usually not found in full adults. Note the adult-type uniform flight feathers. Spain, 7.10.2007 (DF)

990. Eleonora's Falcon *Falco eleonorae*. Presumed 3ʳᵈ cy bird of the dark morph. See comment on **989**. Cyprus, 17.10.2014 (DF)

991. Eleonora's Falcon *Falco eleonorae*, adult male of the light morph, breeding. The intensity of the colour as well as the streaking of the underbody varies individually, but light morph birds always show a Eurasian Hobby-like head. Note uniformly dark underwings, typical of full adults. The sex is revealed by the yellow eye-ring and cere. Cyprus, 17.10.2014 (DF)

992. Eleonora's Falcon *Falco eleonorae*, adult female of light morph, breeding. The rather intensely marked underwing-coverts and tail may suggest a younger age for this adult female. This bird is sexed by its bluish eye-ring and cere. Spain, 8.10.2007 (DF)

993. Eleonora's Falcon *Falco eleonorae*, adult male of light morph, breeding. Some (old?) adults can look very dark below. Spain, 7.10.2007 (DF)

994. Eleonora's Falcon *Falco eleonorae*, adult female of the light morph, breeding. Females are clearly broader-winged and heavier-bodied than males. Cyprus, 17.10.2014 (DF)

995. Eleonora's Falcon *Falco eleonorae*, adult male of the light morph, breeding. Cyprus, 17.10.2014 (DF)

996. Eleonora's Falcon *Falco eleonorae*, adult male of the light morph, breeding. Males are lighter and greyer above than females. Cyprus, 17.10.2014 (DF)

997. Eleonora's Falcon *Falco eleonorae*, adult male of the light morph, breeding (same as **996**). Cyprus, 17.10.2014 (DF)

998. Eleonora's Falcon *Falco eleonorae*, adult female of the dark morph, breeding. Dark morph adults appear uniform, with strong contrast between dark underwing-coverts and glossy grey flight feathers. Cyprus, 17.10.2014 (DF)

999. Eleonora's Falcon *Falco eleonorae*, adult male of the dark morph, breeding. Compare the proportions with the female in **998**. Spain, 6.10.2007 (DF)

1000. Eleonora's Falcon *Falco eleonorae*, adult male of the dark morph, breeding. More uniform above than adult light morph, but the difference is rather marginal. Spain, 6.10.2007 (DF)

SOOTY FALCON
Falco concolor

VARIATION Monotypic

DISTRIBUTION Rare inhabitant of coastal islands in the Red Sea and along the coasts of Arabia, with isolated inland populations in eastern parts of the Sahara and the Negev Desert.

BEHAVIOUR Highly aerial, feeding on insects for most of the year, but raising young in autumn on migrant songbirds.

SPECIES IDENTIFICATION An elegant and slim falcon, with adults and juveniles clearly different in shape. Adults are the longest-winged of any falcon, but comparatively short-tailed, and in active flapping flight the tips of the sharply bent wings appear to reach beyond the tip of the tail. Juveniles are far less extreme in shape and are in fact proportioned like juvenile Eurasian Hobby or Eleonora's Falcon, with comparatively shorter and broader wings and a longer tail compared to the adults. Juveniles are best told from other similar small falcons by their very dark underwings, with diffusely marked coverts and practically unbarred blackish remiges (individually variable subtle barring is confined to the feather-bases). Adults are uniformly grey, with darker wingtips and tail, and conspicuous yellow feet and cere.

MOULT In juveniles moult commences during the first autumn with a partial body moult, comprising most of the body and some of the upperwing-coverts, but not the underwing-coverts nor the remiges. Body moult is completed during the first winter on the wintering grounds. Adults start their annual complete moult during the breeding season, females earlier than males, but moult is suspended before autumn migration. By Oct, most females show 3–4 moulted median primaries, while most males have moulted only 1–2 primaries, if any. At the same time most 2nd cy birds show on average two new median primaries. Many adult males start their moult only after arriving on the wintering grounds in Nov, at the same time that females resume their suspended moult. Moult is completed in late winter, in good time before the spring passage, by which time the adults are in mint condition.

PLUMAGES In autumn three age-classes can be separated: juvenile, transitional and adult; in spring and summer only immature and adult; occasionally a third plumage can be recognised.

Juvenile

Superficially similar to juvenile Eurasian Hobby, and only marginally bigger. Upperparts are almost identical, while the underparts are distinctly different and diagnostic. The ground colour below is deeper than in juvenile Eurasian Hobby, more tawny-buff, and the darker streaking is typically rather diffuse, which is diagnostic. The underwing-coverts appear rather dark, with subtle and diffuse paler fringes, and the remiges are also very dark overall, with some faint pale barring showing furthest in, next to the coverts. The whole underwing thus appears very dark from below, lacking the contrasting markings of juvenile Red-footed or Eleonora's Falcons, being darker and even more uniform than in juvenile Eurasian Hobby. Also, the undertail lacks distinct barring, with visible barring confined to the inner part of the feathers, leaving the outer part uniformly black. Both flight feathers and tail feathers show distinct pale tips in fresh plumage, which, however, wear off during the first winter.

The head is rather similar to juvenile Eurasian Hobby, although it appears darker, thanks to the broader and blunter moustache.

The upperparts appear dark, with distinct tawny fringes and tips, and with black primaries contrasting with the greyer innerwing. The tail looks uniformly dark above with a broad pale tip and the proximal barring becomes visible only when the tail is fanned to the extreme.

During the winter the pale fringes of the upperparts wear off, and the dark ground colour wears down to a drab greyish-brown. At the same time the body plumage starts to moult, and by spring most of the body has moulted into an adult-like plumage, while the wings and the tail feathers remain juvenile.

Transitional plumage (2nd cy autumn)

Birds resemble adults greatly in this plumage, although only a part of the juvenile plumage has been replaced. Since all the pale fringes of the upperparts, so conspicuous in fresh juveniles, have worn off, the upperparts have become uniform and drab. The upperwing looks uniformly dull brownish-grey, lacking the diagnostic contrast of the adult between pale bluish-grey innerwing and darker tip. The underbody is largely moulted and adult-like, although often appearing somewhat mottled, and a few streaked juvenile feathers may remain, but this varies individually. Although the head has been moulted it differs from adults in showing paler cheeks and chin, and a darker moustache in most. The flight feathers (remiges and tail) are still largely juvenile (see under Moult), but good conditions are required to see the faint barring at the base of the remiges, while the barring in the tail is easier to see. The underwing-coverts are still mostly juvenile, but because of wear they too look uniform and do not differ that much from an adult's.

Third-plumage

Some birds in this age-class can be separated from the older adults. Birds with adult-type wing and tail feathers, lacking any type of barring, but with heavily worn upperparts in late summer–early autumn, are thought to be in their 3rd cy (pers. obs.). They have moulted the upperparts extensively during the previous (2nd cy) summer, replacing the juvenile plumage with the first-adult plumage, and these feathers have not been replaced again since,

explaining their faded and worn condition. Older adults would replace their plumage in winter, and would turn up in perfect condition on the breeding grounds.

Adult

Very uniform in colour, beautiful slate-grey to ashy-grey. On the upperparts the darker outer tail and the dark primaries stand out from afar, a feature that is unique among small falcons, but reminiscent of adult Peregrine. The underside is very plain, with the body usually appearing paler than the uniformly grey underwing with a darker tip. The bare parts are bright yellow to orange-yellow and really stand out against the grey plumage.

SEXING Adult males are clearly slimmer and narrower-winged than adult females and tend to be lighter bluish-grey above, possibly also with more orange-yellow bare parts compared with females. Juveniles cannot, as far as is known, be sexed on silhouette or plumage.

CONFUSION RISKS Adults are rather straightforward to identify given good views, but juveniles and some transitional immatures need care in order to tell them from similar plumages of Eurasian Hobby, and Eleonora's and Red-footed Falcons.

1001. Sooty Falcon *Falco concolor*, recently fledged juvenile. Identified by rich sandy-buff ground colour with diffuse streaking on underbody, broadly black, unbarred outer tail, faintly marked, rather dark underwing remiges, and a Eurasian Hobby-like head. Egypt, 28.10.2008 (DF)

1002. Sooty Falcon *Falco concolor*, recently fledged juvenile. The underwings are much more uniform and rather dark compared to any other juvenile falcon. Note that the pale tail-barring stops well before the tail-tip and that the markings of the underbody and underwing-coverts are diffuse. Egypt, 28.10.2008 (DF)

1003. Sooty Falcon *Falco concolor*, recently fledged juvenile. The upperparts are very similar to juvenile Eurasian Hobby, except for the rich tawny-buff colour of the feather margins and cheeks. Egypt, 28.10.2008 (DF)

1004. Sooty Falcon *Falco concolor*, 2nd cy bird in transitional plumage. The body has largely moulted to grey adult plumage, but the tail and wings are still largely juvenile, except for a few replaced median primaries and secondaries. Egypt, 28.10.2008 (DF)

1005. Sooty Falcon *Falco concolor*, 2nd cy bird in transitional plumage. Note the dark wedge formed by replaced adult-type median primaries, while the pale throat and cheeks are a sign of immaturity shown by many 2nd cy birds. Egypt, 28.10.2008 (DF)

1006. Sooty Falcon *Falco concolor*, 2nd cy bird in transitional plumage. From a distance 2nd cy birds are easily mistaken for adults. Note that both juvenile wing as well as tail feathers are very sparingly barred and that the retained juvenile underwing-coverts appear almost uniform. Egypt, 28.10.2008 (DF)

1007. Sooty Falcon *Falco concolor*, adult male, breeding. Males are slimmer than females, with a thinner body and narrower wings. Note arrested moult, with one new primary, p4; the others will be moulted on the winter quarters. Egypt, 28.10.2008 (DF)

1008. Sooty Falcon *Falco concolor*, adult female, breeding. Females are more thickset and broader-winged. The retained subtle tail-barring and the pale chin may indicate this bird is in its 3rd cy. Egypt, 28.10.2008 (DF)

1009. Sooty Falcon *Falco concolor*, adult female, breeding. Adult Sooty Falcons proportionately have the longest wings of all falcons, but a relatively short tail. Egypt, 28.10.2008 (DF)

1010. Sooty Falcon *Falco concolor*, adult, breeding. The upperparts are a beautiful silvery-grey with clearly contrasting dark tail and hand. Egypt, 28.10.2008 (DF)

1011. Sooty Falcon *Falco concolor*, adult, breeding. The bare parts around the eyes and bill are a deep yellow, well visible from afar. Egypt, 28.10.2008 (DF)

MERLIN
Falco columbarius

VARIATION Two very similar subspecies in Europe: in N Europe ssp. *aesalon*, wintering in C and S Europe south to N Africa and the Middle East, and the slightly larger ssp. *subaesalon* in Iceland, wintering in the British Isles. Breeding birds from Britain, Ireland and the Faeroes are intermediate between these two. N American nominate *columbarius* is a rare vagrant to the Atlantic islands, possibly also to the European mainland, while ssp. *pallidus* from the steppes of C Asia is a scarce winter visitor to the Middle East.

DISTRIBUTION Breeds on moors and heathlands of the British Isles and Iceland and in the boreal taiga zone and tundra of N Europe. Migrating and wintering birds may be found anywhere in the region, mostly in open areas, but hunts also in open woodland.

BEHAVIOUR Mostly identified by very fast and rocketing flight low over the ground when hunting small birds, the main prey. Active flight on migration is a series of fast wing beats, with rather short but very pointed wings, interspersed by short glides, with flapping sequences longer and glides shorter compared to most other small raptors.

SPECIES IDENTIFICATION A very small falcon with comparatively broad-based but very pointed wings and a rather long tail. Could be mistaken for a Eurasian Hobby, but the silhouette is rather diagnostic with its shorter wings and longer tail, and the dashing flight is also different. Lacks strong head-markings in all plumages. The row of bright spots to the rear upper hand is diagnostic, giving the wing a speckled and translucent impression when seen against the light, a feature not found in other falcons. The tail-barring in juveniles and females, with sparse, almost equally wide pale and dark bands, is another diagnostic feature separating Merlins from other small falcons. Adult males show unmistakable bluish-grey upperparts, dark outer wings and a broad black subterminal tail-band.

MOULT Complete annual mouth during the breeding season, between Apr/May and Sep/Oct. Juveniles start to replace their body feathers by late first winter, permitting the reliable sexing of juveniles.

PLUMAGES Two age-classes can be separated throughout the year: adults and juveniles. Adult males

and females are easy to tell apart, while all juveniles are very similar to adult females, particularly in spring when the upperparts are faded and more greyish-brown. Sexing juveniles requires excellent conditions and is often not possible in the field.

Juvenile

Very similar to adult female, but upperparts are more uniform, rufous-brown or may show a greyish bloom, all depending on the angle of the light, compared to the more clearly barred and more greyish-brown upperparts of adult females. The underparts are yellowish-buff with dark brown streaking, changing to heavier and broader markings towards the flanks. It is often not possible to distinguish a juvenile from an adult female in normal fly-by situations.

By late winter juveniles start to moult some body feathers of the upperparts and by then males in particular can be identified by the contrast between old juvenile feathers, which are brown, and fresh grey adult-type feathers, although these may be difficult to discern in flight. Spring juveniles can be very difficult to tell from older females, as the uniformly dark brown upperparts of the fresh juvenile have become greyer-brown through wear, but they still lack the barred upperparts typical of adult females.

Sexing juveniles is sometimes possible. Males are distinctly smaller than females, with narrower wings, but this is rarely of use in the field without direct comparison. The inner pale bands of the uppertail tend to be greyish in males and more rusty-buff in females, but the difference is not constant, as some males show buffish bands.

Adult male

Unmistakable, with bluish-grey upperparts and dark primaries, a broad, black subterminal tail-band and a golden neck-band. Finer dark bars, or partial bars, can be seen on a fanned tail in many individuals. Head markings are mostly ill-defined, except for a clearly darker crown in some. Underparts are rusty-yellow to off-white, with more deeply coloured trousers, variably streaked, with broader and browner streaks in some but fine and black streaking in others, while the remiges are distinctly barred from below in black and white.

Adult female

Very similar to the juvenile, but upperparts are greyer-brown and slightly more variegated, with barring across the mantle, rump and upperwing-coverts, and the rump stands out as greyer and paler than the rest of the upperparts. The underparts are often whiter and the streaks are more drop-shaped, lighter brown and less contrasting compared to juveniles, changing to more distinct barring on the flanks. In summer and early autumn adults are always told from fresh juveniles by signs of wing moult.

CONFUSION RISKS It is possible to confuse Merlins with other small falcons or even Eurasian Sparrowhawk, but flight behaviour and silhouette are distinctive.

IDENTIFICATION OF VAGRANT SUBSPECIES

Ssp. *pallidus* 'Steppe Merlin' from the steppes of C Asia is overall paler and more finely marked below, like a washed-out version of *aesalon*. Adult males are very pale grey above and pale rusty-yellow below, with faint head markings and much finer underwing barring, and with primaries also strongly marked above, unlike in *aesalon*. Adult females and juveniles resemble corresponding plumages of *aesalon*, but are overall paler, with finer dark markings on the underwings and more clearly marked upperwings, making the whole bird look like a bleached *aesalon*.

North American *columbarius* ('Boreal' or 'Taiga' Merlin) is clearly darker than *aesalon* throughout. Adult males are dark grey above, the underparts are strongly marked and the dark grey tail shows two broad dark bands inside the broad subterminal band. Adult females and juveniles are clearly darker, and more uniformly slate-coloured above than *aesalon*, lacking the rich rusty fringes and markings of the latter. The single most important identification feature is the *dark tail with three narrow pale bands ('string of pearls')*, best seen on the uppertail.

1012. Merlin *Falco columbarius*, juvenile, migrant (all images refer to ssp. *aesalon*, unless otherwise stated). Short wings with broad arm and pointed hand, and a rather long tail describe the typical Merlin silhouette. Head markings are less striking than in Eurasian Hobby-like small falcons and the underside colours are yellowish-buffs and browns. Finland, 27.9.2005 (DF)

1013. Merlin *Falco columbarius*, juvenile, migrant. Note diagnostic proportions and unique chestnut underwing-coverts with pale spotting. Finland, 21.9.2014 (DF)

1014. Merlin *Falco columbarius*, juvenile, migrant (same as **1013**). Juveniles are richer buff below with darker brown streaking compared to the whiter and less contrastingly marked adult females. Finland, 21.9.2014 (DF)

1015. Merlin *Falco columbarius*, recently fledged juvenile. Finland, 19.8.2011 (DF)

1016. Merlin *Falco columbarius*, juvenile, migrant, (same as **1013**). The upperparts and inner wings appear uniformly brown in juveniles, while the primaries show diagnostic pale spotting. Finland, 21.9.2014 (DF)

1017. Merlin *Falco columbarius*, adult male, breeding. Adult males show strongly barred remiges, rusty-yellow underparts with variable dark streaking, and a broad subterminal tail-band, while the head-markings are rather indistinct. Norway, 14.7.2011 (DF)

1018. Merlin *Falco columbarius*, adult male, breeding. Norway, 23.5.2013 (DF)

1019. Merlin *Falco columbarius*, adult male, breeding. Adult males are a beautiful bluish-grey above with a rusty neck-band, a broad, dark tail-band, and slightly darker, boldly spotted primaries. Norway, 31.5.2015 (DF)

1020. Merlin *Falco columbarius*, adult female, breeding. Adult females recall juveniles, but the breast markings are drop-shaped rather than streaks, and individual marks are lighter and less contrasting. The tail-barring is diagnostic for females and juveniles, showing sparse barring, with dark bars slightly broader than the light bars. Finland, 1.6.2014 (DF)

1021. Merlin *Falco columbarius*, adult female, breeding. The upperparts are more variegated than in juveniles, with more clearly marked upperwing-coverts in particular. Finland, 15.5.2014 (DF)

1022. Merlin *Falco columbarius*, adult female, breeding (same as **1020**). Note diagnostic tail-barring and heavily marked remiges, and compare with other small falcons. Finland, 1.6.2014 (DF)

1023. Merlin *Falco columbarius*, adult male of ssp. *pallidus*, breeding. Note bleached appearance with diluted head pattern and finer underwing markings compared to ssp. *aesalon*. Kazakhstan, 13.5.2010 (DF)

1024. Merlin *Falco columbarius*, adult male of ssp. *pallidus*, breeding. Note almost whitish head and light grey upperparts with distinctly marked upperwing primaries. Kazakhstan, 13.5.2010 (DF)

1025. Merlin *Falco columbarius*, adult female of ssp. *pallidus*, breeding. Strikingly pale with a whitish head and washed-out fine markings to the underwings and body. Kazakhstan, 11.5.2010 (DF)

1026. Merlin *Falco columbarius*, juvenile of ssp. *columbarius*, Taiga or Boreal Merlin. Note uniformly dark brown upperparts and uppertail with narrow, pale tail-bands. USA, autumn (Jerry Liguori)

LANNER FALCON
Falco biarmicus

VARIATION Five subspecies are usually recognised, with S European ssp. *feldeggii* being the darkest and most heavily marked; ssp. *erlangeri* of NW Africa is smaller and paler, with adults often poorly marked below, while ssp. *tanypterus* of the Middle East and NE Africa is intermediate between the former two. Two more subspecies are found in sub-Saharan Africa, ssp. *abyssinicus* across the continent south of Sahara (very similar in appearance to *tanypterus)*, and the rather distinct nominate ssp. *biarmicus* over most of tropical and S Africa. Subspecific identification, except for the rather distinct-looking adults of *feldeggii* and *biarmicus*, is difficult or even impossible (in the case of juveniles), as the individual and age- and sex-dependent plumage variation within each subspecies is considerable and intermediate types occur. The three subspecies fringing the Sahara, *erlangeri, tanypterus* and *abyssinicus* are all rather alike and one may question whether they are sufficiently distinct to warrant subspecific status.

DISTRIBUTION European *feldeggii* is a rare breeding resident in Italy, the Balkans and Asia Minor, while N African *erlangeri* and Middle Eastern *tanypterus* show somewhat stronger populations. Dispersing juvenile *erlangeri* from Morocco frequently reach Spain.

BEHAVIOUR Hunts mainly medium-sized and small birds, locally also rodents where abundant. Active flight more relaxed than Peregrine's, on longer, more flexible wings and longer tail. Soars freely and makes low swoops barely above ground when hunting rodents or larks. Often hunts in pairs, with male and female performing together.

SPECIES IDENTIFICATION Looking at plumage features adults are most similar to adult Peregrines or Barbary Falcons, but differ in head, underwing and flank patterns and colour of upperparts. Their silhouette, however, differs clearly from birds of the Peregrine-group in having a longer and slimmer-based tail and more evenly broad wings, with a broader and more rounded hand. Juveniles are most similar to juvenile Saker and separating the two can be a real challenge, particularly in the case of *feldeggii*, but details of head, underbody and underwings differ in most. Note that juveniles have a slimmer silhouette than adults, largely owing to narrower wings and

longer tail, and juveniles may resemble juvenile Barbary Falcon or even juvenile Eleonora's Falcon.

MOULT Complete annual moult during the breeding season. Juveniles acquire an adult-type head and upper breast in a partial body moult during the first autumn or winter.

PLUMAGES Juveniles and adults are easily separated. Adult birds continue to change as they age, with the markings of the underbody becoming finer and sparser, and the birds thus turning lighter. Adult males are usually paler and less heavily marked below compared to females of a similar age.

Juvenile

Uniformly brown above, with rufous feather margins of mantle and coverts visible only in fresh plumage, and with the tail appearing slightly paler brown when seen from a distance. Head pattern varies individually, and also depending on subspecies, but is mostly distinct, with contrasting solid, dark, narrow and pointed moustache connected to a similarly dark line through the eye and continuing down the side of the neck. As a rule the head markings of juvenile Lanners are darker and more contrasting compared to juvenile Sakers, but exceptions occur, notably in the more streaked and less diagnostic *feldeggii*, while the lightest birds (Saharan *erlangeri*) may lack head markings almost completely. Contrary to Peregrine/Barbary and Saker, most birds show a dark diadem across the forecrown, leaving the rest of the crown and hindneck orange-buff, with a black, downward-pointing triangle on the nape, but again, juvenile *feldeggii* can look very similar to the aforementioned species. The underparts are creamy with darker streaking confined to the upper breast and flanks, while the lower breast and lower flanks, as well as the vent, trousers and undertail-coverts are contrastingly pale, a good character compared to most Peregrines and Sakers. Lighter-coloured birds are finely streaked below, while more pigmented birds can appear completely dark on the upper breast, as the dark streaks merge with each other. The underwings show a strong contrast between dark coverts and paler grey remiges, similar to Saker but unlike Peregrine/Barbary, and the greater and median coverts are variably pale-spotted. The primaries are finely barred, with extensively dusky tips (narrow and distinct in Peregrine/Barbary), while the secondaries are often darker greyish-brown, with barring visible only proximally, another similarity with Saker, but different from Peregrine/Barbary, which show distinctly barred secondaries. The tail pattern varies individually, but all juveniles show broad white tips to the tail feathers when fresh. The rest of the tail-barring varies as in juvenile Peregrine, from more complete ochre barring to just oval spots confined to the inner vanes of each feather. Usually the pattern is only visible from below, but when the tail is fanned, also from above. The central pair of tail feathers can be uniformly brown, or may show transverse spotting, but the ochre spots are always oval, not round as in Saker. Although most juvenile Lanners already show yellow feet when leaving the nest, birds with pale greenish or even bluish legs do occur.

Adult

Ashy-grey above (not bluish as in many Barbary Falcons) with a lighter uppertail, *lacking the contrast between the distinctly paler rump and darker outer tail of Peregrine/Barbary*. The crown is typically yellowish-ochre. Only at close range is the fine dark barring visible on the upperparts, strongest and darkest in European *feldeggii* and often almost lacking in *tanypterus/erlangeri*. Underparts are buffish-white, with variable dark spotting on the breast and belly, while the flanks can be either broadly barred or spotted, depending on the bird's age and provenance (European *feldeggii* being the most heavily barred and most coarsely marked). The crop area is poorly marked and adult Lanners lack the often clear-cut division between the white crop and barred breast of adult Peregrines. The narrow and pointed dark moustache and the dark eye-line curving down the side of the head are both mostly distinct, but some older birds (particularly in *erlangeri* and *tanypterus*) may show rather faint head markings, thus approaching certain Sakers. The underwing-coverts are pale, with the greater coverts more strongly marked than lesser and median, while the remiges are finely barred, with more extensively dark wingtips compared to Barbary or Peregrine. The silvery-grey uppertail is finely barred from base to tip, lacking Peregrine's and Barbary's dark and strongly barred outer tail.

SEXING Males are clearly smaller than females, but size is not easy to assess on single birds. While juvenile males and females appear similar in plumage, adult males tend to be less boldly marked below than adult females, but the extent and size of the markings of the underbody are also related to age.

CONFUSION RISKS Juveniles can be very similar to juvenile Sakers, in plumage as well as proportions, but they could also be confused with juvenile Peregrine and Barbary Falcons or even juvenile Eleonora's Falcon. Adults are mostly confused with adults of the sympatric Barbary Falcon, but should be readily recognised by the uniformly light grey uppertail, more extensively dusky wingtips below and more languid flight.

1027. Lanner Falcon *Falco biarmicus*, juvenile, presumed ssp. *erlangeri*, vagrant. Even at a distance differs from Peregrine/Barbary by dark breast and contrasting dark underwing-coverts. Interestingly, this bird still shows bluish feet. Spain, 16.9.2005 (DF)

1028. Lanner Falcon *Falco biarmicus*, juvenile, presumed ssp. *tanypterus*. Note broad dark wingtips and distinctly barred primaries contrasting with darker secondaries, a feature shared by juvenile Saker, but differing from Barbary/Peregrine (which see). Juveniles of the different Lanner subspecies are rather similar and individual variation prevents subspecific identification. Oman, 3.1.2005 (Tom Lindroos)

1029. Lanner Falcon *Falco biarmicus*, juvenile, presumed ssp. *erlangeri*, vagrant. Note recently moulted adult-type head and upper beast. Gibraltar, Mar 2015 (Stewart Finlayson)

1030. Lanner Falcon *Falco biarmicus*, juvenile, presumed ssp. *abyssinicus*. A rather average-looking juvenile with typical underwing and with body-streaking confined to the breast leaving the rear underbody pale and practically unmarked. The intensely spotted underwing-coverts are typical of juvenile Lanner and these markings differ from both Saker (more streaked) and Barbary/ Peregrine (barred). Ethiopia, 17.11.2011 (DF)

1031. Lanner Falcon *Falco biarmicus*, juvenile, presumed ssp. *abyssinicus*. A lighter and more loosely marked bird. Possibly, the head has already been moulted into adult-type plumage. Ethiopia, 17.11.2011 (DF)

1032. Lanner Falcon *Falco biarmicus*, juvenile, presumed ssp. *tanypterus* (same as **1028**). The upperparts are plain brown, already quite worn in this bird, and do not differ from juvenile Saker in any way. Oman, 3.1.2005 (Tom Lindroos)

1033. Lanner Falcon *Falco biarmicus*, adult male, ssp. *tanypterus*, resident. Adult males are often very finely marked below. Note finely spotted body, evenly barred tail, extent of dark wingtips and markings of underwing-coverts, and compare with Saker and Peregrine/Barbary. Egypt, 16.11.2008 (DF)

1034. Lanner Falcon *Falco biarmicus*, adult female, ssp. *tanypterus*, resident and paired with **1033**. This adult female is probably a younger bird judging by the heavier markings on the body and underwings. Egypt, 16.11.2008 (DF)

1035. Lanner Falcon *Falco biarmicus*, adult female, ssp. *abyssinicus*, resident. Although a representative of the sub-Saharan subspecies this female does not differ that much from an adult female *tanypterus*. The Gambia, 1.1.2012 (DF)

1036. Lanner Falcon *Falco biarmicus*, adult male, ssp. *erlangeri*, resident. An adult male of the smallest and palest of all Lanner subspecies. Note the fine body markings and faintly marked head, yet typically marked underwings. In desert populations the head markings may all but disappear, and in particular juveniles may look very similar to some pale juvenile Sakers. Morocco, 18.1.2014 (DF)

1037. Lanner Falcon *Falco biarmicus*, adult male, ssp. *erlangeri*, resident (same as 1036). Adult Lanners appear uniformly barred ashy-grey above with a slightly paler tail (compare tail-barring with adult Peregrine/Barbary). Morocco, 18.1.2014 (DF)

1038. Lanner Falcon *Falco biarmicus*, adult male, ssp. *abyssinicus*, breeding. Uniformly grey upperparts with an orange-buff crown are diagnostic of adult Lanners. Ethiopia, 9.2.2012 (DF)

SAKER FALCON
Falco cherrug

VARIATION A highly variable species, with a complex taxonomy. Formerly, several geographically separated subspecies were recognised, but most of these are currently explained as being part of the extensive individual and clinal variation. Occurs in the Western Palearctic as nominate *cherrug*; further east, in Mongolia and China, as ssp. *milvipes*, in which juveniles are similar to the nominate, while adults are heavily barred above (resembling a kestrel) with distinct barring on the flanks.

DISTRIBUTION Saker is a rare breeding bird from E Europe across the plains of S Russia and Asia Minor to C Asia and S Central Siberia. Populations in Europe are slowly increasing, while many areas in C Asia and Mongolia have been severely depleted by falconers, and the species is becoming increasingly rare. Nominate birds winter in the Middle East and N and NE Africa. Adults in particular winter near the breeding grounds, weather permitting.

BEHAVIOUR A skilful hunter, which feeds on medium-sized birds and mammals depending on local supply, with sousliks and feral pigeons being the staple food in many areas. Capable of impressive Peregrine-like stoops, but also indulges in long tail-chases to run down bird prey. Mammals are taken in low swoops close above the ground. Sakers are also skilled food-pirates, stealing prey from other raptors, harriers and buzzards in particular.

SPECIES IDENTIFICATION Many plumages are confusingly similar to Lanner or Gyr Falcon, especially those of juveniles, but also certain adult types. The proportions of juvenile Lanner and Saker are very similar, and correct identification always requires detailed plumage characters. Note also that juveniles are clearly slimmer than adults, with narrower wings and longer-looking tail, while adults look heavier and broader-winged, with proportions that may even suggest a Gyr Falcon. Juveniles of Saker and Gyr Falcon can be told by the underwing barring of the primaries, which is distinct and contrasting in Saker but greyish and muted in juvenile Gyr. Sakers also show rounded and distinct pale spots on the upperparts, compared to the more irregular and marbled markings in juvenile Gyr, but the spotting

may be completely lacking in both depending on the degree of pigmentation.

MOULT Complete moult annually during the breeding season, completed by Sep–Oct.

PLUMAGES A highly variable large falcon, with lighter and darker siblings even occurring in the same brood. Birds with a lighter-coloured head are usually also less heavily marked below and therefore look overall paler, while darker birds, particularly juveniles, can be rather similar to juvenile Gyr or Lanner Falcons. Saker is the only large falcon in which separating juvenile and adult plumages may prove difficult, but normally juvenile and adult plumages are separable by different markings on the underparts: streaked in the juvenile, spotted in older birds. However, birds in first-adult plumage can be rather heavily marked below and can easily be mistaken for a juvenile. The dark spots of adults become smaller with increasing age, and males are more finely marked compared to females of a similar age.

Juvenile

Resembles the juveniles of other large falcons, and great care is needed when identifying any large falcon. The head pattern of darker birds is often contrasting and bold, with a distinct, rather narrow and pointed dark moustache and a dark eye-line curving down the side of the neck. The crown is often dark, sometimes creamy or rusty-orange with dark streaking, divided by a pale supercilium from the dark eye-line. Lighter and less heavily pigmented birds have a paler head with reduced dark markings, leaving just a prominent dark eye standing out. The upperparts are dark brown with rusty margins in fresh plumage (never rusty-coloured in Gyr Falcon), but by spring the margins are worn off and the coverts and mantle appear greyer-brown because of wear. Normally the upperparts lack any pale markings, but if any do exist, they should appear as distinct pale, rounded spots, differing from the irregular pale markings of juvenile Gyrs. The folded tail looks uniformly brown above, but when fanned reveals dense pale barring formed by pale spots or ovals, and the tip is broadly white in fresh plumage. Underparts vary depending on the general pigmentation of the bird and matches the pigmentation of the head: the darker the head, the darker the underparts. Darker birds have an almost uniformly dark breast, as the individual streaks overlap, while lighter birds are more finely streaked below. The thighs are dark in many (then a good character compared to juvenile Lanner), creating dark rear flanks in flight, *but the thighs can also be light*, while the vent and undertail-coverts are always pale cream. The underwings show a marked contrast between dark coverts and paler remiges in most, the darkness of the coverts mirroring the general pigmentation of the bird, while lighter-coloured birds may just show a more intensely marked wing-band. In juvenile Saker the greater coverts are often dark with a pale outer margin, creating a streaked impression, while in Lanner they mostly show light-coloured spots; however, strongly pigmented birds of both species may have very similar, largely dark underwing-coverts. The secondaries are often rather uniformly grey and contrast with the densely and distinctly barred primaries, but a similar pattern is also shown by many juvenile Lanners, but not by Peregrines, where the secondaries are distinctly barred.

The feet of juveniles are pale bluish-grey, and will remain that colour for at least a year, sometimes for much longer, which is a good character compared to juvenile Peregrines and Lanners, in which the legs turn yellow soon after leaving the nest or at the latest during their first autumn.

Adult

The pattern and colour of the head is variable, rather whitish in some, with a poorly defined moustache and eyeline, while others show contrasting head markings and a deep ochre crown, thus resembling adult Lanner, but lacking the dark diadem across the forecrown of Lanner. Upperparts are usually uniformly greyish-brown, sometimes more purely grey with darker barring, and then similar to adult Lanner. Tail is brownish-grey above, with pale markings varying from round spots to more regular bands. Underparts are white with dark round spots, the size and density of the spots decreasing with age. Young adults may be heavily spotted below, with spots almost covering the white ground colour entirely, hence recalling juveniles, while old birds, males in particular, can appear almost

immaculate white below. The trousers are normally (but not always) distinctly darker than the rest of the underbody, either appearing as a dark patch on the rear flanks, or forming a dark V across the lower belly and vent. Underwings vary according to pigmentation of the body plumage, with younger birds showing more heavily marked and darker underwing-coverts than older birds, the coverts appearing streaked compared to the more barred impression in other large falcon species. The remiges are rather distinctly barred below (cf. Lanner and Gyr Falcons), with the primaries often showing reduced barring, particularly in older birds, leaving a diagnostic wide pale window in the central hand.

The rare but distinct *saceroides* form of Saker (not a subspecies), has greyish and barred upperparts and tail, a rusty-ochre crown with dark streaking and a tendency towards heavily barred flanks. It could thus easily be mistaken for an adult Lanner, but the dark thighs, if present, should be diagnostic.

SEXING Not reliably sexed by plumage, although when comparing adults, males tend to be more finely and more sparsely marked below than females. Males are smaller than females, but the size difference is apparent only in direct comparison.

CONFUSION RISKS Young birds can easily be confused with juveniles of Lanner and Gyr Falcons, while certain adults may resemble adult Lanners. Because of the great individual variation in large falcons, and because of the extensive trade in various large falcon hybrids, large falcons should always be scrutinised with the greatest care when attempting to identify them.

NOTE Artificial hybrids are commonly used in falconry, with hybrids and different types of backcross intergrades between Gyr and Saker Falcons being particularly popular. Juveniles of such hybrids/intergrades can be impossible to identify in the field, as they may be practically identical to either of the parental species.

References

Barthel, P. & Fünfstück, H.-J. 2012. Das Problem der Hybriden zwischen Grossfalken *Falco* spp. *Limicola* 26: 21–43.

1039. Saker Falcon *Falco cherrug*, recently fledged juvenile. Dark intermediate birds, such as this individual, are very similar to juvenile Gyr Falcons, but have on average more distinctly barred flight feathers below and a darker wingtip. Mongolia, 12.6.2012 (DF)

1040. Saker Falcon *Falco cherrug*, recently fledged juvenile. Mongolia, 12.6.2012 (DF)

1041. Saker Falcon *Falco cherrug*, juvenile. A rather average-looking juvenile, with contrasting, Lanner-like facial markings. Note streaked impression of primary coverts and blue feet. Hungary, Jun 2009 (Peter Csonka)

1042. Saker Falcon *Falco cherrug*, recently fledged juvenile. Dark juveniles can look similar to juvenile Lanners, but retain the bluish feet for at least one year. Slovakia, 20.7.2011 (DF)

1043. Saker Falcon *Falco cherrug*, 2nd cy immature. Note very pale head and spotted underbody of this worn and moulting immature. Mongolia, 9.6.2012 (DF)

1044. Saker Falcon *Falco cherrug*, adult. Rather similar to juvenile, but note spotted breast, distinctly barred secondaries and yellow feet. Slovakia, Mar 2010 (Peter Csonka)

1045. Saker Falcon *Falco cherrug*, adult male, breeding. Adults vary individually and can be more or less heavily marked, lighter or darker, partly depending on age and sex of the bird. Slovakia, 20.7.2011 (DF)

1046. Saker Falcon *Falco cherrug*, adult male, breeding. Probably an older bird showing reduced markings on body plumage. Mongolia, 5.6.2012 (DF)

GYR FALCON
Falco rusticolus

VARIATION Formerly divided into several subspecies, but now regarded as monotypic, with a range of geographically separated colour morphs. Grey morphs dominate over most of range, in the lower Arctic and boreal taiga zones, while white birds are confined to the High Arctic. Icelandic birds are intermediate between the former two, being basically grey, but lighter milky-grey and more distinctly barred above compared to Scandinavian birds. A very dark form occurs in E Canada.

DISTRIBUTION A circumpolar species restricted to the far north, including northern latitudes of the boreal taiga zone as well as the arctic tundra. Mostly resident, with short distance dispersal after the breeding season, but some juveniles and more rarely also some adults move farther south for the winter. White birds from high arctic Greenland (and Canada?) reach Ireland and Britain fairly regularly, occasionally also the continental coast, but they are very rare elsewhere in the region.

BEHAVIOUR Perches on rocks, or on the ground, for long periods when hunting. Feeds mostly on medium-sized birds, such as ducks, grouse, waders and gulls, which are pursued in a fast level flight, but Gyrs are also capable of vertical stoops, like Peregrine. Usually flies low over the ground, like a huge Merlin, but also hunts on the wing from considerable heights, sometimes soaring high out of sight to the naked eye.

SPECIES IDENTIFICATION Large and heavy falcon, with broad and rather rounded wings and a long, full tail. Most likely to be confused with sympatric Peregrine, and although the majority of adults clearly differ in plumage and silhouette, certain juveniles can appear rather similar in both proportions and plumage. In juvenile plumage the two species are best separated by different underwing patterns, with Gyrs showing paler and diffusely barred primaries and rather uniform secondaries contrasting with darker and distinctly patterned underwing-coverts, while in Peregrines the entire underwing is evenly and distinctly marked.

In juvenile plumage most similar to juvenile Saker, and escaped artificially produced hybrids between Gyr and Saker, popular among falconers, can be impossible to tell from either species. As for other large falcons, juveniles and adults show different proportions, juveniles being slimmer and adults typically more compact and broader-winged.

MOULT Annual complete moult during the breeding season, from Apr–Sep, but moult is arrested for part of the nestling period.

PLUMAGES Adults and juveniles are normally distinctive, but birds of the white morph are more similar and may be difficult to age by plumage. Ageing is always possible by the colour of the feet and cere, being bluish in juveniles and yellow in adults.

Juvenile

Generally more uniform above, the uppertail being concolorous with the rest of the upperparts, except when the tail is fanned, when the barring becomes visible. Creamy underbody with dark streaking, changing to broader arrowheads towards the flanks. Feet, cere and eye-ring pale blue regardless of colour morph.

Grey morph Greyish-brown above, but appears often surprisingly grey in flight. Upperwing-coverts and mantle show pale creamy tips and individually variable pale but faint transverse markings. As a rule, darker birds are less marked above than paler birds. The primaries are typically rather softly barred below compared to other large falcon species, making the heavily streaked underwing-coverts stand out as darker and much more clearly marked. Head markings vary depending on general pigmentation of the bird, with darker birds appearing almost dark-hooded, while intermediates may show almost Peregrine-like heads. Paler birds show a faint eye-line and moustache on an otherwise largely light-coloured head.

White morph Basically white with variable dark markings above and below. The white markings of the upperparts appear as spots, compared to a more clearly barred pattern in adults. In fresh plumage the dark markings appear almost black, but fade to brown during the course of the winter, while adults remain more contrasting, black and white.

Dark morph Rather similar to grey morph juveniles, but general impression is very dark: uniformly dark slate above with dark head, while heavy streaking on the underparts merges to create an all-dark breast.

Adult

Clearly barred above, and the tail is pale grey with fine darker barring, while the underparts are spotted and coarsely barred on the flanks. Feet, cere and eye-ring bright yellow.

Grey morph Upperparts appear rather uniformly grey from a distance, but closer views reveal paler grey barring across the entire upper surface, with the uppertail appearing paler than the rest, lacking the dark outer tail of adult Peregrine. Underparts are creamy white, with individually variable (age- and sex-related) dark spotting on the mid-breast, changing to coarse barring on the flanks. Undertail-coverts barred. Remiges are grey below, mostly with ill-defined barring compared to other large falcons, and contrasting with whiter and distinctly barred or streaked underwing-coverts. Head pattern varies individually from almost completely dark-headed birds to individuals with a narrow moustache, largely whitish cheeks and broad supercilium, and a contrasting dark crown, some birds looking not that different from Peregrines.

White morph Similar to white morph juvenile, but markings darker and more contrasting, and upper-parts more clearly barred compared to juveniles. Note also difference in colour of bare parts between young and adult birds.

Dark morph Similar to dark juvenile, but tail may be grey with fine dark barring and the undertail-coverts are heavily barred. Juveniles and adults also show different coloration of the bare parts: blue in young birds and yellow in adults.

SEXING Plumages rather similar, but when comparing adult males and females of breeding pairs males tend to show a smarter head pattern and are more finely marked below and more clearly ashy-grey above compared to females, which are more heavily marked below and duller in upperparts colour, but plumage differences are also age-related, with size and extent of markings decreasing with age, also in females. Despite a marked difference in body weight, females and males can be difficult to tell apart in the field, even when comparing them directly.

CONFUSION RISKS Adults are rather distinctive, but dark-headed grey morph males could be mistaken for adult Peregrine. Juveniles can be very similar to juvenile Saker, but differ by slightly bulkier build and softer underwing barring.

1047. Gyr Falcon *Falco rusticolus*, recently fledged juvenile male. Compare general colour and type of markings of underwings and body with juvenile Peregrine and Saker. Norway, 20.7.2014 (DF)

1048. Gyr Falcon *Falco rusticolus*, juvenile male. This typical Scandinavian bird shows broad wings and a rather long tail; strongly marked underwing-coverts contrasting with diffusely marked flight feathers are typical of most juvenile Gyrs (cf. Peregrine). Norway, 22.8.2011 (DF)

1049. Gyr Falcon *Falco rusticolus*, juvenile male (same as **1048**). The upperparts appear very uniformly greyish-brown, possibly with a slightly lighter tail if anything. The narrow and faint creamy feather edges of the upperparts are normally not seen in the field. Norway, 22.8.2011 (DF)

1050. Gyr Falcon *Falco rusticolus*, juvenile. Although the wings may appear pointed from certain angles they always appear clearly broader but also shorter compared to Peregrine. Norway, 14.7.2011 (DF)

1051. Gyr Falcon *Falco rusticolus*, juvenile. The soft marbling of the remiges is also a good feature for separating juvenile Gyrs from more distinctly barred juvenile Sakers. Norway, 14.7.2011 (DF)

1052. Gyr Falcon *Falco rusticolus*, juvenile. Norway, 14.7.2011 (DF)   **1053**. Gyr Falcon *Falco rusticolus*, juvenile. Norway, 14.7.2011 (DF)

1054. Gyr Falcon *Falco rusticolus*, adult male, breeding. Except for boldly barred greater underwing-coverts, the underwing is diffusely marked. Norway, 6.6.2011 (DF)

1055. Gyr Falcon *Falco rusticolus*, adult female, breeding. A more heavily marked, probably younger, breeding female. Norway, 2.6.2014 (DF)

1056. Gyr Falcon *Falco rusticolus*, adult male, breeding (paired to **1055**). The sparse markings on this male indicate a greater age. Norway, 2.6.2014 (DF)

1057. Gyr Falcon *Falco rusticolus*, adult female. breeding. This female is known to be at least five years old. Norway, 23.5.2013 (DF)

1058. Gyr Falcon *Falco rusticolus*, adult male, breeding (same as **1054**). Adult Scandinavian Gyrs are ashy-grey and densely barred over the entire upperparts. The uppertail is evenly barred from base to tip and lacks the dark outer tail of adult Peregrines. Norway, 6.6.2011 (DF)

1059. Gyr Falcon *Falco rusticolus*, adult female, breeding. Norway, 2.6.2014 (DF)

1060. Gyr Falcon *Falco rusticolus*, juvenile, white morph. Unmistakable. During the winter the dark markings will fade to become clearly browner compared to more black-and-white adults. Greenland, Sep 2011 (Bruce Mactavish)

1061. Gyr Falcon *Falco rusticolus*, juvenile males, white morph. The amount of dark markings varies considerably between birds. Greenland, Sep 2011 (Bruce Mactavish)

1062. Gyr Falcon *Falco rusticolus*, juvenile male (left) and female, white morph. The size difference between the sexes is obvious here, but is soon lost without direct comparison. Greenland, Sep 2011 (Bruce Mactavish)

PEREGRINE FALCON
Falco peregrinus

VARIATION Several subspecies occur in the region, but complex taxonomy, intergradation between subspecies and considerable individual variation, even within subspecies, make subspecific identification of single individuals challenging. Nominate *peregrinus* breeds over most of Europe and the temperate zone of Asia, with northern populations being migratory, wintering as far south as the Mediterranean basin and Africa. The long-winged arctic ssp. *calidus* from the Russian tundra east of the White Sea is a long-distance migrant to temperate and tropical parts of Asia and Africa, even reaching S Africa, passing through Europe on migration. Of these two subspecies only the most typical individuals should be identified, as *peregrinus* and *calidus* seem to intergrade over a wide area. For instance, some birds from Taimyr, which should be *calidus* on breeding range, may look like *peregrinus*, while many Finnish breeding birds (ssp. *peregrinus* by definition) approach the phenotype of *calidus*. Ssp. *calidus* may in fact represent the extreme in a cline, in which Peregrines gradually get paler and less heavily marked from C Europe towards the north-east. Resident ssp. *brookei* from S Europe, the Mediterranean and N Africa is smaller and more compact, usually also darker and more strongly coloured below than *peregrinus*. The picture is further complicated by Barbary Falcon (which see), which in some areas appears to interbreed with Peregrines (W Morocco and Canary Islands) with hybrid offspring not always separable from either *brookei* or *pelegrinoides*. A similar situation of neighbouring taxa intergrading occurs in N Spain, where birds of both *peregrinus* and *brookei* phenotypes coexist and interbreed, resulting in offspring with intermediate features (Zuberogoitia *et al.* 2009). The plumage of the endemic Cape Verde Peregrine ssp. *madens* looks superficially similar to the Barbary Falcons of the Canary Islands, and is probably another hybrid form, awaiting further research. The Nearctic ssp. *tundrius* of N America and Greenland is rather similar to *calidus*, but is smaller on average, in both sexes, with a slightly more contrasting plumage on average, both as a juvenile and as an adult. It may well reach the Atlantic islands and the western seaboard of Europe as an autumn vagrant, but proving this would be challenging owing to the great similarity to *calidus*. The entire Peregrine complex is still in need of further taxonomic research (White *et al.* 2013).

DISTRIBUTION A cosmopolitan species, with a vast worldwide range. After severe large-scale crashes in the mid-20th century populations have largely recovered, and in many places Peregrines are now more common than ever. Occurs across the region on sea cliffs, mountain ranges and isolated rocks, and now also in cities, breeding on buildings. Birds from the taiga and tundra zones frequently nest on the ground or in tree-nests of other birds.

BEHAVIOUR Hunts mostly medium-sized birds in flight, taking them in mid-air, often after a long dive or a vertical stoop. Active flight is powerful, with shallow wing-beats on rather stiff-looking wings, usually interspersed with short glides, but the proportion between glides and flapping sequences varies depending on purpose of flight.

Perches high or low, from tops of aerial masts and church towers to telegraph poles and fence posts, often even on the flat ground, but always chooses places with open views.

SPECIES IDENTIFICATION Peregrines are medium-sized or large falcons, with considerable geographical variation in size and plumage. They are best told from other large falcon species by uniformly marked underwings, lacking strong contrasts between the coverts and remiges, while the dark tips of the primaries *form a narrow but distinct dark trailing edge to the hand* (cf. other large falcons).

Typical adults are easy to identify on muscular appearance, strong flight and diagnostic plumage, with *white upper breast contrasting with dark head and finely barred underparts*, while on the upperside *the contrasting pale grey rump and inner tail* is a good character. Juveniles are more difficult to tell from other large falcons, particularly the paler and large, long-winged northern birds, but they differ from Lanners and Sakers by their stiffer, more powerful wing-action in flight, their uniformly marked underwings and, as in other Peregrine forms, most individuals (but not all) show barred undertail-coverts.

Subspecific identification

Given good views, typical representatives of each subspecies can be identified, at least tentatively, but individuals with intermediate features occur commonly and not every individual can be assigned to a certain subspecies.

Adults

Nominate *peregrinus*: Upperparts are dark slate-grey, sometimes with a bluish cast, and head is black, with a broad moustache, sometimes covering the cheeks and giving the bird an almost hooded appearance. Underparts often show a cream or even light salmon-pink wash on the breast while the barring on the flanks is distinct and blackish, contrasting strongly with the white upper breast.

Ssp. *brookei*: The smallest and most compact subspecies of the region. Upperparts vary from blackish to bluish-grey, the head is black with a broad moustache, often with rufous patches on the nape and a rufous wash to the pale cheeks. Underparts often with deep salmon-pink upper breast and greyish flanks, but flank barring variable and may be fine and even reduced in some apparently old males, approaching Barbary Falcon in this respect. Interbreeds with ssp. *peregrinus* in the W Pyrenees (Spain) where intermediate-type birds are frequent (Zuberogoitia *et al.* 2009), and also with Barbary Falcon further south (which see).

Ssp. *calidus*: Longer-winged and larger than nominate *peregrinus*, but differences are subtle and rarely of use in the field. Adults are on average paler grey above than ssp. *peregrinus*, with a darker head, crown often greyish and contrasting with black moustache and eye-line. Moustache narrower compared to previous forms, with white cheeks typically extensive and gleaming white, almost reaching the eye in many. Underparts pure white, sometimes with a creamy hue, with sparse and fine dark markings; entire underbody, including finely barred underwings, looks very pale in the field.

Ssp. *tundrius*: Very similar to *calidus*, and owing to individual variation and extensive overlap in characters single birds perhaps not separable in the field.

Juveniles

Nominate *peregrinus*: Highly variable, but typical birds dark brown above, with dark head and pale cheeks, the latter often with darker streaking. Underparts rather deep ochre with variably broad dark streaking, the streaks mostly broader than in *calidus*. Underwings heavily barred, looking generally rather dark. However, much lighter birds, approaching *calidus* in plumage, may hatch from typical *peregrinus* parents.

Ssp. *brookei*: Rather similar to juvenile *peregrinus*, but smaller and on average overall darker, more deeply rusty-ochre below, but again rather variable with many birds much lighter and not dissimilar to juvenile *peregrinus*, or even young Barbary Falcon. Many birds show rufous markings on the nape and sides of neck. Starts to moult body plumage by first autumn, much sooner than any other subspecies, with adult-type feathers emerging on the head, upper breast and upperparts as early as Sep–Oct.

Ssp. *calidus*: On average lighter brown above with conspicuous pale ochre feather margins, creating a contrast on the upperwing between lighter and browner coverts and darker remiges. Head typically appears pale with reduced dark markings, often recalling juvenile Saker or Lanner at first glance. Most birds show a well-developed pale supercilium between the darker crown and dark eye-line, with the narrow moustache often the darkest feature of the head. Underparts look very white, often pale cream in fresh plumage, with fine brown streaking, revealing more of the white ground colour compared to *peregrinus* or *brookei*, while the underwings are lighter, owing to finer barring, than in either *peregrinus* or *brookei*.

Ssp. *tundrius*: Very similar to *calidus*, and probably not always possible to separate. This subspecies is a smaller bird, which anyway would be difficult to confirm in the field, and possibly the head markings and the streaking below are on average darker and more contrasting than in *calidus*, but extensive variation in both makes identification of single individuals questionable. In fresh plumage juvenile *tundrius* is on average more deeply buffish below than *calidus*, but during the winter the fading of the plumage makes the two even more similar.

MOULT Entire plumage is moulted annually, but differences in moulting strategies and timing allow separation of the different populations. Resident populations moult between Mar and Sep/Oct. Long distance migrants, such as *calidus* and northern *peregrinus* only moult a few primaries in the breeding areas, before the moult is suspended for the autumn migration. Moult is then resumed in the winter quarters, where it is completed well before the spring migration in Jan–Mar. In 2nd cy birds the moult is usually more advanced compared to adults at the same time. Immatures of *calidus* frequently retain juvenile feathers in the plumage until the late 2nd cy autumn–3rd cy spring.

The body moult starts in juvenile *brookei* early in first autumn (Sep–Oct), when the long-distance migrants still appear in fresh juvenile plumage.

PLUMAGES Normally only two plumage classes can be separated, juvenile and adult. Birds in their 2nd cy can be identified by retained juvenile feathers until moult is completed; in long-distance migrants such as *calidus* and northern *peregrinus*, with a protracted moult, ageing may be possible until the early 3rd cy spring.

Juvenile

Upperparts are usually uniformly dark brown, with narrow rufous fringes. Head is dark, with a broad, rounded moustache and a variable lighter cheek-patch, which may or may not be streaked. Many birds show some pale markings on the supercilium, nape and lower neck, regardless of subspecies. Underparts vary from ochre to whitish, with variable heavy streaking on breast and belly, the markings changing to diagnostic arrowheads or rhomboids on the flanks, while the undertail-coverts are barred (cf. Gyr, Saker and Lanner Falcons). *Underwings are evenly and distinctly barred and there is usually no clear colour contrast between the flight feathers and coverts* (cf. juvenile Gyr, Saker and Lanner). The only notable character of the underwing is *the dark tips of the primaries*, which *form a distinct dark trailing edge to the hand*, another good feature compared to juveniles of Lanner and Saker. Although the feet are usually yellow at fledging, some birds, particularly of northern populations, may still show rather colourless pale greenish or even bluish feet in late autumn/winter.

Transitional plumage

Immatures can be identified by retained juvenile feathers until they are moulted, which in the long-distance migrant populations doesn't happen until late in the 2nd cy or early in the 3rd cy. *These immatures can be identified by retained brownish juvenile-type flight feathers (best seen on the upperwing) and retained brown underwing- and upperwing-coverts.* The head is often the first to attain adult plumage, in resident populations as early as the first winter, but in long-distance migratory populations not until the 2nd cy spring–summer.

Adult

Upperparts vary from dark slate-grey to pale bluish-grey, depending on sex, age and geographical provenance, but the rump and inner tail always contrast as paler compared to the outer tail and mantle. The head looks black, often appearing darker than the mantle, with a variable white cheek-patch, while the moustache is black, broad and rounded at the tip. The upper breast appears unmarked and white to pale pink (cleaner and more spotless in males than in females), while the flanks are densely barred with black, changing to dark spotting towards the mid-breast. The underwings are finely yet distinctly barred, looking very uniform from a distance, save for the *distinct dark tips of the primaries*.

SEXING In direct comparison males are clearly smaller and slimmer than females, but the size can be difficult to assess on single birds. Males appear smaller-bodied, bigger-headed and larger-footed than females, which appear bulkier but with a comparatively smaller head and broader wings. On plumage males are more finely marked below than females of a similar age, but this feature also depends on the age of the bird, with birds becoming more finely marked as they age. On average males tend to show a clean white upper breast and lighter and more bluish upperparts, while females are distinctly spotted on the crop and tend to be darker above.

CONFUSION RISKS Small males could be mistaken for a Eurasian Hobby in poor light, but different flight, more muscular proportions and a broader arm should reveal the mistake. The real challenge is to separate the different large falcon species from each other, although Peregrine is one of the easier, especially the distinctive adult birds. Juveniles are best told from other large falcons by their uniformly patterned underwings, lacking the marked contrast between coverts and remiges of Gyr, Lanner and Saker Falcons, and by their barred undertail-coverts. They also have bright yellow feet and cere from very early on, while other large falcons do not, although northern juvenile Peregrines may show washed-out pale greenish feet. Until recently juveniles of the *calidus*-type were frequently misidentified as Gyrs, Sakers or Lanners when they turned up in Europe in Sep–Oct. For separation from Barbary Falcon, see that species.

References

White, C. M., Cade, T. J. & Enderson, J. H. 2013. *Peregrine Falcons of the World.* Lynx Edicions.
Zuberogoitia, I., Azkona, A., Zabala, J., Astorkia, L.,

Castillo, I., Iraeta, A., Martínez, J. A. & Martínez, J. E. 2009. Phenotypic variations of Peregrine Falcon in subspecies distribution border. In: J. Sielicki and T. Mizera (eds.). *Peregrine Falcon populations – status and perspectives in the 21ˢᵗ century*, pp.295–308. European Peregrine Falcon Working Group.

1063. Peregrine Falcon *Falco peregrinus*, juvenile male. A recently fledged juvenile from the *peregrinus/calidus* intergradation zone in N Scandinavia, at lat. 70°N. Norway, 22.8.2011 (DF)

1064. Peregrine Falcon *Falco peregrinus*, juvenile female, sibling to (**1063**). A typical juvenile Peregrine. Note the uniformly marked underwings and the barred undertail-coverts. Norway, 22.8.2011 (DF)

1065. Peregrine Falcon *Falco peregrinus*, juvenile female. The mint condition of the plumage this late in the season indicates a bird of boreal or arctic origin with a late fledging date, thus it is either a northern *peregrinus* or a *calidus*. Oman, 20.11.2014 (DF)

1066. Peregrine Falcon *Falco peregrinus*, juvenile female, migrant. The lightness of this bird in combination with its reduced dark markings indicates a northern/arctic provenance, possibly a *calidus*. Note the bluish feet of this individual. Israel, 10.11.2009 (DF)

1067. Peregrine Falcon *Falco peregrinus*, juvenile female, migrant. Identified as a *calidus*-type bird by its overall whiteness below, reduced dark head markings and fine streaking on the underparts; this bird is yellow-footed. Finland, 27.9.2005 (DF)

1068. Peregrine Falcon *Falco peregrinus*, juvenile. This small *brookei*-type bird is identified by its small size and rather richly coloured underparts with heavy dark markings. Note also the rufous nape-patches. Juvenile Barbary Falcons could look very similar, but are as a rule lighter sandy-buffish below with finer and lighter brown markings. Morocco, 13.9.2012 (DF)

1069. Peregrine Falcon *Falco peregrinus*, juvenile (same as **1068**). Upperparts are similar to other Peregrines, but owing to an earlier breeding season southern populations are already rather worn above by early autumn. Morocco, 13.9.2012 (DF)

1070. Peregrine Falcon *Falco peregrinus*, juvenile female, migrant/ wintering. This light bird of obvious northern origin has already started to moult in adult-type feathers on the head, crop and thighs, although the general impression is still very juvenile. Israel, 21.4.2009 (DF)

1071. Peregrine Falcon *Falco peregrinus*, 2ⁿᵈ cy female, migrant. This *peregrinus/calidus*-type bird in its first adult plumage still retains a fair number of juvenile feathers in its plumage, including wing and tail feathers and underwing-coverts. Finland, 7.10.2012 (DF)

1072. Peregrine Falcon *Falco peregrinus*, 2ⁿᵈ cy female, migrant (same as **1071**). The retained brown juvenile feathers are often more easily seen from above, when they contrast against fresh grey adult-type feathers. This bird still retains its innermost and outer two primaries and several outer secondaries, as well as scattered coverts on the upperwings and scapulars. Finland, 7.10.2012 (DF)

1073. Peregrine Falcon *Falco peregrinus*, adult female, breeding. This *peregrinus*-type female shows heavy barring below and a rather dark head with a broad moustache. It has only just started its moult and has dropped its first primaries, p4 and p5. Finland, 4.6.2014 (DF)

1074. Peregrine Falcon *Falco peregrinus*, adult female, migrant. This *peregrinus/calidus*-type female is less heavily marked. It still retains some brown underwing-coverts, indicating that it might be in its 3rd cy. Finland, 4.5.2012 (DF)

1075. Peregrine Falcon *Falco peregrinus*, adult female, migrant/wintering. This *calidus*-type bird is still only at the beginning of its primary moult, with new p4, growing p5, missing p3 and p6, and all the rest are retained old feathers. Note the grey, not black head of this bird, a typical *calidus* feature. Israel, 8.10.2009 (DF)

1076. Peregrine Falcon *Falco peregrinus*, adult male, breeding. This bird from the intergradation zone between *peregrinus* and *calidus* is a typical representative of northern breeding birds. Finland, 20.5.2015 (DF)

1077. Peregrine Falcon *Falco peregrinus*, adult male, breeding and father to **1063** and **1064**. This N Scandinavian male has only just started its primary moult, by dropping p4. Note the whiteness of the underparts and the fine markings of this bird from the *peregrinus/calidus* intergradation zone. Norway, 22.8.2011 (DF)

1078. Peregrine Falcon *Falco peregrinus*, adult, migrant. This is a classic *calidus* in all respects, with white and finely marked underparts and a grey rather than black head. Greece, 26.4.2007 (DF)

1079. Peregrine Falcon *Falco peregrinus*, adult female ssp. *brookei*, resident. This colourful female shows the typical salmon-buff breast and grey, densely barred flanks. It will have completed its moult in a few days, once p10 is fully grown. Cyprus, 18.10.2014 (DF)

1080. Peregrine Falcon *Falco peregrinus*, adult male ssp. *brookei*, resident (paired to **1079**). Slightly less colourful than its mate, this male will finish its primary moult a week or so later than her. Cyprus, 18.10.2014 (DF)

1081. Peregrine Falcon *Falco peregrinus*, adult female ssp. *brookei*, resident. This adult from Mallorca shows slightly more heavily marked underparts, but its moult follows exactly the same timing as in its Cyprus relatives. Spain, 9.10.2007 (DF).

1082. Peregrine Falcon *Falco peregrinus*, adult female, wintering. This *peregrinus*-type female shows rather dark upperparts, but with a contrasting lighter grey back, rump and inner tail. Although moult is nearly completed, p10 is not yet fully grown. Morocco, 16.1.2014 (DF)

1083. Peregrine Falcon *Falco peregrinus*, adult female, breeding (same as **1073** but here six weeks later). This *peregrinus*-type female from the *peregrinus/calidus* intergradation zone is lighter above than the female in **1082**. Note that the primary moult has been arrested after replacing only p4 and p5. Finland, 18.7.2014 (DF)

1084. Peregrine Falcon *Falco peregrinus*, adult male, breeding (paired with **1083**). This classic *calidus*-type male from the *peregrinus/calidus* intergradation zone in N Finland (lat. 69°N) shows typically pale grey upperparts and even grey head, and contrasting dark outer tail and wingtips. Finland, 4.6.2014 (DF).

BARBARY FALCON
Falco peregrinus pelegrinoides

VARIATION The taxonomic status of this form is debated; it is either regarded as a full species or a subspecies of Peregrine. If regarded as a full species, two subspecies are recognised, with nominate *pelegrinoides* in the west, replaced by paler and rufous-headed *babylonicus* (Red-naped Shaheen) in the deserts from the Caspian Sea eastwards to C Asia and Pakistan, W India and possibly Mongolia (although some records may relate to misidentified eastern Saker Falcons *Falco cherrug milvipes*, which may look superficially similar but are much bigger). There is an urgent need for further taxonomic studies, particularly in the western part of its Palearctic range, where Peregrine and Barbary Falcons are said to be sympatric. Birds from the Canary Islands and coastal W Morocco are highly variable, with many birds looking very similar to *brookei* Peregrines from the European continent, while others are more similar to Barbary Falcons of the Middle East. Mixed breeding pairs have been noted, where one bird looks like a Peregrine and the other like a typical Barbary (Rodriguéz *et al.* 2011 and pers. obs. in Morocco 2014–2015). The considerable individual variation found within the Moroccan and Canarian populations possibly indicates a large-scale exchange of genes between Peregrine and Barbary. The disputed '*atlantis*' Peregrines from the Moroccan Atlantic coast (Schollaert & Willem 2000) are perhaps best seen as a stable hybrid population between Barbary and Peregrine. Pending further genetic studies and given the extensive apparent hybridisation with Peregrine, Barbary Falcon is here treated as a subspecies of Peregrine.

DISTRIBUTION In Western Palearctic confined to dry desert-like environment from the Canary Islands, throughout N Africa (reaching the Sahel zone south of the Sahara) to the Middle East and Arabia. Further east blends gradually into the eastern subspecies *babylonicus*, from Iraq and Iran to C Asia, Mongolia, W India and Pakistan. Birds looking like Barbary Falcons have been found breeding with typical ssp. *brookei* Peregrines in Spain.

BEHAVIOUR Similar to Peregrine, although takes comparatively larger prey, like doves and pigeons, despite smaller size.

SPECIES IDENTIFICATION Very similar to some ssp. *brookei* Peregrines, and although most birds are separable some are confusingly similar, possibly owing to extensive intergrading in certain areas (Canaries, Morocco, perhaps even Spain). Small and compact with body and feet comparatively large in relation to the small wings. Often soars with wings in a shallow V, possibly owing to high wing-loading (heavy body in relation to small surface area of the wings). Adults in flight are said to show a narrower hand compared to Peregrine (White *et al.* 2013).

MOULT Complete annual moult during the breeding season from Feb to Oct, earlier in the western part of the range than in the eastern part. Juveniles undergo a partial body moult during the first winter, replacing the feathers of the head and upper breast, which become adult-like by spring. The different timing of the moult and the state of plumage wear are important features when separating Barbary Falcon and northern migratory Peregrine populations, which may have overlapping ranges in winter.

PLUMAGES Adults and juveniles are easily separated. In certain conditions first-year birds can also be identified by retained juvenile feathers until the first moult is completed, late in the 2nd cy.

Juvenile

Largely similar to juvenile nominate Peregrine, but the underparts have a distinct sandy-buffish ground colour, deeper tawny-buff when fresh, and the streaking is finer, more similar to *calidus* Peregrine, but the coloration is different: rather light brown fine streaks on a tawny background, while *calidus* is much more contrasting with dark streaks on a whiter background. Upperparts lighter brown compared to juvenile Peregrine, with broader rufous feather margins, the latter bleaching and wearing off during the first winter. Head pattern similar to many juvenile *calidus* Peregrines, with pronounced pale supercilium and forecrown, and pale markings on sides of neck, but these markings are ochre or rufous instead of creamy or whitish, as in *calidus*, while other juveniles can be much darker-headed, more similar to Peregrine. It is important to note that juveniles are more similar in silhouette to other large falcons, showing the 'general falcon-shape', while adults are the more specialised and diagnostic 'Peregrine-shape'.

Transitional plumage

Immatures can be identified by retained juvenile feathers until they are replaced by the 2nd cy autumn. The last juvenile feathers to be replaced are the outer primaries, the outer secondaries and the underwing-coverts. Occasionally the upperwing-coverts may also show some retained brown juvenile feathers after the moult is completed.

Adult

The adult plumage has been over-simplified in field guides and is actually rather variable. Two basic types occur: dark birds, which are very similar to *brookei* Peregrines, when it comes to head and upperparts, and pale birds, which are pale bluish or silvery-grey and rather distinctly barred above, including the upperwings, with a largely rufous hindcrown and extensive rufous nape-patches. In both types the pale cheeks and the breast and belly show a diagnostic rich sandy or orange-buffish wash, with the throat usually distinctly paler, while the underparts are finely marked (finer than in Peregrines), with faint barring mostly confined to the greyish flanks, somewhat more distinctly barred in females than in males, while the mid-breast and belly can be largely unmarked. Birds from the Canary Islands and Morocco are more similar to Peregrines, with coarser underside barring, even more distinct and more complete in old birds, whereas birds from the Middle East are much more finely marked, often appearing almost unmarked on the underbody, particularly males. All adults show a distinctly pale grey rump and innertail compared to the rest of the upperparts, as in Peregrine, with black tail-barring increasing in width towards the tip of the tail.

SEXING Reversed sexual size dimorphism is considerable, with females clearly bigger than males, but this is usually difficult to assess in the field. The small males often appear remarkably big-headed and large-footed in flight, which may be worth noting.

CONFUSION RISKS Barbary Falcons and certain *brookei* Peregrines can look very similar, owing partly to possible gene-flow between populations, and hence only typical individuals should be identified. Difficult types are becoming more common westwards, where Peregrine and Barbary are sympatric in some areas, whereas in the eastern part of the range sympatric breeding forms of Peregrine are lacking and Barbary Falcons look more typical.

References

Brosset, A. 1986. Les populations du Faucon Pélerin *Falco peregrinus* Gmelin en Afrique du Nord: un puzzle zoogéographique. *Alauda* 51: 1–14.

Rodriguéz, B., Siverio, F., Siverio, M. & Rodriguéz, A. 2011. Variable plumage coloration of breeding Barbary Falcons *Falco (peregrinus) pelegrinoides* in the Canary Islands: do other Peregrine Falcon subspecies also occur in the archipelago? *Bull. Brit. Orn. Club* 131: 140–153.

Schollaert, V. & Willem, G. 2000. Taxonomy of the Peregrine *Falco peregrinus*/Barbary Falcon *F. (peregrinus) pelegrinoides* complex in Morocco. *Bull. African Bird Club* 7: 101–103.

White, C. M., Cade, T. J. & Enderson, J. H. 2013. *Peregrine Falcons of the World*. Lynx Edicions.

1085. Barbary Falcon *Falco peregrinus pelegrinoides*, juvenile. Overall impression is pale buffish with fine and inconspicuous breast streaking. Israel, 24.3.2010 (DF)

1086. Barbary Falcon *Falco peregrinus pelegrinoides*, juvenile. Note sandy-buffish ground colour below with fine brown streaking. Very similar to some juvenile *brookei* Peregrines, and possibly not always separable. Oman, 11.11.2005 (Tom Lindroos)

1087. Barbary Falcon *Falco peregrinus pelegrinoides*, adult. A typical adult of the Middle Eastern type, pale sandy-buffish and practically lacking markings on the underbody. Israel, 22.3.2012 (DF)

1088. Barbary Falcon *Falco peregrinus pelegrinoides*, adult, presumed female. A slightly more heavily marked bird of the Middle Eastern type. This individual has just started its complete moult by dropping primary p4. Israel, 16.3.2013 (DF)

1089. Barbary Falcon *Falco peregrinus pelegrinoides*, adult male, breeding. A typical finely marked adult male with Peregrine-type head markings. Note typical colour of the underparts, including rufous cheeks, and distinctly darker outer tail; compare with ssp. *brookei* Peregrines. Israel, 26.3.2007 (DF)

1090. Barbary Falcon *Falco peregrinus pelegrinoides*, adult female, breeding, paired to **1089**. Shows typical sandy-buff overall colour, with more distinctly marked underbody compared to its mate. Note the rufous colour of the neck and moustache, and how the outertail-bars are clearly wider and darker than the inner ones. Israel, 26.3.2007 (DF)

1091. Barbary Falcon *Falco peregrinus pelegrinoides*, adult female, breeding (same as **1090**). Note uniformly light ashy-grey upperparts and extensive rufous areas of the nape, neck and cheeks, differing from any other Peregrine. Although Barbary Falcons are on average much lighter above than Peregrines, the variation is considerable and birds darker than this dominate. Israel, 26.3.2007 (DF)

1092. Barbary Falcon *Falco peregrinus pelegrinoides*, adult, presumed female. The birds of the Canary Islands are by definition Barbary Falcons, but many show intermediate features between Peregrine and Barbary. This female is clearly heavier and broader-winged compared to Middle Eastern Barbary Falcons, and also more heavily marked below. The Canary Islands population is probably best treated as a hybrid swarm of intergrades between Peregrines and Barbary Falcons. The extreme phenotypic variation within this population also supports the hybrid population theory. Lanzarote, Canary Islands, Spain, 17.12.2011 (DF)

1093. Presumed intergrade between Peregrine and Barbary Falcon, adult, from the west coast of Morocco. These coastal birds have been referred to as *'atlantis'* Peregrines, but are in fact more likely to be a hybrid population featuring a mix of Barbary and Peregrine characters, similar to the situation in the Canary Islands. They could easily be identified as one or the other species depending on the dominant features of each individual. Morocco, 21.1.2014 (DF)

1094. Presumed intergrade between Peregrine and Barbary Falcon, adult, from the west coast of Morocco. Although superficially looking like a Barbary Falcon from this angle, this bird shows a mix of characters indicating hybrid ancestry. Morocco, 16.1.2014 (DF)

1095. Possible intergrade between Peregrine and Barbary Falcon, juvenile. This bird cannot be identified with certainty, and could either be a juvenile *brookei* Peregrine, a juvenile Barbary Falcon or a hybrid/intergrade between the two. Morocco, 16.1.2014 (DF)

REFERENCES

Barthel, P. & Fünfstück, H.-J. 2012. Das Problem der Hybriden zwischen Grossfalken *Falco* spp. *Limicola* 26: 21–43.

Bernis, F. 1975. Migración de falconiformes y *Ciconia* spp. por Gibraltar II. Análisis descriptive del verano-otoño 1972. *Ardeola* 21: 489–580.

Blanc, J.-F., Sternalski, A., & Bretagnolle, V. 2013. Plumage variability in Marsh Harriers. *British Birds* 106:145–158.

Brosset, A. 1986. Les populations du Faucon Pélerin *Falco peregrinus* Gmelin en Afrique du Nord. un puzzle zoogéographique. *Alauda* 51: 1–14.

Campora, M. & Cattaneo, G. 2005. Ageing and sexing Short-toed Eagles. *British Birds* 98: 370–376.

Clark, W. S. 2005. Steppe Eagle *Aquila nipalensis* is monotypic. *Bull. Brit. Orn. Club* 125: 149–153.

Corso, A. 2009. Successful mixed breeding of Atlas Long-legged Buzzard and Common Buzzard on Pantelleria, Italy, in 2008. *Dutch Birding* 31: 224–226.

Corso, A. & Forsman, D. 2008. A hybrid Imperial Eagle x Golden Eagle in Romania. *Birding World* 21: 304–305.

Corso, A. & Monterosso, G. 2000. Ein unbeschriebene dunkle variante des Baumfalken *Falco subbuteo* und ihre Unterscheidung vom Eleonorenfalken *F. eleonorae*. *Limicola* 14: 209–215.

Corso, A. & Monterosso, G. 2004. Further comments on dark Hobbies in southern Italy. *British Birds* 97: 411–414.

Crochet, P.-A. 2008. A Taiga Merlin on the Azores: an overlooked vagrant to Europe? *Birding World* 21: 114–116.

van Duivendijk, N. 2011. Steppenbuizerd in Nederland. herziening, status en determinatie. *Dutch Birding* 33: 283–293.

Elorriaga, J. & Muñoz, A.-R. 2013. Hybridisation between the Common Buzzard *Buteo buteo buteo* and the North African race of Long-legged Buzzard *Buteo rufinus cirtensis* in the Strait of Gibraltar: prelude or preclude to colonization? *Ostrich* 84: 41–45.

Faveyts, W. 2010. Ringvangst van een donkere Boomvalk te Brecht in mei 2009. *Natuur.oriolus* 76: 109–112.

Faveyts, W., Valkenburg, M. & Granit, B. 2011. Crested Honey Buzzard: identification, western occurrence and hybridisation with European Honey Buzzard. *Dutch Birding* 33: 149–162.

Fefelov, I. V. 2001. Comparative breeding ecology and hybridisation of Eastern and Western Marsh Harriers *Circus spilonotus* and *C. aeruginosus* in the Baikal region of Eastern Siberia. *Ibis* 143: 587–592.

Forsman, D. 1993. Hybridising Harriers. *Birding World* 6: 313.

Forsman, D. 1994. Field identification of Crested Honey Buzzard. *Birding World* 7: 396–403.

Forsman, D. 1995. Male Pallid and female Montagu's Harrier raising hybrid young in Finland in 1993. *Dutch Birding* 17: 102–106.

Forsman, D. 1998. *The Raptors of Europe and the Middle East – A Handbook of Field Identification*. T. & A. D. Poyser.

Forsman, D. 2005. Rüppell's Vultures in Spain. *Birding World* 18: 435–438.

Forsman, D. & Lämsä, E. 2007. Successful interbreeding between Common Buzzard and Rough-legged Buzzard in Finland. *Linnut* 42: 36–37 (in Finnish).

Forsman, D. & Nye, D. 2007. A hybrid Red Kite x Black Kite in Cyprus. *Birding World* 20: 480–481.

Forsman, D. & Peltomäki, J., 2007. Hybrids between Pallid and Hen Harrier – A New Headache for Birders? *Alula* 13: 178–182.

Forsman, D. 2009. Hybrid harriers on the move. *Birding World* 22: 469–470.

Forsman, D. & Erterius, D. 2012. Pallid Harriers in northwest Europe and the identification of presumed Pallid Harrier x Hen Harrier hybrids. *Birding World* 25: 68–75.

Fransson, T. & Pettersson, J. 2001. *Swedish Bird Ringing Atlas*. Vol 1. Stockholm.

García, V., Moreno-Opo, R. and Tintó, A. 2013. Sex differentiation of Bonelli's Eagle *Aquila fasciata* in Western Europe using morphometrics and plumage colour patterns. *Ardeola* 60: 261–277.

Garner, M. 2002. Identification and vagrancy of American Merlins in Europe. *Birding World* 15: 468–480.

Gjershaug, J. O., Forset, O. A., Woldvik, K. & Espmark, Y. 2006. Hybridisation between Common Buzzard *Buteo buteo* and Rough-legged Buzzard *B. lagopus* in Norway. *Bull. Brit. Orn. Club* 126: 73–80.

Gonzales, J. M. & Margalida, A. (eds.) 2008. *Conservation biology of the Spanish Imperial Eagle* (Aquila adalberti). Organismo Autónomo Parques Nacionales. Ministerio de Medio Ambiente y Medio Marino y Rural. Madrid.

Houston, D. C. 1975. The moult of the White-backed and Rüppell's Griffon Vultures *Gyps africanus* and *G. rueppellii. Ibis* 117: 474–488.

Johnson, J. A., Watson, R. T., & Mindell, D. P. 2005. Prioritizing species conservation: does the Cape Verde Kite exist? *Proc. Royal Soc. London* B 272: 1365–1371.

Lontkowski, J. & Maciorowski, G. 2010. Identification of juvenile Greater Spotted Eagle, Lesser Spotted Eagle and hybrids. *Dutch Birding* 32: 384–397.

Maciorowski, G., Lontkowski, J. & Mizera, T. 2014. *The Spotted Eagle – Vanishing Bird of the Marshes*. Poznan.

Malmhagen, B. & Larsson, H. 2013. Fältbestämning av subadult östlig och spansk kejsarörn. *Vår Fågelvärld* 3: 47–49 (in Swedish).

McCarthy, E. M. 2006. *Handbook of Avian Hybrids of the World*. Oxford University Press.

Mullarney, K. & Forsman, D. 2011. Identification of Northern Harriers and vagrants in Ireland, Norfolk and Durham. *Birding World* 23: 509–523.

Mundy, P., Butchart, D., Ledger, J. and Piper, S. 1992. *The Vultures of Africa*. Academic Press.

Pfander, P. & Schmigalew, S. 2001. Umfangreiche Hybridisierung der Adler, *Buteo rufinus* Cretz. und Hochlandbussarde *B. hemilasius* Temm. et Schlegel. *Ornithol. Mitteil.* 53: 344–349.

Porter, R. F. & Kirwan, G. M. 2010. Studies of Socotran birds VI. The taxonomic status of the Socotra Buzzard. *Bull. Brit. Orn. Club* 130: 116–131.

Ristow, D. 2004. Exceptionally dark-plumaged Hobbies or normal Eleonora's Falcons? *British Birds* 97: 406–411.

Rodriguéz, B., Siverio, F., Siverio, M. & Rodriguéz, A. 2011. Variable plumage coloration of breeding Barbary Falcons *Falco (peregrinus) pelegrinoides* in the Canary Islands: do other Peregrine Falcon subspecies also occur in the archipelago? *Bull. Brit. Orn. Club* 131: 140–153.

Rodriguez, G, Elorriaga, J. & Ramirez, J. 2013. Identification of Atlas Long-legged Buzzard and its status in Europe. *Birding World* 26: 147–173.

Schollaert, V. & Willem, G. 2000. Taxonomy of the Peregrine *Falco peregrinus*/Barbary Falcon *F. (peregrinus) pelegrinoides* complex in Morocco. *Bull. African Bird Club* 7: 101–103.

Strandberg, R. 2013. Ageing, sexing and subspecific identification of Osprey, and two WP records of American Osprey. *Dutch Birding* 35: 69–87.

Väli, U. & Lõhmus, A. 2004. Nestling characteristics and identification of the Lesser Spotted Eagle *Aquila pomarina*, Greater Spotted Eagle *A. clanga* and their hybrids. *J. Ornithol.* 145: 256–263.

Väli, U., Dombrovski, V., Treinys, R., Bergmanis, U., Daróczi S. J., Dravecky, M., Ivanovsky, V., Lontkowski, J., Maciorowski, G., Meyburg, B.-U., Mizera, T., Zeitz, R. & Ellegren, H. 2010. Widespread hybridisation between the Greater Spotted Eagle *Aquila clanga* and the Lesser Spotted Eagle *Aquila pomarina* (Aves. Accipitriformes) in Europe. *Biol. J. Linnean Soc.* 100: 725–736.

Vaurie, C. 1965. *The Birds of the Palearctic Fauna. Non-passeriformes*. Witherby, London.

Wheeler, B. K. 2003. *Raptors of Eastern North America*. Princeton University Press, Princeton, New Jersey.

White, C. M., Cade, T. J. & Enderson, J. H. 2013. *Peregrine Falcons of the World*. Lynx Edicions.

Yosef, R., Helbig, A. J., & Clark, W. S. 2001. An intrageneric *Accipiter* hybrid from Eilat, Israel. *Sandgrouse* 23: 141–144.

Yosef, R., Lontkowski, J., Stawarczyk, T. & Fehervari, P. 2002. Biometric variation in three migratory *Accipiter* species at Eilat, Israel. *Sandgrouse* 24: 55–57.

Zuberogoitia, I., Azkona, A., Zabala, J., Astorkia, L., Castillo, I., Iraeta, A., Martínez, J. A. & Martínez, J. E. 2009. Phenotypic variations of Peregrine Falcon in subspecies distribution border. In. J. Sielicki and T. Mizera (eds.). *Peregrine Falcon populations – status and perspectives in the 21ˢᵗ century*, pp.295–308. European Peregrine Falcon Working Group.

INDEX

A

Accipiter badius 287
Accipiter brevipes 282
Accipiter gentilis 270
Accipiter nisus 275
Aegypius monachus 192
African Fish Eagle 138
African White-backed Vulture 174
American Kestrel 464
American Swallow-tailed Kite 119
Amur Falcon 475
Aquila adalberti 370
Aquila chrysaetos 352
Aquila clanga 395
Aquila fasciata 425
Aquila heliaca 359
Aquila nipalensis 373
Aquila pennata 432
Aquila pomarina 409
Aquila rapax 388
Aquila verreauxii 421
Aquila wahlbergi 442
Atlas Long-legged Buzzard 335

B

Bald Eagle 135
Barbary Falcon 535
Bateleur 215
Bearded Vulture 184
Black Eagle 421
Black Kite 81
Black Vulture 192
Black-eared Kite 91
Black-shouldered Kite 115
Black-winged Kite 115
Bonelli's Eagle 425
Booted Eagle 432
Buteo buteo buteo 304
Buteo buteo vulpinus 315
Buteo lagopus 343
Buteo rufinus cirtensis 335
Buteo rufinus rufinus 327
Buzzard, Atlas Long-legged 335
   Common 304
   Eastern Honey 67

   Eastern Long-legged 327
   Eurasian Honey 56
   Rough-legged 343
   Steppe 315

C

Cinereous Vulture 192
Circaetus gallicus 206
Circus aeruginosus aeruginosus 222
Circus cyaneus 234
Circus hudsonius 242
Circus macrourus 246
Circus pygargus 254
Clanga clanga 395
Clanga pomarina 409
Common Buzzard 304
Common Kestrel 446
Crested Honey-buzzard 67

D

Dark Chanting Goshawk 293

E

Eagle, African Fish 138
   Bald 135
   Black 421
   Bonelli's 425
   Booted 432
   Eastern Imperial 359
   Golden 352
   Greater Spotted 395
   Lesser Spotted 409
   Pallas's Fish 131
   Pallas's Sea 131
   Short-toed 206
   Short-toed Snake 206
   Spanish Imperial 370
   Steppe 373
   Tawny 388
   Verreaux's 421
   Wahlberg's 442
   White-tailed 121
   White-tailed Sea 121
'Eastern Black Kite' 95
Eastern Honey Buzzard 67
Eastern Honey-buzzard 67
Eastern Imperial Eagle 359

Eastern Long-legged Buzzard 327
Eastern Red-footed Falcon 475
Egyptian Vulture 143
Elanoides forficatus 119
Elanus caeruleus 115
Eleonora's Falcon 489
Eurasian Griffon 157
Eurasian Griffon Vulture 157
Eurasian Hobby 481
Eurasian Kestrel 446
Eurasian Marsh Harrier 222
Eurasian Sparrowhawk 275
European Honey-buzzard 56

F

Falco amurensis 475
Falco biarmicus 509
Falco cherrug 514
Falco columbarius 503
Falco concolor 498
Falco eleonorae 489
Falco naumanni 455
Falco peregrinus 526
Falco peregrinus pelegrinoides 535
Falco rusticolus 519
Falco sparverius 464
Falco subbuteo 481
Falco tinnunculus 446
Falco vespertinus 46
Falcon, Amur 475
    Barbary 535
    Eastern Red-footed 475
    Eleonora's 489
    Gyr 519
    Lanner 509
    Peregrine 526
    Red-footed 467
    Saker 514
    Sooty 498

G

Gabar Goshawk 297
Golden Eagle 352
Goshawk 270
    Dark Chanting 293
    Gabar 297
    Northern 270
Greater Spotted Eagle 395
Griffon, Eurasian 157

Himalayan 179
    Rüppell's 166
Griffon Vulture 157
Gypaetus barbatus 184
Gyps africanus 174
Gyps fulvus 157
Gyps himalayensis 179
Gyps rueppellii 166
Gyr Falcon 519

H

Haliaeetus albicilla 121
Haliaeetus leucocephalus 135
Haliaeetus leucoryphus 131
Haliaeetus vocifer 138
Harrier, Eurasian Marsh 222
    Hen 234
    Marsh 222
    Montagu's 254
    Northern 242
    Pallid 246
    Western Marsh 222
Hawk, Marsh 242
    Rough-legged 343
Hen Harrier 234
Hieraaetus fasciatus 425
Hieraaetus pennatus 432
Hieraaetus wahlbergi 442
Himalayan Griffon 179
Himalayan Griffon Vulture 179
Hobby, Eurasian 481
Honey Buzzard, Oriental 67
    Western 56
Honey-buzzard, Crested 67
    Eastern 67
    European 56
    Oriental 67
    Western 56
Hooded Vulture 153

K

Kestrel, American 464
    Common 446
    Eurasian 446
    Lesser 455
Kite, American Swallow-tailed 119
    Black 81
    Black-eared 91
    Black-shouldered 115

Black-winged 115
'Eastern Black' 95
Red 107
Swallow-tailed 119
Western Black 82
Yellow-billed 99

L

Lammergeier 184
Lanner Falcon 509
Lappet-faced Vulture 199
Lesser Kestrel 455
Lesser Spotted Eagle 409
Levant Sparrowhawk 282

M

Marsh Harrier 222
Marsh Hawk 242
Melierax metabates 293
Merlin 503
Micronisus gabar 297
Milvus aegyptius 99
Milvus migrans 81
Milvus migrans lineatus 91
Milvus migrans migrans 82
Milvus milvus 107
Monk Vulture 192
Montagu's Harrier 254

N

Necrosyrtes monachus 153
Neophron percnopterus 143
Northern Goshawk 270
Northern Harrier 242

O

Oriental Honey Buzzard 67
Oriental Honey-buzzard 67
Osprey 49

P

Pallas's Fish Eagle 131
Pallas's Sea Eagle 131
Pallid Harrier 246
Pandion haliaetus 49
Peregrine Falcon 526
Pernis apivorus 56
Pernis ptilorhyncus 67

R

Red Kite 107
Red-footed Falcon 467

Rough-legged Buzzard 343
Rough-legged Hawk 343
Rüppell's Griffon 166
Rüppell's Griffon Vulture 166
Rüppell's Vulture 166

S

Saker Falcon 514
Shikra 287
Short-toed Eagle 206
Short-toed Snake Eagle 206
Sooty Falcon 498
Spanish Imperial Eagle 370
Sparrowhawk 275
Eurasian 275
Levant 282
Steppe Buzzard 315
Steppe Eagle 373
Swallow-tailed Kite 119

T

Tawny Eagle 388
Terathopius ecaudatus 215
Torgos tracheliotos 199

V

Verreaux's Eagle 421
Vulture, African White-backed 174
Bearded 184
Black 192
Cinereous 192
Egyptian 143
Eurasian Griffon 157
Griffon 157
Himalayan Griffon 179
Hooded 153
Lappet-faced 199
Monk 192
Rüppell's 166
Rüppell's Griffon 166

W

Wahlberg's Eagle 442
Western Black Kite 82
Western Honey-buzzard 56
Western Marsh Harrier 222
White-tailed Eagle 121
White-tailed Sea Eagle 121

Y

Yellow-billed Kite 99